Virtual Reality in Tunnel Engineering

Editors:

Yuxin Zhang
Department of Urban Management Kyoto University
Kyoto, Japan

Yi Rui
Department of Geotechnical Engineering
Tongji University, Shanghai, China

Xinyan Huang
Department of Building Environment and Energy Engineering
The Hong Kong Polytechnic University, Hong Kong

Foreword by
Prof Hehua Zhu

CRC Press
Taylor & Francis Group
Boca Raton London New York

CRC Press is an imprint of the
Taylor & Francis Group, an **informa** business
A SCIENCE PUBLISHERS BOOK

First edition published 2027
by CRC Press
2385 NW Executive Center Drive, Suite 320, Boca Raton FL 33431

and by CRC Press
4 Park Square, Milton Park, Abingdon, Oxon, OX14 4RN

CRC Press is an imprint of Taylor & Francis Group, LLC

Library of Congress Cataloging-in-Publication Data (applied for)

ISBN: 978-1-032-82374-4 (hbk)
ISBN: 978-1-032-84686-6 (pbk)
ISBN: 978-1-003-51448-0 (ebk)

DOI: 10.1201/9781003514480

Typeset in Times New Roman
by Prime Publishing Services

Preface

Urban growth, climate change pressures, and the expansion of transport and utility networks have increased the use of underground space. Tunnels now serve not only road and rail corridors, but also metro systems, underground stations, utility corridors, and multi-function underground hubs. At the same time, tunnelling projects continue to face fundamental constraints, including geological uncertainty, limited workspace, complex logistics, strict safety requirements, and high consequences of design or operational errors. These conditions have increased the demand for better planning, clearer cross-disciplinary communication, safer training, and more reliable operation and management.

Virtual Reality in Tunnel Engineering was prepared to address this demand by treating virtual reality (VR) (including augmented reality and extended reality) as engineering instruments rather than visualization products, see Fig. 1. The scope covers the full tunnel lifecycle, including early-stage design and construction coordination, environmental perception and operational performance, inspection and maintenance, emergency response, evacuation behaviour, and the integration of sensing and robotics for safer and more efficient management. A unifying concept throughout the volume is that when field testing is expensive, disruptive, or unsafe, immersive simulation can provide a controlled, repeatable environment for evaluating design options, training personnel, and studying decisions and behaviours under realistic constraints.

The volume adopted a human-centred engineering perspective. Tunnel performance depends on structural and mechanical systems, but it also depends on how workers, drivers, passengers, and responders perceive information and make decisions. Visibility, lighting transitions, noise, guidance and communication, fatigue, stress, and social influence can alter behaviour and can directly affect both safety outcomes and operational efficiency. Immersive methods enable systematic and repeatable experiments, allowing these factors to be quantified, compared, and linked to design and management decisions without exposing participants to real hazards.

VR applications in tunnels are increasingly integrated with digital twins, Building Information Modeling, sensing systems, monitoring data, and robotic platforms. Such integration strengthens situational awareness, supports faster interpretation of complex data, and improves the testability of construction and operational strategies. In this role, immersive environments function as an interface

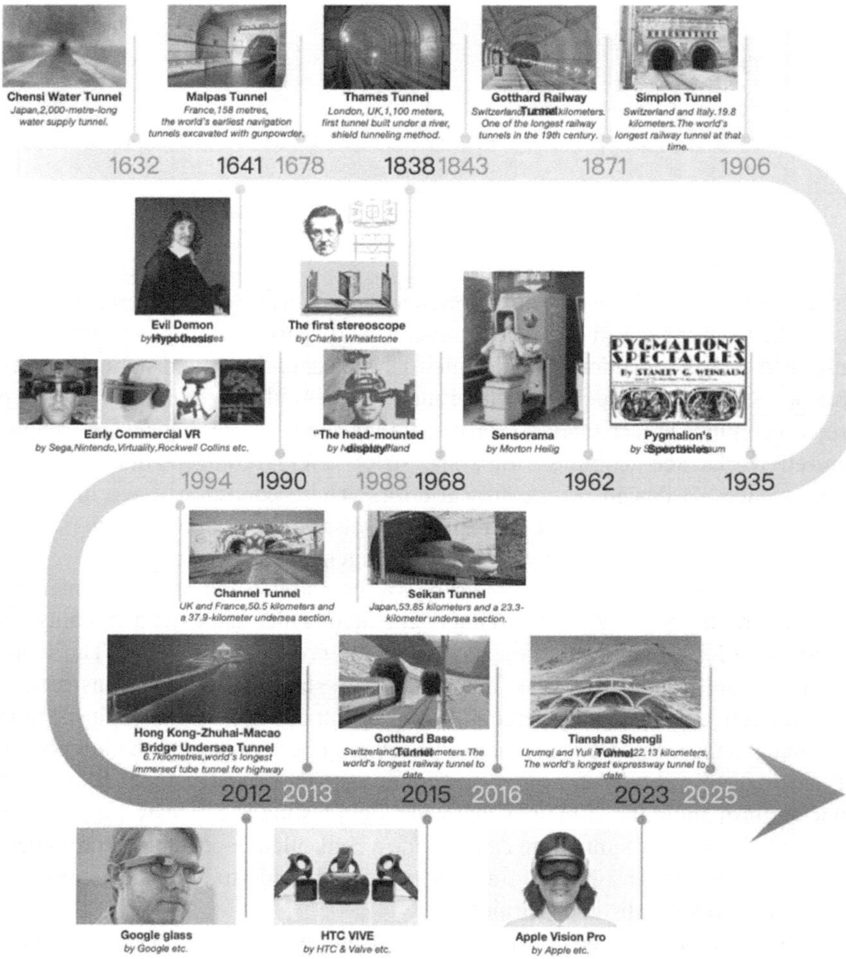

Fig. 1 The coupling history of VR development and tunnelling engineering

that connects models and data with human decision-makers, supporting both technical analysis and coordinated action.

This edited volume brings together contributions from tunnelling engineering, safety science, human factors, lighting and environmental research, and intelligent systems, see Fig. 2. The editors structured the book to provide both technical foundations and application-oriented studies, so that readers can understand the main concepts and also see how immersive methods have been implemented for tunnel problems. The intended audience includes graduate students and researchers seeking methods and research directions, practicing engineers and managers interested in implementable workflows, and stakeholders involved in training, safety planning, and operational governance.

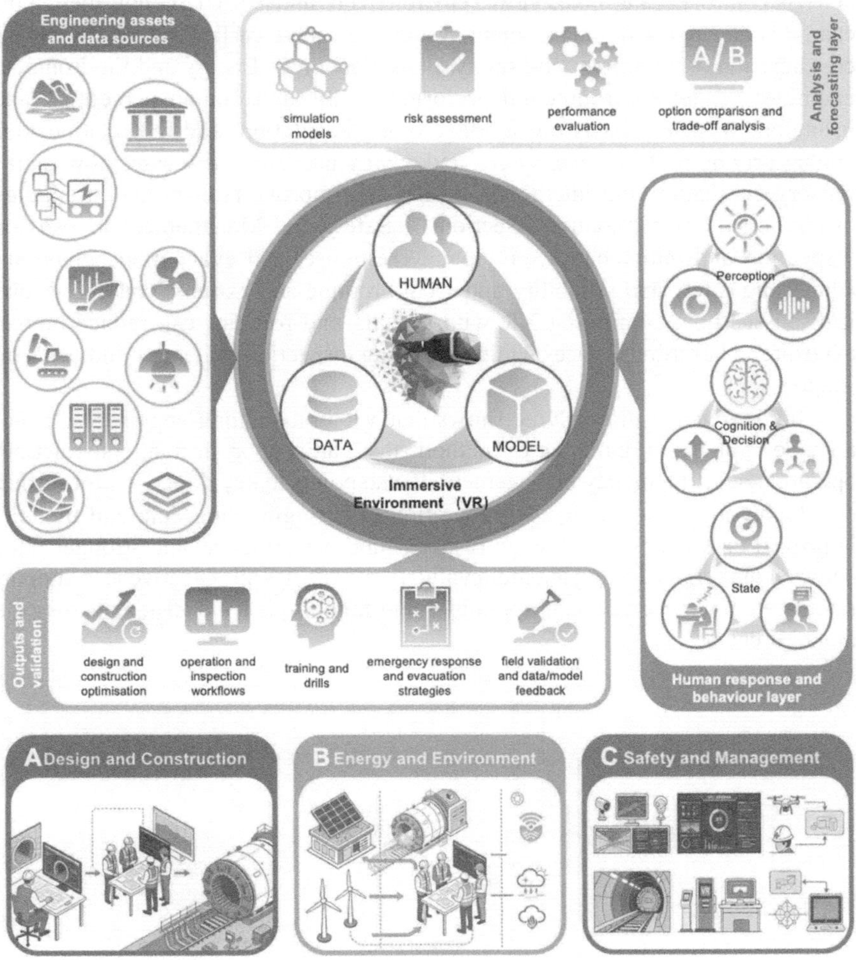

Fig. 2 The diagram and components of this book

The book is organized into two foundation chapters followed by three thematic sections, moving from core concepts to engineering applications. The first foundation chapter introduces the technical basis of VR and explains how immersive systems have progressed from visualization tools to platforms for decision support, training, and system testing in complex underground work. The second foundation chapter introduces core concepts in tunnelling engineering, providing essential background on methods, constraints, and terminology needed for the later application chapters.

After the foundation chapters, the book is divided into three sections. Section A: Design and Construction presents how immersive methods support design review and construction planning, including collaboration, progress representation, and risk control in data-linked virtual environments, and it discusses how these tools can be connected with management systems to enable earlier testing and lower-cost iteration of workflows and safety plans. Section B: Energy and Environment examines tunnel environmental performance through human perception and behaviour, with emphasis on lighting, visual adaptation, and the relationships among driving performance, safety, and energy use, and it discusses how multi-sensory simulation and interaction design can improve realism and strengthen evaluation of user responses. Section C: Safety and Management focuses on inspection and maintenance, emergency response, and evacuation behaviour, addressing safety and reliability under both routine and extreme conditions, and it discusses how combining XR with sensing and robotics can reduce human exposure in hazardous spaces while improving inspection efficiency and decision support.

Taken together, the volume shows that VR in tunnelling engineering is not a single application, but a set of methods that can support design, construction, operation, and emergency management. It also points to key directions for further development, including tighter integration with engineering data and models, improved measurement of human response and performance, and stronger links between immersive testing and field validation. The overall objective is to support safer systems, more efficient operation, and higher-quality decisions across the tunnel lifecycle.

Foreword

IMMERSIVE TECHNOLOGY IS REDEFINING HOW ENGINEERS THINK ABOUT UNDERGROUND SPACE.

Underground space has always resisted observation. Geology is invisible until opened, and the consequences of poor engineering decisions are often irreversible. For generations, tunnel engineers worked at the boundary of what could be known. This volume shows how virtual reality has begun to move that boundary.

The contributions here demonstrate that immersive methods are instruments of engineering reasoning, not visualization products. Design options can be evaluated before ground is broken. Workers can be trained under controlled and repeatable simulated conditions. Human behaviour under stress, in noise, and in degraded visibility can be measured and connected directly to design and management decisions.

When integrated with digital twins, building information models, and live sensor data, immersive environments become an operational interface across the full tunnel lifecycle. This volume brings together tunnelling engineering, safety science, human factors, and intelligent systems to lay the foundations for that integration. It will be a valuable resource for researchers, engineers, and practitioners for years to come.

Zhu Hehua is Honorary Chair Professor at Tongji University and Academician of the Chinese Academy of Engineering. He is Director of the State Key Laboratory of Disaster Reduction in Civil Engineering and Chair of the ISSMGE Technical Committee on Data Standards. A leading figure in digital underground space and infrastructure intelligence, he has been recognized by international peers as a pioneer in information integration for urban infrastructure planning, design, construction, and maintenance.

Contents

List of Contributors

Axel Mossberg
Bengt Dahlgren, Fire Research, Sweden.

Chao Chen
Department of Geotechnical Engineering, Tongji University, Shanghai, 200092, China.

Chen Zhou
School of Psychology and Neuroscience, University of Glasgow.

Furui MAN
Department of Building and Real Estate, Faculty of Construction and Environment, The Hong Kong Polytechnic University, Hong Kong

Guanlin Huang
Department of Geotechnical Engineering, Tongji University, Shanghai, 200092, China.

Haihang Hu
Research Center for Urban Fire Safety Engineering, Zhejiang University, Hangzhou, China.

He Shiyong
State Key Laboratory of Mountainous Bridge and Tunnel Engineering, Chongqing Jiaotong University.

Hongda Lin
Research Center for Urban Fire Safety Engineering, Zhejiang University, Hangzhou, Chin.

Hung-Lin CHI
Department of Building and Real Estate, Faculty of Construction and Environment, The Hong Kong Polytechnic University, Hong Kong

Junyu CHEN
Department of Building and Real Estate, Faculty of Construction and Environment, The Hong Kong Polytechnic University, Hong Kong

Liang Bo
School of Civil Engineering, Chongqing Jiaotong University, Chongqing 400074.

Liu Hao
School of Civil Engineering, Chongqing Jiaotong University, Chongqing 400074.

Max Kinateder
National Research Council, Canada.

Mun On Wong
Department of Civil and Environmental Engineering, Faculty of Science and Technology, University of Macau, Macau SAR, China.

Niu Jia An
School of Civil Engineering, Chongqing Jiaotong University, Chongqing 400074.

Peinan Li
College of Environmental Science and Engineering, Donghua University, Shanghai 201620, China.

Rong Deng
Research Center for Smart Urban Resilience and Firefighting, The Hong Kong Polytechnic University, Hong Kong.

Saizhe Ding
Research Center for Smart Urban Resilience and Firefighting, The Hong Kong Polytechnic University, Hong Kong.

Shaobin Zhou
Department of Civil and Environmental Engineering, Faculty of Science and Technology, University of Macau, Macau SAR, China.

Tianhang Zhang
Research Center for Urban Fire Safety Engineering, Zhejiang University, Hangzhou, Chin.

Wenyu XU
Department of Building and Real Estate, Faculty of Construction and Environment, The Hong Kong Polytechnic University, Hong Kong

Xiaojun Li
Department of Geotechnical Engineering, Tongji University, Shanghai, 200092, China.

Xin Huang
Department of Geotechnical Engineering, Tongji University, Shanghai, 200092, China.

Xinyan Huang
Department of Building Environment and Energy Engineering, The Hong Kong Polytechnic University, Hong Kong.

Xinyu Kou
Department of Geotechnical Engineering, Tongji University, Shanghai, 200092, China.

Yi Rui
Department of Geotechnical Engineering, Tongji University, Shanghai, 200092, China.

Yifei Ding
Department of Building Environment and Energy Engineering, The Hong Kong Polytechnic University, Hong Kong.

Yishu Wang
Department of Geotechnical Engineering, College of Civil Engineering, Tongji University, Shanghai 200092, China.

Yishuo Jiang
Department of Systems Engineering, Cornell University, Ithaca, USA.

Yueyue Zhu
No.1239, Siping Road, Yangpu District, Shanghai; Department of Geotechnical Engineering, College of Civil Engineering, Tongji University, Shanghai 200092, China.

Yuxin Zhang
Department of Urban Management Kyoto University, Kyoto, Japan.

Zeyu Dai
Department of Geotechnical Engineering, College of Civil Engineering, Tongji University, Shanghai 200092, China.

1 The Development and Application of Virtual Reality

Chen Zhou[1*] and Yuxin Zhang[2]

Virtual Reality (VR) has dramatically reshaped the landscape of scientific research in both cognitive science and engineering over the past decade, by offering immersive, interactive, and precisely controlled environments for behavioural experiments and simulation. This chapter begins by tracing the historical development of VR, from early philosophical thought experiments to its present technological sophistication, by highlighting major advances in hardware and software. We then discuss how VR resolves the long-standing tension between experimental control and ecological validity, allowing researchers to systematically manipulate perceptual and social variables while capturing multimodal behavioural, physiological, and neural data in real time. In cognitive science, VR has enabled research in vision, spatial navigation, self-consciousness, and social cognition, supporting the quantification and modelling of human behaviour within realistic yet programmable settings. In engineering, VR has become a critical tool for digital prototyping, human-machine integration, real-time system simulation, and training in high-risk or constrained environments. We further present a novel methodological framework for translating abstract psychological concepts into computational models and socially aware algorithms, with direct applications in robotics and virtual agents. Altogether, these developments suggest that VR is not only a ground-breaking research technology, but also an integrative platform for quantifying and understanding human behaviour, designing intelligent systems, and advancing human-technology interaction.

1. The History of VR: from Early Concepts to Modern Products

Despite physiological limitations and the immutable laws of the physical world, humans haven't stopped pursuing the creation of realistic experiences within virtual environments. The fantasy of immersive experience in an artificial world was deeply rooted in the history of human philosophy, science, and technology. In

1 School of Psychology and Neuroscience, University of Glasgow.
2 Department of Urban Management Kyoto University, Kyoto, Japan.
Email: yx.zhang@polyu.edu.hk[2]
* Corresponding author: chen.zhou@glasgow.ac.uk

DOI: 10.1201/9781003514480-1

1641, René Descartes introduced the Evil Demon Hypothesis in his masterpiece, *Meditations on First Philosophy* [1], proposing that a powerful, evil demon might manipulate his senses, producing a false reality. This thought experiment was designed to challenge the reliability of sensory experience and the certainty of knowledge derived from it, a line of inquiry that later inspired the famous thought experiment "brain in a vat" [2] and many science fiction works closely related to the concept. Beyond its philosophical underpinning, it also inspired scientists to develop equipment capable of systematically replacing or overriding human sensory inputs and perceptions with digitally constructed stimuli. One of the earliest attempts at creating 3D vision by actively presenting different images to each eye to produce binocular vision can be traced back to the 1830s, when Sir Charles Wheatstone first introduced the idea of stereopsis [3]. Based on this principle, he developed the first stereoscope for 3D visualization, see Fig. 1.

However, the concept of the modern Virtual Reality headset was widely believed to be first introduced in a science fiction short story, *Pygmalion's Spectacles* by Stanley Weinbaum in 1935 [4], in which he described a pair of goggles that can alter the sensation of the wearer. The idea was proposed decades before VR was technically possible, making it a milestone in the birth of modern VR equipment. With the rapid development of computer graphics in the 1960s, the prototypes of VR headsets, such as the Sensorama by the filmmaker Morton Heilig [5], were invented and tested in laboratory environments, an immersive film player integrating 3D vision, stereo sound, haptic stimuli, simulated smell, etc. In 1968, Ivan Sutherland invented "the head-mounted display" [6], a display that can track the positions and orientations of the wearer and update the visual image accordingly in real-time. From the 1970s to the 1990s, virtual reality (VR) steadily advanced alongside advancements in computer graphics and was progressively used in a variety of fields, such as academia, gaming, the military, and aviation. Notably, the US Air Force developed a simulated flight system to train pilots with HMDs [7], and NASA further improved VR to simulate a space environment and train astronauts in the lab with digital gloves that can interact with the virtual objects [8]. The commercial products of VR, such as the Sega concept VR headset, Nintendo Virtual Boy, Virtuality Visette, Rockwell Collins ProView SR80, etc., were first released in the 1990s. However, due to limited graphic performance at the time and high price, these products were either not officially released or not commercially successful, which was a big setback for the field.

After a long period of technological maturation, in 2012, Palmer Luckey launched *Oculus Rift* via Kickstarter, leading to Facebook acquiring Oculus for $2 billion in 2014, which ignited the entire sector and sent shockwaves through the VR industry. Since then, leading tech companies around the world joined the competition one after another, including HTC (Vive), Sony (PSVR), Samsung (HMD Odyssey), Microsoft (HoloLens), and Apple (Vision Pro), by releasing their VR/AR products in the coming years. The commercial VR headsets were then dramatically improving and continuing to evolve on a yearly, even monthly,

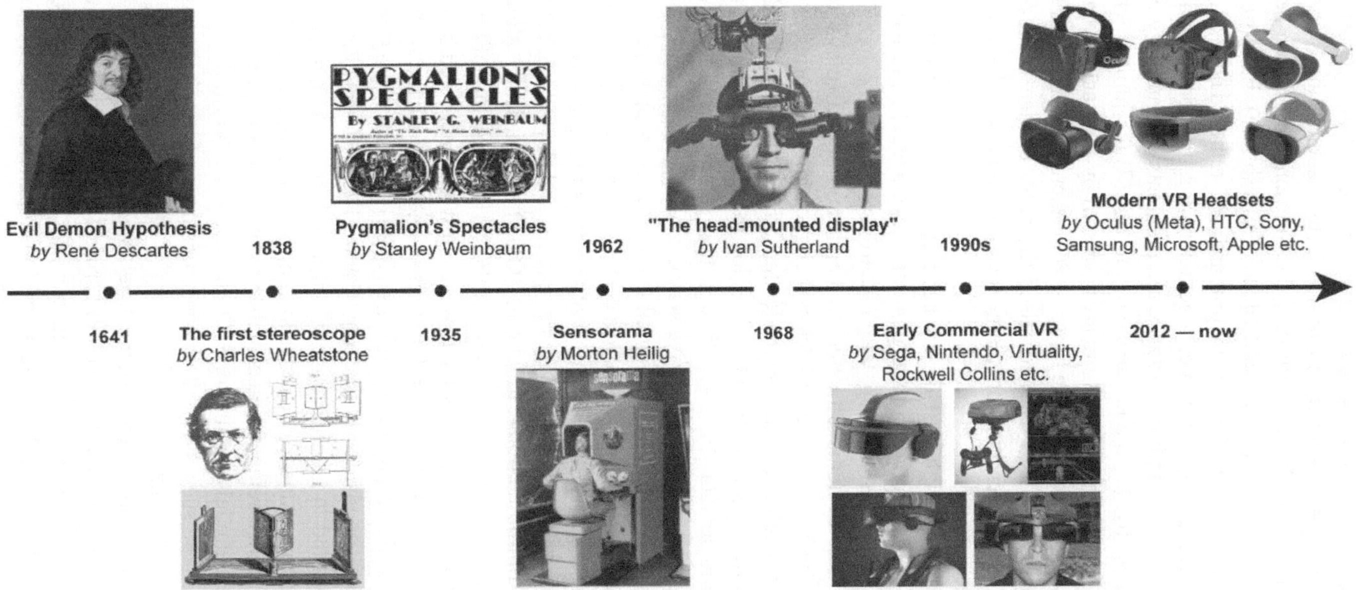

Fig. 1 The development of VR from concepts to products. VR has developed from philosophical concepts to modern products, with seven notable milestones in history.

basis. The performance of these headsets finally began to approach the levels of immersion once dreamed of in fiction, with a relatively affordable price.

In the first five years after the historical Oculus acquisition, wired headsets were the mainstream of the VR industry, focusing on the graphic performance (resolution), field of vision (FOV), and latency (refresh rate). Early models, such as Oculus Rift CV1 and HTC Vive, were primarily PC-tethered headsets, requiring a high-performance gaming computer together with a high-end graphics card to operate. Besides the demanding hardware requirements, the mobility of the users is largely restricted by: (1) the existence of the PC, (2) the limited length of the data cable, and (3) the limited tracking area covered by the tracking base stations (positional tracking sensors). These flaws in PC-tethered VR headsets have made them much more difficult to use and raised the entry barrier, reducing them to professional tools for scientific study, large toys for die-hard techies, and things to display at business events.

The mobility and high cost of VR headsets have been major obstacles in popularizing VR technology among general consumers. In the early years of cognitive science research using VR, researchers employed hybrid inertial/ultrasonic tracking systems attached to head-mounted displays to localize participants and enable real-time updates of their movements. During experiments, an experimenter would manually manage the tethered cable by holding it and following the participant throughout the session to ensure freedom of movement and prevent entanglement [9]. Even though the technical challenge was partly addressed by the release of the HTC Vive Wireless Adapter in late 2018, which used Intel WiGig technology to transmit a signal wirelessly between the headset and PC. This has allowed users to move freely in a designated area without being disturbed by the cable. The area that the tracking base stations and signal transmitters cover, however, still limits this option. Some companies and developers have invented VR-compatible omnidirectional treadmills (ODT) that basically allow users to walk around on a treadmill that can simulate the veridical feeling of locomotion in the virtual environment. These products were not widely accepted by consumers, mainly because of the high price and motion sickness associated with them. The game-changer was the Oculus Quest, released in 2019, which was a standalone VR headset with a built-in inside-out tracking system. The all-in-one headset allowed users to move around in the theoretically unlimited space. Although it sacrificed graphical performance due to the reduced computational power of mobile chips, the overall experience remained acceptable at an affordable price. The product, followed by other models, including PICO headsets (PICO Immersive Pte. Ltd., China), became major successes in both consumer and commercial markets, giving rise to the robust VR content ecosystem.

Beyond hardware advancements, the development of VR-supporting software, particularly game engines, has been equally critical and should not be overlooked. Before the release of Unity 5 by its main rival, Unity Technologies, in 2015, Epic Games published Unreal Engine 4, the fourth generation of its open-source game engine, concurrently with the acquisition of Oculus by Facebook in 2014.

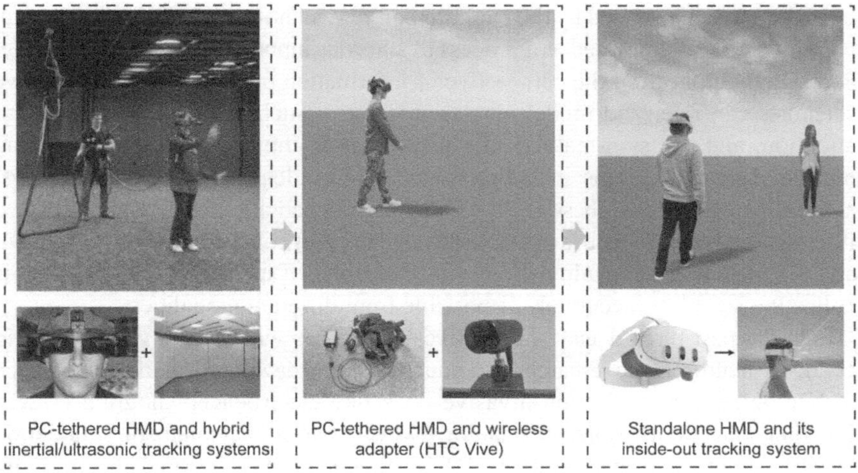

Fig. 2 Advancements in VR Mobility. VR headsets evolved from tethered, costly systems to affordable, standalone headsets, greatly improving user freedom and popularizing the technology.

These two dominant game engines have played a pivotal role in shaping VR content creation, lowering development barriers, and driving industry adoption. With their integrated native VR support, these game engines addressed VR-specific challenges, such as performance optimization and motion interaction. This empowerment allowed small teams to develop VR content independently and significantly lowered the entry barrier for behavioral researchers designing virtual environments and scripting VR experiments. More importantly, these game engines have driven industrial standards in VR content creation. Recently, the plugins of the game engines have provided substantial resources for creating high-quality content, such as the well-known Metahuman Creator of Unreal Engine 5, which generates realistic virtual humans with many built-in animations and functions for communication. Together, the game engines and their associated toolboxes are pushing the boundaries of VR applications.

Today, with the development of both hardware and software, VR is no longer a fantasy. It is an evolving tool for the application in many industries, including gaming, healthcare, education, scientific research, and engineering. Without a doubt, it will shape the future of human communication and the way of interacting with the outside world, shedding new light on the research of human consciousness and the philosophical question of sensory experiences.

2. VR as a Game-changer for Cognitive Science Research: The Revolution from Both Input and Output

Although the development of VR has significantly impacted a lot of research fields and industries, cognitive science is definitely one of the research fields that has been dramatically shaped by the evolution of VR technology [10]. In brief, human cognition can be simplified as a process of stimulus input, information processing, and signal output from the perspective of information processing [11]. Cognitive scientists can infer and model mental processes—such as perception, memory, attention, and decision-making—by manipulating input stimuli and collecting multimodal output signals, including behavior, physiological responses, and neural activity. To achieve the best utility of this process, cognitive scientists have two different approaches. One is trying to achieve the best control of the stimulus input and the most precise manipulation of the variables of interest, while the other is collecting signals as accurately as possible from different modalities.

In the past three decades, cognitive science has witnessed a vast growth in cognitive neuroscience research [12], due to the rapid development of neuroscience technologies, especially non-invasive ones like electroencephalogram (EEG), functional magnetic resonance imaging (fMRI), transcranial magnetic stimulation (TMS), Magnetoencephalography (MEG), functional near-infrared spectroscopy (fNIRS) and transcranial direct current stimulation (tDCS), which give rise to the golden age of neuroimaging studies [13]. Advances in measuring brain outputs have greatly enhanced cognitive scientists' ability to record neural signals, whether from electrical activity or blood oxygen levels. By examining how these signals change in response to varying stimulus intensities, researchers can infer and model the cognitive processes and functional organization of the human mind. [14].

VR, on the other hand, is the new wave of revolution from both the input and output ends [15]. The technical specifications of VR offer several significant advantages for cognitive science research in ways that would be difficult or impossible in traditional experimental methodologies and real-world settings.

The first and one of the most compelling advantages of VR in cognitive science is its unique ability to reach a balance between sophisticated experimental control and ecological validity, which is regarded as a challenge that has long divided laboratory-based experimental research and real-world observation and statistics [16]. Traditional laboratory research tends to abstract scientific questions from real life and investigate them in well-controlled environments. This approach offers better control over variables but often sacrifices realism. On the contrary, field studies were mostly based on the methodology of observation and questionnaire, which provides naturalistic contexts but limits the precision of manipulation and measurement. VR bridges this gap by creating immersive, interactive environments that are realistic to human participants, yet remain entirely programmable and repeatable. Researchers can easily manipulate every aspect of the virtual setting, from environmental variables to the features of the virtual humans (age, gender, ethnicity, facial expression, hand gesture, body posture, voice, walking trajectory, etc.), while maintaining perfect consistency across trials and participants. At the same time, VR allows for naturalistic tasks like communicating with a virtual human or navigating social scenarios, which more closely reflect everyday social behaviours of humans. This amazing balance between ecological immersion and experimental rigor delivered by VR makes it a powerful tool for studying human perception and action, cognition, and decision making in contexts that are both scientifically valid and behaviourally meaningful.

On the output end, VR offers a rich, multimodal framework for capturing human responses, going far beyond traditional behavioural measures [17] (Martin et al., 2022). In addition to collecting trial-by-trial behavioural judgments, such as decisions, reaction times, or accuracy, VR environments can be seamlessly integrated with tools that track bio-signals like heart rate, skin conductance response, and neural signals (through fNIRS, EEG, fMRI, etc.), providing insight into participants' emotional and physiological states [18]. Moreover, built-in sensors and external tracking systems of VR equipment can record facial expressions, allowing researchers to detect subtle affective reactions in real time [19]. VR also enables precise monitoring of eye movements through integrated eye trackers [20] (Adhanom et al., 2023) (HTC Vive Pro Eye powered by Tobii, FOVE, Apple Vision Pro), revealing where attention is directed and how visual information is processed in real-time. Furthermore, full-body motion capture [21] can be used to analyze gait, posture, gesture, and other motor behaviours as participants interact naturally with their environment or virtual humans. This combination of high-resolution real-time behavioural and physiological data within a controlled, immersive environment makes VR a uniquely powerful tool for collecting multimodal human behavioural responses and bio-signals to understand cognitive processes in the human mind.

In summary, just as the maturation of neuroimaging technologies in the late 1990s sparked a revolutionary shift in cognitive science by investigating the neural underpinnings of mental processes, VR now represents the next revolutionary wave in the field. By bridging the long-standing divide between experimental control and ecological validity, especially in social science studies, VR enables researchers to simulate complex, life-like environments with precision and consistency. Its capacity to deliver immersive experiences while collecting rich, real-time multimodal data, spanning behaviour and physiology, makes it an unparalleled tool for exploring human cognition in the complex social world. Therefore, VR is not merely a methodological improvement, but a paradigm shift redefining how we study the human mind in a complex context.

3. The Application of VR in Cognitive Science

Since VR has been useful in controlling experimental environments, continuously manipulating experimental stimuli, and collecting multimodal human responses and bio-signals, the technology is being widely used in many different research fields in cognitive science and plays a crucial role in pushing the boundaries of our understanding of human cognition.

In the field of vision science, William Warren has been a pioneering figure in using VR to investigate the fundamental mechanisms and dynamics of human perception and action. His research has focused on how visual information guides behaviour in dynamic, realistic virtual environments, particularly in the domains of visual control of locomotion [9; 22; 23; 24; 25] and spatial navigation [26]. In

2001, one of his most influential studies used VR to displace visual optic flow from the direction of walking, which was always consistent in real life and bounded by the laws of optics [9]. VR, in this case, successfully created visual stimuli that were not possible to be realized or seen in nature, thus allowing vision scientists to study how human locomotor control is modulated by the combination of visual cues through precisely manipulating the visual stimuli and recording walking trajectories of participants. By further combining immersive virtual environments with sophisticated experimental control of psychophysics, Warren has been able to design complex locomotor tasks such as obstacle avoidance [22], spatial navigation [26], and pedestrian interactions [24]. His pioneering work using VR not only deepens our understanding of the dynamics of perception and action, but also showcases the unique power of VR to study human behaviours in ecologically valid yet scientifically rigorous manners.

VR has also been widely adopted in neuroscience studies, mainly on the topics of self-consciousness, body ownership, and spatial navigation. Olaf Blanke has been at the forefront of research on self-consciousness and body ownership, using VR in combination with neuroimaging technologies to explore how the brain constructs a sense of self [27; 28; 29; 30]. His work was influential in shaping our understanding of the critical philosophical question of self-representation. He has shown that bodily self-consciousness—our sense of owning a body and being located in space—can be manipulated through multisensory conflicts. In pioneering VR experiments, Blanke and his team used visuo-tactile stimulation to induce out-of-body experiences and illusory ownership of virtual bodies, indicating that self-identification and self-location are not fixed, but can be experimentally altered [27]. By further integrating VR with cutting-edge neuroimaging technologies, his research has helped identify key brain regions involved in bodily self-consciousness [30]. These findings have major implications not only for neuroscience and cognitive science but also for clinical contexts, such as understanding disorders of the self in conditions like schizophrenia [31]. Christian Doeller, as another researcher who has successfully combined VR and advanced neuroimaging technologies, has made significant contributions to understanding the neural basis of human spatial navigation [32, 33; 34;]. Inspired by findings from rodent studies, his research has focused on how the human brain encodes spatial information through grid cells, place cells, and head-direction cells, which were widely studied in animal models. Using immersive VR environments that allow participants to freely navigate virtual spaces while undergoing fMRI or MEG scanning, Doeller and colleagues have identified grid-like activity patterns in the human entorhinal cortex, providing powerful evidence that similar neural coding mechanisms support spatial navigation in humans [32–34](Doeller et al.,. His work has further shown that these spatial representations are not only used for navigation but are also involved in episodic memory and conceptual cognitive maps, suggesting a more general coding principle in human cognition [35]. Doeller's research clearly demonstrated how VR can empower neuroscience research with immersive virtual environments to accurately simulate naturalistic visual experiences, in order to capture how the human brain represents and processes real-world stimuli.

The most influential application of VR in the field of cognitive science is in social cognition [36]. Mel Slater, as a well-known pioneer in the intersection of VR and psychology, has made groundbreaking contributions in integrating VR in social psychology studies on how immersive virtual environments influence human perception and social behaviours [37]. His work has demonstrated the power of VR to induce body ownership illusions (e.g., the rubber hand illusion), showing how virtual embodiment can alter self-perception, implicit biases, and social attitudes [37-39]. Slater also explored prosocial behaviour in VR, revealing how virtual experiences can foster empathy and reduce prejudice by allowing users to embody avatars with different genders and races [40]. His studies on social presence laid the foundations for understanding how virtual interactions mirror real-world social dynamics, including conformity, proxemics, and other nonverbal social cues [15; 37; 41]. Through sophisticated experimental designs, his work bridges psychological theory with VR technology, leading to transformative applications in therapy and education. Jeremy Bailenson, as the founding director of the Virtual Human Interaction Lab at Stanford University, is a leading figure in exploring how VR can be used to study human social interaction. His pioneering research on the function of nonverbal social cues of virtual humans during social interaction, such as gaze [42], proximity[43], and mimicry [44], has demonstrated the ability of virtual humans in influencing actual human social behaviours, indicating the human-likeness of virtual humans. More importantly, his research has inspired the methodological development of using virtual humans to investigate human social interaction in a better-controlled environment [45]. Recently, virtual humans in the virtual environment have been widely adopted in the research of human social behaviours by social psychologists [36]. For example, researchers used VR to study human heroic behaviors when witnessing crime, which effectively bypassed the ethical constraints of conducting similar studies in real-life situations [47].

My previous research used VR to precisely manipulate social variables, such as interpersonal distance, orientation, and postural openness of virtual humans, to measure human perception of social interaction using psychophysical methods [48]. Based on the quantified results, we introduced the Social Interaction Field Model (SIFM) by borrowing the concept of field theory from Newtonian physics to describe the distribution of social impact around the human body. Analogous to the gravitational field around objects, the strength of social interaction between individuals could be calculated from the distribution of the social interaction field, both conceptually and mathematically. This work demonstrated the feasibility of quantifying human social behaviour using classical psychophysical techniques within a controlled virtual environment, offering new insights into the integration of cutting-edge VR technology with rigorous psychophysical measurement.

Building on these findings, we investigated how the perception of social interactions and the identification of social groups guide human navigation through crowds [49]. Participants were asked to walk within immersive virtual social environments while wearing a VR HMD, during which their walking trajectories were recorded under systematically manipulated social settings. We subsequently developed a social locomotion model based on these behavioural results to predict human path selection and walking patterns in complex dynamic social scenes. This model was then implemented into a socially aware navigation algorithm for a service robot. Robots utilizing the social locomotion algorithm exhibited more human-like behaviour and received significantly higher comfort ratings compared to those using previous social navigation algorithms. Overall, this research illustrates a novel approach for translating human behavioural findings into robotic engineering by combining VR-based behavioural quantification with computational modeling to derive predictive mathematical models.

4. The Methodological Framework to Translate Psychological Theories into Computational Models and Algorithms

Methodologically, abstract psychological concepts are complex and thus challenging to manipulate and model in a quantifiable way. Therefore, current psychological theories cannot be directly translated into explainable computational models and algorithms that meet industrial requirements, leaving a significant knowledge gap. To bridge this gap, I have established an innovative methodological framework to translate abstract social signals into computational models and algorithms that are both understandable and explainable. First, with cutting-edge VR technology, I create well-controlled virtual environments, manipulate social variables in a quantified way, and use classic psychophysics, which offers rigorous methods to measure human social behaviours, including with social stimuli. Next, using computational modelling, I build basic mathematical functions to describe human social behaviours. These basic functions can be generalized to complex social scenes by mathematical deduction. Then, I transform the model into algorithms that can be applied to virtual agents and social robots. Finally, the algorithm performance is validated and refined in human-machine interaction experiments to meet industrial requirements. This framework has successfully translated human perception of social groups to socially-aware navigation algorithms by first quantifying human locomotion behaviours in a virtual environment facing virtual humans. Then, I built a Social Locomotion Model based on the perception of other people's social influence to model the human path selection behaviours and walking trajectories. The model was then verified and modified in complex social scenarios. Finally, the computational model was transformed into algorithms that were further tested and refined in human-robot interaction experiments.

Technologically, the critical challenge is that human social behaviours are multi-modal and can only be measured in realistic social contexts (vs the laboratory). However, traditional methods used in social psychology, such as questionnaires and observation, cannot be used to precisely measure human social behaviours for computational modelling. To achieve this, I have developed a cutting-edge VR

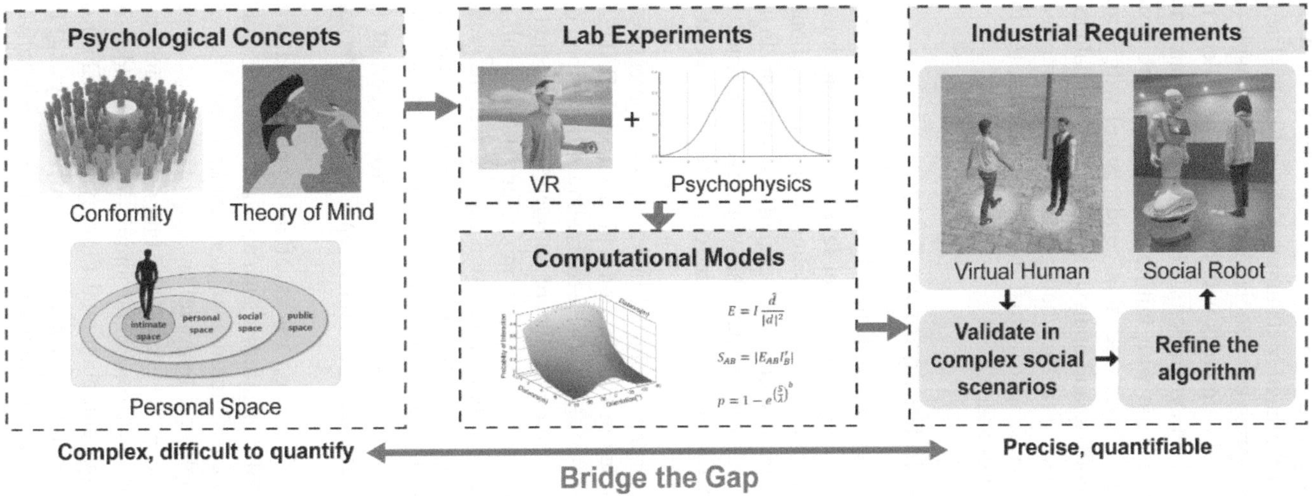

Fig. 3 The methodological framework to translate psychological concepts into algorithms that meet industrial requirements. Complex psychological concepts are first manipulated and quantified in VR with psychophysical methods in lab experiments. Computational models are then built based on the behavioural results and transformed into algorithms that can be applied to virtual agents and social robots, to validate and refine the model in real-life social scenarios.

platform that can precisely manipulate the realistic appearances and behaviours of virtual humans by leveraging the leading-edge game engine from Unreal Engine 5 (Epic Games) and its associated Metahuman Creator Plugin. The platform can thus produce continuous variables on social dimensions that enable (1) collection of trial-by-trial behavioural judgments with hand-held controllers, recording (2) eye movements with built-in eye trackers, (3) facial expressions with structured light cameras, (4) body motion with full-body motion capture system, and (5) multiple biomarkers (e.g., heart rate, skin conductance) with a multi-channel physiological signal recording and analysis system. The standalone VR headset enables participants to view realistic social stimuli while immersed in realistic social environments, interact with virtual humans, and track (6) their walking trajectories in boundaryless space.

5. VR Application in Engineering Fields

Virtual Reality (VR) has become an increasingly integral tool in various branches of engineering, where it supports system-level simulation, human-machine integration, and real-time environment modeling. Unlike conventional simulation tools, VR allows engineers to interact with digital environments in an immersive and spatially coherent way, enabling both the subjective experience and objective measurement of engineering processes under controllable yet ecologically valid conditions.

In mechanical and product design engineering, VR is integrated into digital prototyping pipelines, often in combination with parametric CAD systems (e.g., SolidWorks, CATIA, Siemens NX). These tools allow direct import of 3D models into immersive VR environments, where engineers can conduct ergonomics assessment, reachability analysis, and assembly simulation at full scale. Head-mounted displays (HMDs) paired with motion capture systems (e.g., OptiTrack, Vicon) enable spatial mapping between the user's body and the virtual model, allowing quantification of joint angles, reach vectors, and biomechanical stress during task simulation. These features are particularly important for assessing human-in-the-loop scenarios such as maintenance operations or cockpit layout evaluation [50].

In construction engineering, VR is often synchronized with Building Information Modeling (BIM) platforms such as Autodesk Revit and Navisworks to support spatial-temporal simulation of construction sequences, scaffolding logistics, and safety compliance. Temporal layering of construction phases can be visualized in 4D simulations (3D geometry + time), allowing for clash detection and schedule validation. Environmental simulations embedded in VR, such as airflow modeling using computational fluid dynamics (CFD) or structural deformation under load conditions using finite element analysis (FEA), provide multi-layered feedback for design validation [51]. These quantitative models can be parameterized and adapted dynamically based on user input during the VR session, creating closed-loop design-experiment iterations.

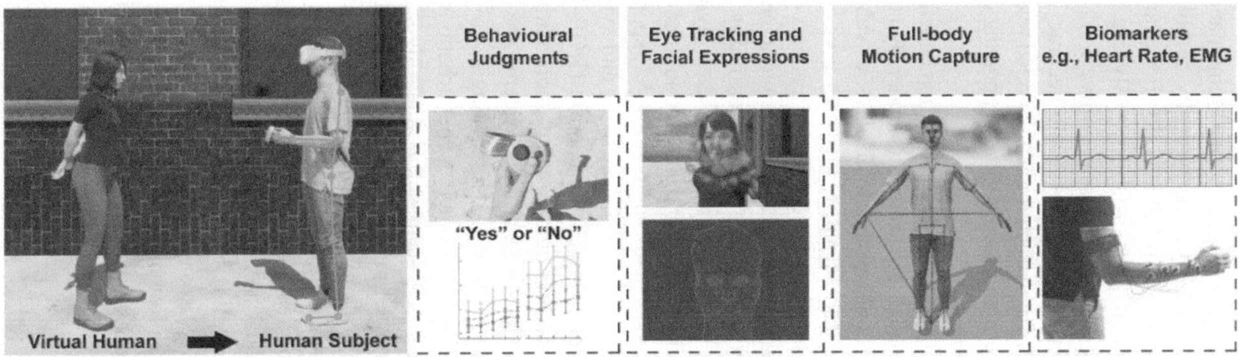

Fig. 4 The cutting-edge VR experimental platform. The platform can manipulate the personal attributes and behaviours of virtual humans from the input end, and integrate behavioural judgments, eye-tracking, facial expressions, body motions, and biomarkers from the output end to achieve accurate measurement of the social behaviours and physiological signals of human participants in social contexts.

Training and operational rehearsal in engineering domains make extensive use of VR, particularly where real-world exposure is risky, expensive, or logistically limited. VR-based simulators are deployed in nuclear engineering (e.g., reactor control system training), aerospace (e.g., engine maintenance procedure), and chemical process industries. These systems typically combine virtual scenario scripting with sensor-based assessment frameworks. For example, physiological responses during training (heart rate variability, galvanic skin response) are captured using integrated biosensors to assess cognitive workload or stress level. Metrics such as task completion time, tool usage accuracy, and adherence to standard operating procedures are extracted via embedded telemetry. This allows fine-grained behavioural performance evaluation, supporting quantitative models of learning and skill acquisition.

Human factors engineering and interface evaluation benefit significantly from VR's capacity to simulate control environments under dynamic task demands. Virtual prototypes of control rooms, wearable displays, or vehicle dashboards are evaluated through immersive task scenarios in which user attention, workload, and error rates are measured. Integrated eye-tracking systems (e.g., Tobii Pro VR, Varjo XR-3) provide real-time heatmaps of gaze behaviour, enabling quantitative assessment of interface saliency and information accessibility. Full-body tracking and gesture recognition can be used to model postural dynamics and micro-movements related to physical workload and interaction fluency.

In robotics and autonomous systems engineering, VR supports the simulation of agent-environment interactions using digital twins and synthetic sensor data. This includes simulation of multi-modal perception (RGB-D, LiDAR, IMU), robot kinematics and dynamics, and human-agent coordination in collaborative tasks. Frameworks like ROS (Robot Operating System) are commonly used in tandem with game engines (e.g., Unity3D, Unreal Engine) to render and control robotic behaviour within VR. Engineers can inject perturbations (e.g., sensor noise, unexpected obstacles) into the virtual scene and evaluate algorithm robustness and failure recovery [52]. Furthermore, human subjects can be introduced into these simulations to study shared autonomy protocols, handover strategies, and safety margins during co-located task execution. [53].

Teleoperation systems and remote inspection applications also leverage VR interfaces, where real-time data streams from sensors (e.g., 3D SLAM, stereo vision) are rendered in VR to guide human control of distant actuators. Haptic interfaces and force-feedback devices (e.g., SensAble Phantom, HaptX gloves) are integrated into the VR loop to allow bi-directional interaction. These systems are used in hazardous environments such as subsea infrastructure inspection, disaster response robotics, or extraterrestrial exploration, where conventional physical presence is not feasible.

In engineering education and laboratory training, VR creates virtual labs where learners can interact with parameterized physical simulations. For example, fluid flow visualization using Navier-Stokes simulation, structural resonance under dynamic loading, or thermodynamic cycle analysis can be explored using interactive, embodied methods [54; 50]. Virtual labs are instrumented to log behavioural data such as exploration patterns, hypothesis generation, and interaction sequences, which can then be analyzed to model cognitive learning processes and instructional effectiveness.

In conclusion, VR provides engineering with a high-fidelity, multimodal experimental platform that combines perceptual immersion with precise control and measurement. Its integration with biomechanical modeling, computational simulation, physiological sensing, and robotic control systems transforms it from a visualization tool into a full-scale experimental framework. As game engines continue to incorporate native support for physics-based modeling, real-time data streaming, and AI-assisted behaviour generation, VR will become a critical infrastructure for engineering research and practice, particularly in areas requiring close integration between human behaviour, physical systems, and computational models [55].

Looking ahead, VR stands at the intersection of human civilization. It is a technology that will fundamentally reshape how humans interact with each other, with machines, and with the environment around them. Beyond its current applications in cognitive science research and engineering, VR offers an unprecedented platform for bridging the inner mind and the outer physical world, enabling new forms of social relationships and interaction with the environment. As hardware and software continue to evolve in the coming years, the boundaries between the physical and virtual worlds will become increasingly fluid, pushing new frontiers for scientific research, industrial applications, and the understanding of human consciousness itself. Just as Descartes once questioned the reliability of sensory experience through his Evil Demon Hypothesis, VR now challenges the very definition of reality and perception, offering new possibilities to explore the nature of human existence. In this unfolding future, VR is not merely a tool; it is a new frontier that will redefine human civilization.

References

[1] Descartes, R. (1641). *Meditations on First Philosophy: With Selections from the Objections and Replies* (J. Cottingham, Ed.; Second Edition). Cambridge University Press. https://doi.org/10.1017/9781107416277.

[2] Putnam, H. (1981). Brains in a vat. In *Reason, Truth and History* (pp. 1–21). Cambridge University Press.

[3] Wheatstone, C. (1838). XVIII. Contributions to the physiology of vision.—Part the first. On some remarkable, and hitherto unobserved, phenomena of binocular vision. *Philosophical Transactions of the Royal Society of London, 128*, 371–394. https://doi.org/10.1098/rstl.1838.0019.

[4] Weinbaum, S. G. (1935). *Pygmalion's Spectacles*. Floating Press.

[5] Heilig, M.L. (1962). *Sensorama simulator* (United States Patent No. US3050870A). https://patents.google.com/patent/US3050870/en.

[6] Sutherland, I.E. (1968). A head-mounted three dimensional display. *Proceedings of the December 9-11, 1968, Fall Joint Computer Conference, Part I*, 757–764. https://doi.org/10.1145/1476589.1476686.

[7] Furness, III, T.A. (1986). The Super Cockpit and its Human Factors Challenges. *Proceedings of the Human Factors Society Annual Meeting, 30*(1), 48–52. https://doi.org/10.1177/154193128603000112.

[8] Fisher, S.S., McGreevy, M., Humphries, J. and Robinett, W. (1987). Virtual environment display system. *Proceedings of the 1986 Workshop on Interactive 3D Graphics*, 77–87. https://doi.org/10.1145/319120.319127.

[9. Warren, W.H., Kay, B.A., Zosh, W.D., Duchon, A.P. and Sahuc, S. (2001). Optic flow is used to control human walking. *Nature Neuroscience*, *4*(2), 213–216. https://doi.org/10.1038/84054.

[10] Cipresso, P., Giglioli, I.A.C., Raya, M.A. and Riva, G. (2018). The Past, Present, and Future of Virtual and Augmented Reality Research: A Network and Cluster Analysis of the Literature. *Frontiers in Psychology*, *9*. https://doi.org/10.3389/fpsyg.2018.02086.

[11] Anderson, J.R. (1990). *Cognitive psychology and its implications, 3rd ed* (pp. xvi, 519). W H Freeman/Times Books/ Henry Holt & Co.

[12] Gazzaniga, M.S. (2014). *Handbook of Cognitive Neuroscience*. Springer.

[13] Friston, K.J. (2009). Modalities, Modes, and Models in Functional Neuroimaging. *Science*, *326*(5951), 399–403. https://doi.org/10.1126/science.1174521.

[14] Friston, K.J. (2005). Models of Brain Function in Neuroimaging. *Annual Review of Psychology*, *56*(1), 57–87. https://doi.org/10.1146/annurev.psych.56.091103.070311.

[15. Sanchez-Vives, M.V. and Slater, M. (2005). From presence to consciousness through virtual reality. *Nature Reviews Neuroscience*, *6*(4), 332–339. https://doi.org/10.1038/nrn1651.

[16] Parsons, T.D. (2015). Virtual Reality for Enhanced Ecological Validity and Experimental Control in the Clinical, Affective and Social Neurosciences. *Frontiers in Human Neuroscience*, *9*. https://doi.org/10.3389/fnhum.2015.00660.

[17] Martin, D., Malpica, S., Gutierrez, D., Masia, B. and Serrano, A. (2022). Multimodality in VR: A Survey. *ACM Comput. Surv.*, *54*(10s), 216:1-216:36. https://doi.org/10.1145/3508361.

[18] Kuai, S.-G., Liang, Q., He, Y.-Y. and Wu, H.-N. (2021). Higher anxiety rating does not mean poor speech performance: Dissociation of the neural mechanisms of anticipation and delivery of public speaking. *Brain Imaging and Behavior*, *15*(4), 1934–1943. https://doi.org/10.1007/s11682-020-00387-3.

[19] Cha, H.-S., Choi, S.-J. and Im, C.-H. (2020). Real-Time Recognition of Facial Expressions Using Facial Electromyograms Recorded Around the Eyes for Social Virtual Reality Applications. *IEEE Access*, *8*, 62065–62075. https://doi.org/10.1109/ACCESS.2020.2983608.

[20] Adhanom, I.B., MacNeilage, P. and Folmer, E. (2023). Eye Tracking in Virtual Reality: A Broad Review of Applications and Challenges. *Virtual Reality*, *27*(2), 1481–1505. https://doi.org/10.1007/s10055-022-00738-z.

[21] Li, X., Fan, D., Feng, J., Lei, Y., Cheng, C. and Li, X. (2025). Systematic review of motion capture in virtual reality: Enhancing the precision of sports training. *Journal of Ambient Intelligence and Smart Environments*, *17*(1), 5–27. https://doi.org/10.3233/AIS-230198.

[22] Fajen, B.R., and Warren, W.H. (2003). Behavioral Dynamics of Steering, Obstacle Avoidance, and Route Selection. *Journal of Experimental Psychology: Human Perception and Performance*, *29*(2), 343–362. https://doi.org/10.1037/0096-1523.29.2.343.

[23] Warren, W.H. (2006). The dynamics of perception and action. *Psychological Review*, *113*(2), 358–389. https://doi.org/10.1037/0033-295X.113.2.358.

[24] Rio, K.W., Dachner, G.C. and Warren, W.H. (2018). Local interactions underlying collective motion in human crowds. *Proceedings of the Royal Society B: Biological Sciences*, *285*(20180611), 20180611. https://doi.org/10.1098/rspb.2018.0611.

[25] Baxter, B.A., and Warren, W.H. (2020). Route selection in barrier avoidance. *Gait and Posture*, *80*(December 2019), 192–198. https://doi.org/10.1016/j.gaitpost.2020.04.009.

[26] Warren, W.H., Rothman, D.B., Schnapp, B.H., and Ericson, J.D. (2017). Wormholes in virtual space: From cognitive maps to cognitive graphs. *Cognition*, *166*, 152–163. https://doi.org/10.1016/j.cognition.2017.05.020.

[27] Lenggenhager, B., Tadi, T., Metzinger, T. and Blanke, O. (2007). Video Ergo Sum: Manipulating Bodily Self-Consciousness. *Science*, *317*(5841), 1096–1099. https://doi.org/10.1126/science.1143439.

[28] Blanke, O. and Metzinger, T. (2009). Full-body illusions and minimal phenomenal selfhood. *Trends in Cognitive Sciences*, *13*(1), 7–13. https://doi.org/10.1016/j.tics.2008.10.003.

[29] Slater, M., Spanlang, B., Sanchez-Vives, M.V., and Blanke, O. (2010). First Person Experience of Body Transfer in Virtual Reality. *PLOS ONE*, *5*(5), e10564. https://doi.org/10.1371/journal.pone.0010564.

[30] Blanke, O. (2012). Multisensory brain mechanisms of bodily self-consciousness. *Nature Reviews Neuroscience*, *13*(8), 556–571. https://doi.org/10.1038/nrn3292.

[31] Sass, L.A. and Parnas, J. (2003). Schizophrenia, Consciousness, and the Self. *Schizophrenia Bulletin*, *29*(3), 427–444. https://doi.org/10.1093/oxfordjournals.schbul.a007017.

[32] Doeller, C.F., King, J.A. and Burgess, N. (2008). Parallel striatal and hippocampal systems for landmarks and boundaries in spatial memory. *Proceedings of the National Academy of Sciences*, *105*(15), 5915–5920. https://doi.org/10.1073/pnas.0801489105.

[33] Doeller, C.F., Barry, C. and Burgess, N. (2010). Evidence for grid cells in a human memory network. *Nature*, *463*(7281), 657–661. https://doi.org/10.1038/nature08704.

[34] Bellmund, J.L., Deuker, L., Navarro Schröder, T. and Doeller, C.F. (2016). Grid-cell representations in mental simulation. *eLife*, *5*, e17089. https://doi.org/10.7554/eLife.17089.

[35] Bellmund, J.L.S., Gärdenfors, P., Moser, E.I. and Doeller, C.F. (2018). Navigating cognition: Spatial codes for human thinking. *Science*, *362*(6415), eaat6766. https://doi.org/10.1126/science.aat6766.

[36] Pan, X. and Hamilton, A.F. de C. (2018). Why and how to use virtual reality to study human social interaction: The challenges of exploring a new research landscape. *British Journal of Psychology*, *109*(3), 395–417. https://doi.org/10.1111/bjop.12290.

[37] Slater, M. (2009). Place illusion and plausibility can lead to realistic behaviour in immersive virtual environments. *Philosophical Transactions of the Royal Society B: Biological Sciences*, *364*(1535), 3549–3557. https://doi.org/10.1098/rstb.2009.0138.

[38] Banakou, D., Groten, R. and Slater, M. (2013). Illusory ownership of a virtual child's body causes overestimation of object sizes and implicit attitude changes. *Proceedings of the National Academy of Sciences*, *110*(31), 12846–12851. https://doi.org/10.1073/pnas.1306779110.

[39] Maister, L., Slater, M., Sanchez-Vives, M.V. and Tsakiris, M. (2015). Changing bodies changes minds: Owning another body affects social cognition. *Trends in Cognitive Sciences*, *19*(1), 6–12. https://doi.org/10.1016/j.tics.2014.11.001.

[40] Peck, T.C., Seinfeld, S., Aglioti, S.M. and Slater, M. (2013). Putting yourself in the skin of a black avatar reduces implicit racial bias. *Consciousness and Cognition*, *22*(3), 779–787. https://doi.org/10.1016/j.concog.2013.04.016.

[41] Slater, M. and Sanchez-Vives, M.V. (2016). Enhancing Our Lives with Immersive Virtual Reality. *Frontiers in Robotics and AI*, *3*. https://doi.org/10.3389/frobt.2016.00074.

[42] Bailenson, J.N., Blascovich, J., Beall, A.C. and Loomis, J.M. (2001). Equilibrium Theory Revisited: Mutual Gaze and Personal Space in Virtual Environments. *Presence: Teleoperators and Virtual Environments*, *10*(6), 583–598. https://doi.org/10.1162/105474601753272844.

[43] Bailenson, J.N., Blascovich, J., Beall, A.C. and Loomis, J.M. (2003). Interpersonal Distances in Virtual Environments. *Personality and Social Psychology Bulletin*, *29*(7), 819–833. https://doi.org/10.1177/0146167203253270.

[44] Bailenson, J.N. and Yee, N. (2005). Digital Chameleons. *Psychological Science*, *16*(10), 814–819. https://doi.org/10.1111/j.1467-9280.2005.01619.x.

[45] Blascovich, J., Loomis, J., Beall, A., Andrew, C., Swinth, Kimberly R., Hoyt,Crystal L. and Bailenson, J.N. (2002). TARGET ARTICLE: Immersive Virtual Environment Technology as a

Methodological Tool for Social Psychology. *Psychological Inquiry*, *13*(2), 103–124. https://doi.org/10.1207/S15327965PLI1302_01.

[46] Jin, K., Wu, J., Zhang, R., Zhang, S., Wu, X., Wu, T., Gu, R. and Liu, C. (2024). Observing heroic behavior and its influencing factors in immersive virtual environments. *Proceedings of the National Academy of Sciences*, *121*(17), e2314590121. https://doi.org/10.1073/pnas.2314590121.

[47] Zhou, C., Han, M., Liang, Q., Hu, Y.F. and Kuai, S.G. (2019). A social interaction field model accurately identifies static and dynamic social groupings. *Nature Human Behaviour*, *3*(8), 847–855. https://doi.org/10.1038/s41562-019-0618-2.

[48] Zhou, C., Miao, M.C., Chen, X.R., Hu, Y.F., Chang, Q., Yan, M.Y. and Kuai, S.G. (2022). Human-behaviour-based social locomotion model improves the humanization of social robots. *Nature Machine Intelligence*. https://doi.org/10.1038/s42256-022-00542-z.

[49] Zhang, Y., Kinateder, M., Templeton, A. and Warren, W. H. (2025). A Virtual Reality Experiment on Visual and Auditory Guidance for Egress in Road Tunnel Fires. *Fire Technology*. https://doi.org/10.1007/s10694-024-01668-9.

[50] Höllerer, T., Interrante, V., Lecuyer, A., Suma, E., IEEE Computer Society. Technical Committee on Visualization and Graphics, IEEE Computer Society., & Institute of Electrical and Electronics Engineers. (n.d.). *2016 IEEE Virtual Reality (VR) : proceedings : Greenville, South Carolina, USA, 19-23 March 2016.*

[51] De Fino, M., Tavolare, R., Bernardini, G., Quagliarini, E. and Fatiguso, F. (2023). Boosting urban community resilience to multi-hazard scenarios in open spaces: A virtual reality – serious game training prototype for heat wave protection and earthquake response. Sustainable Cities and Society, 99. https://doi.org/10.1016/j.scs.2023.104847.

[52] Engelbrecht, H., Lindeman, R.W., and Hoermann, S. (2019). A SWOT Analysis of the Field of Virtual Reality for Firefighter Training. In *Frontiers in Robotics and AI* (Vol. 6). Frontiers Media S.A. https://doi.org/10.3389/frobt.2019.00101.

[53] Kinateder, M., Wirth, T.D., and Warren, W.H. (2018). *Crowd Dynamics in Virtual Reality*. 15–36. https://doi.org/10.1007/978-3-030-05129-7_2.

[54] Fan, C., Zhang, C., Yahja, A. and Mostafavi, A. (2021). Disaster City Digital Twin: A vision for integrating artificial and human intelligence for disaster management. *International Journal of Information Management*, *56*. https://doi.org/10.1016/j.ijinfomgt.2019.102049.

[55] Zhou, C., Miao, M.C., Chen, X.R., Hu, Y.F., Chang, Q., Yan, M.Y. and Kuai, S.G. (2022). Human-behaviour-based social locomotion model improves the humanization of social robots. *Nature Machine Intelligence*. https://doi.org/10.1038/s42256-022-00542-z.

2 Introduction to Tunnelling Engineering

Yi Rui[1*] and Yueyue Zhu[2]

With the compression of urban living spaces and the increasing population, the development and utilization of underground space play a crucial role in enhancing human living quality. Tunnel engineering occupies a significant position in the development of underground space. Currently, tunnel engineering has undergone long-term historical development and technological accumulation, resulting in a variety of tunnel usage types, cutting-edge construction techniques, and typical engineering cases. This chapter outlines the basic concepts of tunnel engineering by defining and categorizing tunnels and clarifying the history and development of tunnel engineering through the introduction of typical ancient and modern projects, as well as revolutionary technologies. Subsequently, it systematically introduces the construction techniques and key issues faced in tunnel engineering from three aspects: construction methods, key technologies, and risk management and prevention. Finally, it summarizes some classic engineering cases in the history of world tunnel construction and briefly introduces and analyzes some emerging applications of intelligent and digital technologies in recent years. This chapter's systematic introduction of the concepts, history, technology, and engineering of tunnel engineering lays the theoretical foundation and knowledge system for readers to understand the key focus of this book, namely, virtual reality technology in tunnel engineering.

1. Basic Concepts and Developmental History

1.1 Overview of Tunnel Engineering

Tunnel Definition

Tunnels are a form of human utilization of underground space, structures buried in the ground layers [1]. In 1970, the Tunnel Conference convened by the Organization for Economic Cooperation and Development comprehensively defined tunnels as: a strip-shaped cavity with a cross-sectional area greater than 2 m^2, constructed in a prescribed shape and size for a specific purpose underground by any means. Its core

[1,2] No.1239, Siping Road, Yangpu District, Shanghai; Department of Geotechnical Engineering, College of Civil Engineering, Tongji University, Shanghai 200092, China.

* Corresponding author: ruiyi@tongji.edu.cn

DOI: 10.1201/9781003514480-2

features include: (1) a clear purpose (such as transportation, water conservancy, etc.); (2) being located underground, constructed by artificial methods (such as excavation or drilling); (3) having a prescribed shape and size, with a cross-sectional area greater than 2 m^2.

A tunnel is a passage built in rock, soil, or underwater with entrances at both ends, used for vehicles, pedestrians, pipelines, cables, water flow, logistics, etc. The length of a railway tunnel is the distance between the outer surface of the portal wall at the entrance and exit and the elevation intersection of the track's top surface. The length of a highway tunnel is the distance between the outer surface of the portal wall at the entrance and exit and the point where the intersection of the road surface and the centerline of the road occurs.

Classification System

Currently, many different types of tunnels have been constructed in the project. Tunnels are classified into 7 categories as shown in Table 1, and a brief introduction is provided for each type of tunnel.

Table 1 Classification of tunnels.

No.	*Based on*	*Category*
1	Tunnel length	Short tunnel, medium tunnel, long tunnel, super long tunnel
2	Geological conditions of tunnels	Soil tunnel, rock tunnel
3	Tunnel cross-sectional area	Ultra-small cross-section tunnel, small cross-section tunnel, medium cross-section tunnel, large cross-section tunnel, extra-large cross-section tunnel
4	Excavation span of tunnels	Small span tunnel, medium span tunnel, large span tunnel, extra-large span tunnel
5	Tunnel burial depth	Shallow-buried tunnel, deep-buried tunnel
6	Tunnel purposes	Transportation tunnel, hydraulic tunnel, municipal tunnel, mining tunnel
7	Tunnel location	Mountain tunnel, underwater tunnel, urban tunnel

Classification according to tunnel use

1. Traffic tunnel

Traffic tunnels are a type of architectural structure, usually located underground in mountainous areas or cities, that connect two places for vehicles or pedestrians to pass through [2]. As shown in Fig. 1(a). Depending on different construction purposes and usage, traffic tunnels can be divided into several types, including highway tunnels, railway tunnels, water supply tunnels, subway tunnels, and underwater tunnels. Traffic tunnels are an important architectural structure, with different types serving various construction and usage goals, necessitating the study of various variables and concerns. When constructing traffic tunnels, it is necessary to strictly adhere to relevant regulations and standards to ensure their safety, reliability, and efficiency.

2. Water conveyance tunnel

The water tunnel is a tunnel excavated in the mountains or underground for water passage, as shown in Fig. 1(b). Water tunnels can be used for irrigation, power generation, flood discharge, water conveyance, construction diversion, and navigation [3]. An unpressurized tunnel has free water flowing through it; a pressurized tunnel has water throughout its cross-section and exerts a specific amount of water pressure on its wall. Water tunnels can be divided according to their functions into water diversion and water conveyance tunnels, diversion and flood discharge tunnels, tailwater tunnels, and sediment flushing tunnels . Water tunnels generally include three main parts: the inlet structure, tunnel body, and outlet structure.

(a) (b)

Fig. 1 (a) Traffic tunnel [4]; (b) Hydraulic tunnel [5].

3. Municipal tunnel

The municipal tunnel is built underground in the city to house various municipal facilities and underground pipelines. The number of underground pipelines for municipal facilities is rising due to the necessity for additional industrial development and the enhancement of city dwellers' quality of life. Municipal tunnels are divided into drainage tunnels, water supply tunnels, gas tunnels, heating and hot water pipeline tunnels, electrical and cable tunnels, and mixed tunnels according to their intended use. The tunnel can usually be constructed by the open-cut method, pipe jacking method, shield method, or mining method.

4. Mine tunnel

The mine tunnel is mainly used for mining and logistical transportation, which is the construction structure that passes roads or railways through mountains or underground in the mining area, as shown in Fig. 2(a) [6]. The construction of mine tunnels needs to consider factors such as geological conditions, mineral resources, and miner safety in the mining area, while also considering ventilation, lighting, drainage, and emergency evacuation. The construction of mine tunnels needs to strictly comply with relevant regulations and standards to ensure the safety and efficiency of mining operations.

Classified by the position of the tunnel

1. Mountain tunnel

Mountain tunnels refer to tunnels built under mountains or hills to shorten the distance and avoid steep slopes for vehicle passage. The construction of mountain tunnels, as shown in Fig. 2(b), is an important part of modern transportation development, promoting economic growth, strengthening connections, and enhancing communication between different regions. In addition, the construction of mountain tunnels must fully consider construction techniques and design requirements, and also consider its impact on the environment and ecology.

(a) (b)

Fig. 2 (a) Mine Tunnel [7]; (b) Mountain tunnel [8].

2. Underwater tunnel

A tunnel constructed beneath the sea or underwater to link two coasts or islands is referred to as an underwater tunnel. Its primary purposes are electricity transmission, water supply, communication, and transportation. They are usually used to connect two areas separated by water bodies, providing a more convenient and stable passage than bridges. The main construction methods used for building underwater tunnels are the diaphragm wall method, the compressed air caisson method, the shield method, and the immersed tube method. The main part of the underwater tunnel is located in the rock and soil layers beneath the river or seabed. It is constantly below the groundwater level, bearing the full water pressure from the water surface to the depth of the tunnel. Therefore, underwater tunnels face severe waterproofing problems from construction to operation.

3. Urban tunnel

Urban tunnels refer to buildings constructed underground in cities to accommodate the passage of vehicles through the city for railway needs. Urban tunnels face complex geological and environmental conditions during construction, requiring special attention to soil deformation and the protection of nearby buildings and underground pipelines.

Classified by the length of the tunnel

Due to the differences in the usage scenarios of tunnels, there are differences in the classification of tunnel lengths for railway tunnels and road tunnels [9]. The specific length classification standards for both are as follows.

1. Short tunnel

- Railway tunnel: with a total length of 500 m or less.
- Road tunnel: with a total length of 500 m or less.

This type of tunnel is quite common in urban traffic, mainly used to connect different areas of the city and provide convenient passage routes.

2. Middle tunnel

- Railway tunnel: with a total length of over 500 m to 3000 m, including 3000 m.
- Road tunnels: with a total length of over 500 m up to 1000 m, including 1000 m.

This type of tunnel is usually used to cross mountainous areas or connect two important transportation nodes, which is of great significance for improving transportation efficiency.

3. Long tunnel

- Railway tunnel: with a total length of over 3000 m up to 10000 m, including 10000 m.
- Road tunnel: with a total length of over 1000 m up to 3000 m, including 3000 m.

The function and effect of this kind of tunnel are similar to those of the central tunnel.

4. Extra-long tunnel

- Railway tunnel: over 10,000 m in total length.
- Road tunnel: over 3000 m in length.

These tunnels are usually used to cross complex geological environments, such as deep mountain valleys or underwater, and they are the most challenging in terms of engineering difficulty and technical requirements.

Classified by the cross-sectional area of the tunnel

According to the classification criteria of tunnel cross-sectional area size recommended by the International Tunneling Association (ITA), tunnels can be divided into 5 categories, as shown in Table 2.

Table 2 Tunnel classified by cross-sectional area.

No.	*Name*	*Classification*
1	Extra-large cross-section tunnel	Tunnel with a clear area larger than 100.0 m^2
2	Large cross-section tunnel	Tunnel with a clear area of 50.0 m^2~100.0 m^2
3	Medium cross-section tunnel	Tunnel with a clear area of 10.0 m^2~50.0 m^2
4	Small cross-section tunnel	Tunnel with a clear area of 3.0 m^2~10.0 m^2
5	Ultra-small cross-section tunnel	Tunnel with a clear area of less than 3.0 m^2

Tunnel burial depth

1. Shallow-buried tunnel

Shallow-buried tunnels refer to a type of tunnel with a small thickness of the covering layer at the top, where the supporting structure bears the soil pressure and has a low correlation with the burial depth, topography, and surface environment of the tunnel. As shown in Fig. 3(a). The specific criteria for determination vary depending on the type of engineering. The core characteristic of shallow buried tunnels in highway tunnels is that the supporting structure bears the soil pressure without significant influence from the burial depth, topography, or surface environment. In railway tunnels, the determination is based on the comprehensive assessment of the rock mass grade and burial depth. Urban subway tunnels use the span-depth ratio for assessment. In addition, some codes divide the tunnels based on the diameter multiple, where tunnels with a burial depth less than 2 times the tunnel diameter are called shallow buried tunnels [10].

2. Deep-buried tunnel

The deep-buried tunnel refers to the vertical distance from the top of the tunnel excavation section to the natural ground, with a small impact on the ground during excavation, and does not even affect the ground tunnel. As shown in Fig. 3(b). The depth is usually determined by whether a “natural arch” can be formed. The surrounding rock pressure of deep-buried tunnels mainly comes from the loosening pressure of the surrounding rock, with the value based on the height of the construction collapse. The stability of deep-buried tunnels is relatively high, with small deformation of the surrounding rock after excavation, and minimal impact on the ground [11,12].

(a) (b)

Fig. 3 (a) Shallow buried tunnel [4]; (b) Deep buried tunnel [4].

Geological conditions of the tunnel

1. Soil tunnel

Soil tunnels refer to tunnels built in soil layers, as shown in Fig. 4(a). These tunnels are usually constructed in weak geological environments, such as clay or sandy soil, and special attention needs to be paid to the stability of the strata during their construction. The construction of soil tunnels is relatively difficult due to the need to deal with possible collapses and water seepage issues.

2. Rock tunnel

Rock tunnels are tunnels constructed in rock formations, as shown in Fig. 4(b). These tunnels are usually built in geological conditions where the rock is stable and of high strength. Precision blasting techniques and effective support measures are required during the construction of rock tunnels to ensure safety. They serve as important transportation means connecting two locations, and are widely used in transportation construction projects such as highways, railways, and subways.

(a) (b)

Fig. 4 (a) Soil tunnel [4]; (b) Rock tunnel [4].

Classify according to the excavation span of tunnels

1. Small-span tunnel

Small-span tunnels generally refer to tunnels with spans below 10m, characterized by low construction difficulty, and usually do not require specialized construction techniques, such as road underpass tunnels, municipal pipeline tunnels, etc.

2. Medium-span tunnel

Medium-span tunnels typically refer to tunnels with spans ranging from 10 m to 30 m, characterized by increasing construction difficulty and technical requirements, taking into account various factors such as geological conditions, environmental factors, etc. Such tunnels include railway tunnels, highway tunnels, etc.

3. Large-span tunnel

The large-span tunnel usually refers to tunnels with a span between 30 m and 100 m, which are characterized by difficult construction, requiring specialized construction techniques and equipment, such as spiral tunnels, long railway tunnels, long highway tunnels, etc.

4. Extra-large span tunnel

Extra-large span tunnels typically refer to tunnels with spans of over 100m, characterized by extremely difficult construction and requiring advanced construction techniques and equipment, such as underwater tunnels and high-speed railway tunnels.

Characteristics and Components of Tunnels

1. Characteristics of a tunnel

(1) Tunnels are an effective method for transportation routes to pass through natural barriers.
(2) The geological conditions in the area where the crossing is located are complex and constantly changing, requiring adjustments to engineering positioning, design, and construction methods.
(3) The construction site is narrow, with restrictions on labor and equipment, and high requirements for industrialization and mechanization of construction.
(4) Expensive to build.

2. Composition of the tunnel

The main structure of the tunnel consists of the tunnel body and the tunnel portal. At the tunnel entrance or exit section, if the surface rock and soil on the slope are severely weathered and prone to collapse, the tunnel body will be extended, or a cut-and-cover tunnel entrance will be added to ensure driving safety.

The ancillary buildings of the tunnel include: pedestrian walkways (or escape tunnels) and drainage facilities. For long or very long tunnels, there are also ventilation ducts, ventilation rooms, power supply, lighting, signaling, fire protection, communication, and rescue, as well as other auxiliary facilities for measurement, monitoring, etc.

1.2 Tunnel Engineering: History and Development

Ancient Tunnels

Humans have long known how to use natural caves as dwellings. When society developed tools for digging, human-made tunnels were discovered. During this period, excavation relied mainly on the "fire burning method" and primitive tools such as iron hammers and steel drills, making construction very difficult. Tunnel construction was still in an experiential stage, all based on the long-term experience accumulated by builders, with no theoretical guidance.

Typical ancient tunnel cases include:

(1) Between approximately 2180 and 2160 BC, the pedestrian tunnel built under the Euphrates River in the city of Babylon is the earliest known tunnel used for transportation. The tunnel is a brick structure.

(2) The Poisilip Tunnel, built between Naples and Pozzuoli in 36 BC, might have been the largest ancient tunnel structure. The tunnel was excavated in tuff without a lining.

(3) The Shimen Tunnel, located in Hanzhong County, Shaanxi Province, China, was built in 66 AD. This is China's earliest mountain-piercing tunnel, constructed using the "fire-burning and water-pouring" method, and also the world's earliest artificial traffic tunnel. The width and height of the tunnel's inner wall are both over 4 meters.

Development of Modern Tunnels

Tunnel construction in modern times originated in the canal era. With the use of explosives, the excavation efficiency was greatly improved, marking the beginning of modern tunnel engineering. Among them, the Malpas Canal Tunnel in France may be the earliest tunnel excavated with gunpowder. The tunnel was built between 1666 and 1681, with a total length of 157 m. Around 1830, there were more and more railway tunnels, starting from the economically developed countries in Europe, followed by the United States and Japan after the Meiji Restoration.

1. Explosion drilling technique

In the 20th century, the rotary drilling method was adopted, and the rock drill was invented. The blasting technology has also developed from the traditional method of igniting with fuses and detonators to millisecond detonation with electrical detonators, achieving smooth blasting and pre-split blasting. With the continuous deepening of people's understanding of tunnel engineering and the accumulation of engineering experience, many mature construction techniques and methods have been formed, such as the New Austrian Tunnelling Method, the New Norwegian Method, and the Norwegian Method, etc.

2. The origin and development of shield tunneling

In 1847, the shield tunneling machine was invented, which is a specialized equipment with a shield. The shield tunneling machine uses the pre-installed lining blocks at the rear as a fulcrum to advance, cutting through the soil with a cutter head, while removing the excavated soil and assembling the prefabricated concrete lining blocks behind. The first tunnel excavated using shield tunneling technology was the Thames Tunnel. Subsequently, in 1869, the British engineer James Henry Greathead first developed the circular shield tunnel. In 1886, the compressed air method was first combined with shield tunneling in underground construction in London. The first slurry shield tunneling machine, which uses a cutting head to excavate soil and hydraulic slurry removal, was put into use in Japan in 1967. In 1963, the Japanese company Sato Kogyo first developed the earth pressure balanced shield tunneling machine, which was first used in tunnel construction projects in Tokyo, Japan, in 1974.

3. Origin and development of the Tunnel Boring Machine (TBM) method

TBM, also known as Full Face Rock Tunnel Boring Machine, is used for tunnel excavation in medium and hard rock formations [13]. This method was first successfully trialed in 1883. Hard rock TBMs use the rolling cutters on the rotating cutter head to squeeze, shear, and break the rock. The rock chips are picked up by the bucket teeth on the rotating cutter head, dropped onto the main machine conveyor belt for rearward transportation, and then transported to the outside of the tunnel via a muck car or a tunnel continuous conveyor belt. Currently, three types of TBMs have been developed: shieldless TBM, single shield TBM, and double shield TBM, suitable for different rock conditions and excavation methods.

4. Origin and development of the pneumatic caisson method

The immersion tube method is a construction method for building tunnels at the bottom of the water. The immersed tube tunnel is a method of constructing an underwater tunnel by floating several prefabricated sections to the site at sea (or river) level, and sequentially sinking and installing them in a well-prepared trench [14]. The first successful construction using the immersion tube method was the Shirley Drainage Tunnel in Boston, United States, in 1894.

Modern Milestone Projects

1. English Channel Tunnel

The tunnel, which was completed and opened to traffic in 1994, took more than 8 years to build and cost approximately 10 billion pounds. The tunnel was constructed using the TBM (Tunnel Boring Machine) method and has a total length of 50.5 km.

2. Qinghai-Tibet subsea tunnel

Constructed by the Japanese in 1988, it is 53.85 km long, with 23.3 km under the sea. After completion, it became the longest underwater tunnel in the world.

3. Gotthard Base Tunnel in Switzerland

The tunnel, which was opened to traffic in 2016 and has a total length of 57.1 km, set a new record for the length of traffic tunnels.

4. Hong Kong-Zhuhai-Macau Bridge immersed tube tunnel

The undersea section, with a total length of 5.6 km, was opened to traffic in 2018, making it the world's longest road-immersed tube tunnel.

Development Trend of Tunnel Engineering

In the 21st century, with the rapid development of technology and the global population exceeding 6 billion, the scarcity of land resources has become an increasingly common concern among countries. The utilization of underground space will also be given more attention. Looking ahead at the development trends of tunnel engineering, it mainly revolves around several aspects:

(1) Tunnel construction mechanization, including high-precision collaborative positioning of equipment clusters, long-distance control transmission, and rapid and fine reconstruction of equipment operating scenarios.
(2) Tunnel structure prefabrication, with key technologies such as standardized selection and production control of structural components and full-life-cycle manufacturing, construction, control, and traceability.
(3) Tunnel construction informatization and intelligence, including tunnel structure health monitoring identification systems, intelligent traffic safety risk control systems, and a transportation intelligent big data sharing platform.

2. Construction Techniques and Key Issues

2.1 Tunnel Construction Methods

Tunnel construction is the general term for construction methods, construction techniques, and construction management involved in building tunnels and underground chambers. The tunnel construction process usually includes: excavating soil and rock in the formation of a tunnel that conforms to the designed cross-section, carrying out necessary support and lining, controlling the deformation of the surrounding rock of the tunnel, and ensuring the safety of tunnel construction and long-term use. According to the different geological conditions through which the tunnel passes and the current development of engineering technology, tunnel construction methods mainly include the following types, each of which will be detailed.

Cut-and-Cover Method

The cut-and-cover method refers to a construction method in which the ground is first excavated, the lining is built in open situations, and then underground engineering construction is carried out by covering and backfilling. This method is mostly used for shallow tunnels and is the most basic and commonly used construction method in soft soil underground engineering. The advantages of the cut-and-cover method are simple construction technology, fast speed, economy, good overall structural stress conditions, etc. When there are no conditions limiting ground traffic and the environment, it should be the preferred method. However, its disadvantages are also obvious, such as longer interruption of traffic time, noise, and vibration. The application of the cut-and-cover method is related to many factors, such as environmental conditions around the construction, engineering geology, hydrogeological conditions, burial depth of structures, technical and economic indicators, etc. Therefore, when choosing the cut-and-cover method to construct various underground engineering projects, various factors should be comprehensively considered.

Mining Method (Drilling and Blasting Method)

The mining method refers to the construction method of tunnels by excavating underground tunnels [15]. The mining method is named after its use in ore mining, and it includes traditional methods and the New Austrian Tunneling Method (NATM). Its basic principle is that after tunnel excavation and blasting, the rock mass fractures and forms a loosened state, which can collapse at any time. Based on this loosening load theory, the construction method is to excavate in sections one by one and require support while excavating for safety, resulting in complex support structures and high consumption of timber. With the introduction of rock bolting support, the number of sections can be reduced, leading to the development of the NATM.

Shield Tunneling Method

The shield tunneling method is a fully mechanized construction method in underground excavation. It involves the advancement of shield machines in the ground, using the shield casing and segmental lining to support the surrounding rock and prevent collapses into the tunnel. At the same time, the cutting device excavates the soil in front of the face, which is then transported out of the tunnel by excavation machinery, pressurized at the rear using jacks, and assembled with prefabricated concrete segments to form the tunnel structure. The shield machine was invented in 1847 and is a specialized equipment with a shield. There are two main methods of excavating the shield machine: mechanical and hydraulic, with hydraulic being more common. In a hydraulic shield machine, there is a sealing chamber filled with bentonite slurry at the face. The bentonite slurry is used for balancing earth and groundwater pressures and as a medium for conveying excavated soil.

TBM Method

The Tunnel Boring Machine (TBM) is a construction method that uses a full-face tunneling machine to carry out construction processes such as excavation, support, and muck removal in parallel and continuously, with the advantages of fast tunneling speed, environmental friendliness, and high comprehensive benefits. The full-face tunneling machine, which integrates excavation, support, muck removal, and other construction processes in parallel and continuous operation, is a factory-like assembly line tunnel construction equipment integrating mechanical, electrical, hydraulic, optical, and pneumatic systems, capable of achieving the construction of complex geological and topographical deep-buried long tunnels that are difficult to realize using traditional drilling and blasting methods.

Special Construction Techniques

In addition to the commonly used tunnel construction methods mentioned above, there are also some special construction methods that can be used for tunnel engineering. For example:

(1) Immersed tube method. It involves prefabricated tunnel sections being cast on land and then submerged into an underwater trench, assembled into a whole. This method is suitable for underwater tunnels across rivers or seas.
(2) Freezing method. It injects brine at ground temperature into the formation through freezing pipes to freeze the soil and create a temporary water-stop curtain. This method is suitable for highly unstable formations such as water-rich sand layers, silt, etc.
(3) Jacked box method. It uses hydraulic jacks to push prefabricated box sections into the ground, excavating while pushing. This method is suitable for municipal pipelines and short-distance crossings under railways or buildings.
(4) Pipe roof method. Steel pipes are first driven into the tunnel roof to form a support structure, followed by internal excavation. This method is suitable for shallow-buried excavation in loose formations. The selection of these special construction methods needs to comprehensively consider geological conditions, construction period, costs, environmental protection requirements, and other factors. With technological advancements, the integration of digital tools such as BIM modeling and robotic excavation is driving tunnel construction toward a safer and more efficient direction.

2.2 Construction Key Technologies

Advanced Geological Forecasting

1. Definition of advanced geological forecasting

Using drilling and modern geophysical methods, the geological conditions in front of the excavation face of underground engineering projects such as tunnels, adits, and underground chambers are explored. The aim is to understand the structure, properties, and conditions of the rock and soil mass in front of the construction site, as well as information on the presence of groundwater, gas, geological stress, etc. This information is crucial for providing guidance for further construction, to avoid geological disasters such as water inrush, gas outbursts, rock bursts, and large deformations during construction and operation, ensuring the safety and smooth progress of the construction.

2. Purpose of the advanced geological forecast

(1) Further investigate the geological and hydrogeological conditions ahead of the tunnel excavation work face to guide the smooth progress of the construction work.
(2) Reduce the probability and degree of damage caused by geological disasters.
(3) To provide a geological basis for optimizing engineering design.
(4) Provide geological data for the compilation of completion documents.

3. Main methods and applicable conditions of advanced geological forecasting

The advanced prediction methods of tunnel geology mainly include: geological investigation method, geophysical method, advanced drilling method, advanced pilot tunnel method, hydraulic combined measurement method, TSP, TGP method, or TRT method. The main applicable conditions are as follows:

(1) The geological survey method is applicable to advance the geological prediction of tunnels under various geological conditions, and the survey content should include supplementary geological surveys on the tunnel surface and geological surveys inside the tunnel.
(2) The geophysical method is suitable for advanced geological prediction in long, extra-long tunnels or geologically complex tunnels, and the main methods include the elastic wave reflection method, geological radar method, infrared detection method, transient electromagnetic method, and high-resolution direct current method.
(3) Rich water structures, rich water karst development areas, coal-bearing or oil and gas strata, gas development zones, goaf areas, as well as major geophysical anomaly areas and other geologically complex tunnels and underwater tunnels must use advanced drilling methods to forecast and evaluate the geological conditions ahead.
(4) The advanced pilot tunneling method can be carried out by parallel advanced pilot tunneling method and in-tunnel advanced pilot tunneling method, and two parallel tunnels can predict the geological conditions of the tunnel to be excavated later based on the tunnel excavated in advance.

4. Main contents of the advanced geological forecast

The main contents of advanced geological prediction in tunnel construction:

(1) Location, scale, and nature of faults and their impact zones.
(2) The position, scale, and nature of the weak interlayer.
(3) Location, scale, and characteristics of karst.
(4) Position of interfaces with varying rock types and surrounding rock grades.
(5) The possible locations and scale of engineering geological hazards.
(6) The location, scale, and nature of aqueous structures.

(7) Geological methods survey, geophysical methods, drilling methods, and advanced borehole prediction methods are currently commonly used for advanced geological prediction and forecasting.

Rock Mass Stability Analysis

Tunnel engineering is a complex and important engineering project, with one key factor being the stability of the tunnel surrounding the rock mass [16]. The stability of the surrounding rock mass plays a crucial role in the safety and sustainable operation of tunnels. Therefore, the analysis of rock mass stability in tunnel engineering has become a challenging problem that engineers need to study and solve.

The stability analysis of surrounding rock in tunnel engineering can be divided into two aspects: rock mechanics analysis and numerical simulation. Rock mechanics analysis refers to the experimental testing of the physical and mechanical properties of tunnel surrounding rock through field exploration and sampling, and understanding the strength, deformation characteristics, and failure properties of the surrounding rock through theoretical calculation and analysis. This can provide key parameters and references for tunnel design. When conducting rock mechanics analysis, it is necessary to sample the surrounding rock first. By observing the rock cores and geological surfaces, basic information such as color, structure, and rock type of the surrounding rock can be obtained. Then, rock engineering mechanical tests, such as tensile tests, compressive tests, etc., are used to determine the strength and deformation characteristics of the surrounding rock. In addition, uniaxial and triaxial shear tests are needed to assess the shear strength of the rock. These experimental data can provide a foundation for subsequent numerical simulations.

Numerical simulation is used to simulate the deformation and damage of the surrounding rock in tunnel construction and operation processes using computers. Through numerical simulation, a comprehensive and accurate analysis and prediction of the stability of the surrounding rock can be conducted. In numerical simulation, the finite element method is mainly used for calculations. Firstly, it is necessary to establish the material model and boundary conditions of the surrounding rock based on the experimental data obtained from rock mechanics analysis. Then, the tunnel is modeled, and the rock material model is applied to the simulation. Finally, loads are applied to the surrounding rock, and the deformation and damage of the rock are simulated by computer.

In conclusion, the stability analysis of surrounding rock in tunnel engineering is an important and complex task. It involves both experimental testing and analysis of rock mechanics, as well as the use of numerical simulation methods to simulate and calculate the construction and operation of tunnels. Through the comprehensive study of rock mechanics analysis and numerical simulation, engineers can fully understand the stability of the surrounding rock, predict the deformation and failure of the surrounding rock, and develop corresponding construction and safety measures. Only by ensuring the stability of the surrounding rock can the safe operation of the tunnel be guaranteed.

Groundwater Treatment

The function requirements for waterproofing in tunnel engineering are to use effective and reliable waterproofing materials and technical measures to ensure that the tunnel is free from water intrusion and leakage, to protect the tunnel to have a good, safe operating environment, operating conditions, and service life. By adopting the principle of "combination of draining, isolating, and blocking, tailor-made solutions, comprehensive management," the purpose of reliable and economically reasonable waterproofing is achieved.

1. Drainage measures

Taking drainage measures to lower the groundwater level, or diverting and draining surface water, and artificially setting up drainage systems to drain groundwater through tunnels. Common structural drainage facilities include blind ditches, drainage holes, and drainage ditches. During construction, construction drainage should be considered, including external drainage and internal drainage. In addition, construction drainage should be combined with structural drainage.

2. Water cooling measures

Using impermeable or weakly permeable materials to isolate groundwater outside the building space. In addition to external waterproofing, also known as positive waterproofing, external waterproofing can withstand the hydrostatic pressure of groundwater and transmit it to the structure. Internal waterproofing on the inner side of the structure, also known as negative waterproofing, has a weaker ability to withstand water pressure but can act as a barrier to capillary water and gaseous water. It is not suitable for use when the hydrostatic pressure is high. Structural waterproofing involves using the structure itself as a waterproof layer.

3. Measures for blocking water

Water blocking measures refer to injecting waterproof materials into the damaged part (voids, cracks, etc.) and its vicinity when the waterproof structure is damaged and leaks water, playing a certain repair role, so it is often called water blocking. In addition, sealing measures should be taken at the joints of prefabricated components, which is actually a kind of water-blocking method.

Ventilation and Lighting

Tunnel engineering, as an important part of modern transportation construction, bears the task of vehicle and pedestrian passage [17]. Ensuring traffic safety, tunnel lighting, and ventilation facilities are crucial. Below, the respective requirements in tunnel engineering will be introduced from the aspects of ventilation and lighting.

1. Ventilation requirements

(1) Ventilation Supply: In underground tunnels, Yingyin ensures an adequate supply of fresh air to provide good air quality and alleviate the pollution

caused by traffic emissions. For urban tunnels, the supply of fresh air should meet the requirements of ventilation and air exchange, ensuring that the ventilation volume meets the specified standards.

(2) Exhaust Gas Treatment: The exhaust gas in the tunnel is generated by vehicle emissions and other combustion emissions. To ensure the cleanliness of the air in the tunnel and the smooth operation of vehicles, corresponding ventilation measures need to be taken to timely remove the exhaust gas.

(3) Smoke Emission: In the event of a fire or other emergencies, the tunnel's ventilation system should be able to timely and effectively discharge smoke to ensure a safe evacuation pathway for personnel.

(4) Ventilation equipment: The tunnel ventilation system equipment includes ventilators, fans, smoke exhaust fans, etc. This equipment must undergo strict inspection and testing to ensure stable and reliable operation.

2. Lighting requirements

(1) Intensity of illumination: According to the relevant national standards, tunnels must ensure sufficient intensity of illumination to provide good visual conditions and ensure the safety of drivers and pedestrians. Generally speaking, the average illuminance of urban tunnels should be between 100 and 200 lux, while the illuminance of long tunnels should not be less than 50 lux.

(2) Lighting equipment: The lighting equipment commonly used in tunnels mainly includes fluorescent lamps and LED lamps. These lamps have the characteristics of high efficiency, durability, and stability, providing sufficient light intensity.

(3) Lighting layout: The lighting layout in the tunnel should be reasonable to ensure uniform illumination throughout the tunnel, avoiding uneven light intensity and shadow areas. Generally speaking, the lighting fixtures in the tunnel should be in a state of continuous and uniform illumination to provide a good visual effect.

(4) Lighting Control: The lighting system of the tunnel should have intelligent control functions and be able to adjust the brightness of the lighting according to the changes in external natural light, to achieve energy-saving purposes.

2.3 *Tunnel Risk Management and Prevention*

Tunnels are artificial structures built underground. Tunnel engineering projects have a large scale, long construction periods, high investment, high concealment, and uncertain risks. During the construction and operation of tunnel projects, various potential uncertainties may arise, which may lead to damage to the engineering itself, casualties, and property damage [18]. In order to improve the safety and reliability of tunnel engineering, safeguard public lives and property, and enhance the level of tunnel construction, it is necessary to strictly manage and prevent the risks of tunnel engineering.

Identification of Risk Sources

In tunnel engineering construction, there are many factors that affect its construction safety, which are random and uncertain. Generally speaking, the types of accidents that occur in tunnel construction mainly include collapse, rockfall or support object impact, blasting, vehicle injury, lifting equipment injury, gas explosion, powder explosion, electric shock, fire, scalding and drowning, poisoning and suffocation, and other injuries. However, different construction areas require different tunnel construction techniques due to different geological and environmental characteristics, so there is no fixed experience or construction mode for identifying the sources of risk in construction. The sources of risk vary, and new sources constantly emerge. According to statistics on the types of accidents in tunnel construction over the years, the most common types of accidents are collapses, followed by explosions, poisoning and suffocation, and rockfall or support object incidents. Collapse accidents in tunnel construction are mainly influenced by terrain and climatic conditions, as well as the complex geological environment and narrow working space during tunnel construction. In addition, there are many uncertain factors in tunnel construction, which increase the safety risks. However, excluding geological factors, the main causes of collapse accidents during construction are due to improper arching, excavation not following requirements, insufficient safety distances, negligence of construction personnel, or failure to follow operation norms. Collapse accidents can cause serious injuries, even endanger lives, and require special attention and prevention.

Risk Management System

1. Organization and responsibilities

The tunnel engineering project needs to establish a safety risk management working group, which is responsible for the safety risk management of tunnel engineering. The main responsibilities of this group are:

(1) Responsible for identifying, assessing, and controlling safety risks in tunnel engineering.
(2) Develop safety risk management systems and operating procedures.
(3) Organize safety risk training and education.
(4) Supervise and inspect the implementation of safety risk control measures.
(5) Handle behaviors that violate safety risk management systems.

2. Identification and assessment of safety risks

Tunnel engineering safety risk identification includes three aspects:

(1) Risk identification for tunnel engineering design, construction, operation, etc.
(2) Risk identification for geological conditions, construction methods, mechanical equipment, personnel quality, etc.
(3) Risk identification for factors that may cause accidents.

Tunnel engineering safety risk assessment includes two aspects:

(1) Evaluating the severity and probability of identified risks.
(2) Categorizing risks into four levels: major risk, significant risk, general risk, and low risk according to the assessment results.

3. Safety risk training and education

The tunnel engineering project department regularly organizes safety risk training to enhance employees' safety awareness and skills. Safety risk training is provided for new employees, transferred employees, and personnel engaged in special operations. Special safety training is conducted for employees engaged in high-risk operations.

4. Supervision, inspection, rewards, and punishments

The tunnel engineering project department regularly conducts safety risk inspections to ensure that all measures are properly implemented. Any behavior that violates the safety risk management system will be dealt with according to law and regulations. Units and individuals who perform outstandingly in safety risk control work will be commended and rewarded.

Risk Prevention Measures

During tunnel construction, there are multiple sources of risk . The specific preventive measures are introduced from the aspects of geology, technology, equipment, personnel, and environment [19].

1. Geological risk prevention measures

First, conduct a detailed survey to understand the geological conditions of tunnel engineering, including rock types, geological structures, groundwater conditions, etc., to provide a scientific basis for construction. Based on the survey results, develop corresponding risk response plans, including predicting potential geological problems, implementing preventive measures, etc. During the construction process, carry out geological monitoring to timely grasp the geological changes inside the tunnel, and take corresponding measures.

2. Measures for preventing technical risks

Based on the actual situation of tunnel engineering, a scientific and reasonable technical solution is formulated, including tunnel design, construction methods, and support methods. Before construction, technical disclosure work should be done for the construction personnel, so that each worker understands and masters the relevant construction technology. During the construction process, regular technical training should be conducted for the workers to improve their skill level and reduce technical risks.

3. Equipment risk prevention

(1) In terms of equipment selection, suitable construction equipment such as excavators, drilling rigs, and loaders should be chosen according to the needs

of tunnel engineering. Establish an equipment protection system, regularly inspect, maintain, and repair equipment to ensure normal operation.

(2) Establish an equipment protection system, regularly inspect, maintain, and repair equipment to ensure normal operation.

(3) Equipment operation training is provided to equipment operators to enhance their operational skills and prevent risks caused by improper operation.

4. Personnel risk prevention measures

(1) Establish a strict personnel management system, and strictly manage construction personnel to ensure they comply with relevant regulations and requirements.

(2) Provide safety training for construction workers to enhance their safety awareness and reduce risks caused by improper operations.

(3) Provide legal labor protection equipment for construction workers, such as safety helmets, protective clothing, gloves, etc., to reduce the possibility of personnel injury.

5. Measures for preventing environmental risks

First, conduct an environmental impact assessment before tunnel construction to understand the potential impacts on the environment during the construction process. Based on the assessment results, develop corresponding environmental protection measures, including measures to reduce noise, dust pollution, etc. Strictly implement environmental protection measures during the construction process to ensure that the construction of the tunnel complies with relevant environmental protection regulations.

3. Typical Cases and Modern Technology

3.1 Classic Engineering Cases

Qinling Zhongnanshan Mountain Highway Tunnel

The Qinling Zhongnanshan Highway Tunnel is the world's longest two-tube unidirectional highway tunnel. It runs from Qingcha in the Chang'an District of Xi'an in the north to Yingpan Town, Zhashui County, under Shaanxi Province in the south, with a total length of 18.02 kilometers. Designed for a speed of 80 kilometers per hour, it takes only 15 minutes to drive through the Qinling Mountains, the natural boundary between northern and southern China. This "world's best" was entirely designed and constructed by Chinese people independently, and the design embodies a humanistic concept: special light strips are set in the tunnel, presenting scenes of "blue sky," "white clouds," and "rainbows" through different lights and projection patterns, giving drivers and passengers the feeling of being outdoors and helping to alleviate driving and riding fatigue.

Taiwan Shiqiuling Tunnel

The Taiwan Shiqiuling Tunnel in Taiwan is the earliest mountain railway tunnel in China. In June 1887, the Keelung-Taipei-Hsinchu Railway was constructed from Taipei to Keelung, with a length of 32.186 kilometers. The following year, it was extended from Taipei to the south to Hsinchu, with a length of 67.591 kilometers. In the section from Taipei to Keelung, the railway had to cross the Lion Ball Ridge, requiring the excavation of a 261-meter-long tunnel. The tunnel passes through shale, sandstone, and clay layers, with a maximum burial depth of 61 meters. Due to limited funds and technology at that time, the construction had to be done with simplification, combined with the lack of safety facilities, leading to multiple collapses during construction. Fortunately, when the tunnel was finally excavated, the phenomenon of "sleeve piercing" (misalignment) did not occur, but the elevation at the northern end was unexpectedly 4.27 meters higher than the southern end. Nevertheless, the construction of this tunnel has provided future generations with invaluable construction experience.

The Lærdal Tunnel in Norway

The Lærdal Tunnel in Norway is the world's longest road tunnel with a length of about 24.5 kilometers. It is located in central Norway, running from Lærdal in the east to Aurland in the west, serving as a crucial link between the capital, Oslo, and the second largest city, Bergen. Previously, vehicles traveling between Oslo and Bergen had to take a three-hour ferry across the Sognefjord at Lærdal and navigate steep mountain roads between Lærdal and Aurland, which were closed during the winter due to ice. With the opening of the Lærdal Tunnel, travel time between the two cities was reduced from 14 hours to 7 hours, and vehicles could pass through the tunnel even in winter. The design capacity of the Lærdal Tunnel is 400 vehicles per hour, but due to the low population in Norway, only 1000 cars pass through the tunnel each day, which is just one-tenth of its design capacity.

Saint Gotthard Tunnel

The Saint Gotthard Tunnel in Switzerland is the world's longest road tunnel, measuring 16.32 kilometers. The tunnel passes through the Saint Gotthard Peak in the southeastern Alps of Zurich, serving as an important link between the north and south main roads in Switzerland's national highway system. During the design and construction of the tunnel, numerous technical challenges were encountered. The tunnel had to pass through various geological formations such as granite, gneiss, and schist, with the worst sections consisting of weathered gravel and talc. To facilitate ventilation, drainage, and chipping removal, parallel galleries were excavated 30 meters from one side of the tunnel, with transverse connecting passages every 250 meters linking the galleries to the tunnel. These connecting passages could also be used for the construction of a second tunnel in the future. In terms of fire protection design, the top of the tunnel was divided into inlet and outlet air passages, allowing for compartmentalized ventilation in case of a fire without affecting other ventilation zones. Computer-controlled stations were also placed at both ends of the tunnel to monitor and control traffic flow, as well as provide timely warnings for changes in air cleanliness, fires, and lighting brightness.

3.2 Intelligent and Digital Technology

BIM Technology

BIM technology, namely Building Information Modeling, is a digitized management tool based on three-dimensional digital technology [20]. Currently, this technology is widely used in various industries such as construction, road and bridge, and pipeline engineering. Sharing and transmitting data in the process of engineering design, construction, and management enables all parties involved in the project to more accurately understand the building information, thereby improving production efficiency and reducing costs.

The core of BIM technology lies in its information-sharing and management capabilities [21]. By creating a 3D model containing all tunnel information, participants can access real-time project progress, construction safety plans, and related data. This transparency enables team members to quickly respond to changes and ensure the project is completed on time. Against the backdrop of continued economic development, the construction steel industry is increasingly stringent in controlling project costs. The introduction of BIM technology not only alleviates the industry's demand for efficient project cost management but also enhances data analysis and quantity surveying capabilities. The application of this technology marks a profound transformation in the field of project cost management, driving progress across the entire industry.

New Materials and New Construction Methods

Tunnel engineering, as an important field of infrastructure construction, has seen the emergence of innovative materials and new construction methods with the continuous advancement of technology. These technologies have significantly improved construction efficiency, safety, and environmental friendliness.

1. Application of new materials

(1) High-performance concrete: High-performance concrete significantly improves compressive strength, impermeability, and durability by incorporating silica fume, fly ash, or nanomaterials [22]. This material can be used for tunnel linings to withstand high water pressure and corrosive environments, offering advantages such as reducing lining thickness and extending structural life.

(2) Fiber reinforced composites: Fiber-reinforced composite materials use materials such as carbon fiber and glass fiber to replace traditional steel bars, used for strengthening tunnel linings or temporary supports, solving

the problem of steel bar corrosion [23]. This material has the advantages of being lightweight, high-strength , and corrosion-resistant , making it suitable for complex geological conditions.

(3) Geo-polymer grouting material: The polymer grouting material is made from industrial waste slag (such as slag and fly ash) as raw materials, which forms a cementitious material through an alkali-induced reaction [24]. This material is commonly used for tunnel rock grouting reinforcement and can quickly close cracks.

(4) Polymeric waterproof materials: The main types of polymer waterproofing materials are: bentonite waterproof blankets, EVA/ECB waterproof board, and non-curing rubber asphalt coating. These materials have high extensibility, crack resistance, and adaptability to tunnel deformation and can ensure long-term waterproofing effects [25].

(5) Nanomodified materials: The typical applications of this material include: increasing the density of concrete with nano-silica and enhancing the corrosion resistance of steel with graphene coatings. Currently, this intelligent material (such as self-healing concrete) is still in the experimental stage.

2. Innovative construction methods

(1) Intelligent shield tunneling technology: The shield tunneling machine is equipped with sensors to monitor parameters such as cutter head torque, earth pressure, temperature, etc., and optimize the excavation parameters through big data analysis to predict geological risks (such as rock burst, water inflow, etc.).

(2) The digital upgrade of Neo Azeroth: The combination of BIM and 3D laser scanning in New Aufo monitors the deformation of surrounding rock in real-time, dynamically adjusts the support parameters based on monitoring data, and improves safety and economy.

(3) 3D printing lining technology: 3D printing lining technology uses robotic arms to spray concrete or composite materials, directly printing tunnel lining structures. This method can reduce the use of formwork for lining construction and achieve rapid construction of complex structures [26].

Artificial Intelligence Applications

In recent years, with the development of artificial intelligence technology, its application in tunnel detection has gradually become a hot research topic [27]. Currently, significant progress has been made in the application of artificial intelligence technology in tunnel detection, including computer vision, deep learning , and sensor technology with data fusion. Computer vision technology can automatically monitor tunnel structures and abnormal situations through methods such as image processing, object detection and tracking, image recognition, and classification. Deep learning technology uses models such as convolutional neural networks, recurrent neural networks, and generative adversarial networks to learn and analyze large amounts of data, improving the accuracy and efficiency of tunnel detection [28]. In addition, the application of sensor technology and data fusion can provide diverse data sources and more comprehensive analytical capabilities.

As artificial intelligence technology continues to develop, its technology and applications in tunnel detection will become increasingly important [29]. In the future, artificial intelligence technology will play a greater role in tunnel detection, including improving detection accuracy and efficiency, reducing manual intervention, enhancing data security, and so on. At the same time, we also need to recognize the challenges that artificial intelligence technology faces in tunnel detection, such as data acquisition and quality issues, algorithm performance and reliability, privacy and security issues, etc. Only through continuous in-depth research and exploration can we better leverage the role of artificial intelligence technology in tunnel detection to ensure the safety of tunnel traffic and building structures.

References

[1] Chapman, D.N., Metje, N. and Stark, A. (2017). Introduction to tunnel construction. Crc Press.

[2] Haack, A. (2002). Current safety issues in traffic tunnels. *Tunnelling and underground space technology, 17*(2), 117–127.

[3] Zhu, Y. and Liu, C. (2022). Study on the parameter influence of new composite lining of water conveyance tunnel under high internal water pressure. *Frontiers in Earth Science*, 10, 920230.

[4] Ye, F., He, C., Wang, S.M. and Zhang, J.L. (2012). Landscape design of mountain highway tunnel portals in China. *Tunnelling and Underground Space Technology, 29*, 52–68.

[5] Wang, T.T. and Lee, C.H. (2013). Life-cycle design considerations for hydraulic tunnels: lessons learned from inspection and maintenance cases. *Journal of performance of constructed facilities, 27*(6), 796–806.

[6] Song, Z.P., Yang, T.T., Jiang, A.N., Zhang, D.F. and Jiang, Z.B. (2016). Experimental investigation and numerical simulation of surrounding rock creep for deep mining tunnels. *Journal of the Southern African Institute of Mining and Metallurgy, 116*(12), 1181–1188.

[7] Boutin, M., Benzakour, A., Despins, C.L. and Affes, S. (2008). Radio wave characterization and modeling in underground mine tunnels. *IEEE Transactions on Antennas and Propagation, 56*(2), 540–549.

[8] Liu, C., Zhang, D. and Zhang, S. (2021). Characteristics and treatment measures of lining damage: A case study on a mountain tunnel. *Engineering Failure Analysis, 128*, 105595.

[9] Kuesel, T.R., King, E.H. and Bickel, J.O. (2012). Tunnel engineering handbook. Springer Science & Business Media.

[10] Patil, M., Choudhury, D., Ranjith, P.G. and Zhao, J. (2018). Behavior of shallow tunnel in soft soil under seismic conditions. *Tunnelling and Underground Space Technology, 82*, 30–38.

[11] Chen, L., Jin, A., Wu, S., Chu, C. and Li, X. (2022). Numerical study on spalling failure of rock surrounding deep buried tunnel based on DEM. *Computers and Geotechnics, 145*, 104653.

[12] Sun, Z., Zhang, D., Li, M. and Guo, F. (2024). Large deformation characteristics and the countermeasures of a deep-buried tunnel in layered shale under groundwater. *Tunnelling and Underground Space Technology*, 144, 105575.

[13] Gong, Q., Yin, L., Ma, H. and Zhao, J. (2016). TBM tunnelling under adverse geological conditions: an overview. *Tunnelling and Underground Space Technology, 57*, 4–17.

[14] Dong, S., Li, S., Yu, F. and Wang, K. (2023). Risk assessment of immersed tube tunnel construction. Processes, *11*(4), 980.

[15] Hu, J., He, M., Tao, Z., Cheng, T., Xiao, Y., Li, H. and Li, L. (2024). An overview of potential excavation compensation method for tunnelling in deep rock engineering. *Geomechanics and Geophysics for Geo-Energy and Geo-Resources, 10*(1), 137.

[16] Zhang, Z., Shi, X., Wang, B. and Li, H. (2018). Stability of NATM tunnel faces in soft surrounding rocks. *Computers and Geotechnics, 96*, 90–102.

[17] Musa, M.S., Nallagownden, P., Chiu, K.W. and Sarwar, M.B. (2015, October). Design and development of intelligent adaptive tunnel lighting system. In 2015 IEEE Conference on Energy Conversion (CENCON) (pp. 289–292). IEEE.

[18] Deng, M. (2018). Challenges and thoughts on risk management and control for the group construction of a super-long tunnel by TBM. *Engineering, 4*(1), 112–122.

[19] Zhang, C., Liu, N. and Chu, W. (2016). Key technologies and risk management of deep tunnel construction at Jinping II hydropower station. *Journal of Rock Mechanics and Geotechnical Engineering, 8*(4), 499–512.

[20] Kensek, K. (2014). Building information modeling. Routledge.

[21] Volk, R., Stengel, J. and Schultmann, F. (2014). Building Information Modeling (BIM) for existing buildings—Literature review and future needs. *Automation in construction, 38*, 109–127.

[22] Bharatkumar, B.H., Narayanan, R., Raghuprasad, B.K. and Ramachandramurthy, D.S. (2001). Mix proportioning of high performance concrete. *Cement and concrete composites, 23*(1), 71–80.

[23] Yang, J., Zhang, Y. and Huang, J. (2024). The strengthening theory of steel fiber reinforced concrete and its application in tunnel engineering: A review. *Journal of Engineered Fibers and Fabrics*, 19, 15589250241239242.

[24] Liu, F., Zheng, M. and Ye, Y. (2020). Formulation and properties of a newly developed powder geopolymer grouting material. *Construction and Building Materials, 258*, 120304.

[25] Jiang, Y., He, B., Zhao, J., Pei, H., Liu, J. and Wang, H. (2022). Influence of novel polymer waterproofing membrane on mechanical properties of tunnel lining structure. *Construction and Building Materials, 360*, 129579.

[26] Li, X., Shao, Y., Ma, G. and Wang, L. (2024). A New 3D printing method and similar materials of the tunnel lining for the geomechanical model test. *Construction and Building Materials, 433*, 136724.

[27] Wu, X., Park, Y., Li, A., Huang, X., Xiao, F. and Usmani, A. (2021). Smart detection of fire source in tunnel based on the numerical database and artificial intelligence. *Fire Technology, 57*, 657–682.

[28] Wang, X., Lu, H., Wei, X., Wei, G., Behbahani, S. S. and Iseley, T. (2020). Application of artificial neural network in tunnel engineering: A systematic review. *IEEE Access, 8*, 119527–119543.

[29] Liu, L., Song, Z. and Li, X. (2024). Artificial intelligence in tunnel construction: A comprehensive review of hotspots and frontier topics. *Geohazard Mechanics, 2*(1), 1–12.

3 Designing VR-Based Training Scenarios for Tower Crane Lifting Operations Considering Potential Accidents

Junyu CHEN[1], Furui MAN[2], Wenyu XU[3] and Hung-Lin CHI[4*]

Tower cranes, as critical construction machinery, are essential for boosting construction productivity; however, they also pose a significant risk of injuries or fatalities to work crews during operations. Operators in the closest proximity to potential danger assume the most pivotal role in accidents due to the persistent risk of crane-related falling or crane collapses, which makes their occupational training significantly vital. Recently, VR-based tower crane operation training has been widely regarded as the ideal approach for transitioning operators from hazardous and costly on-site training environments to safer settings. Still, current training solutions are challenging in providing effective outcomes. One of the essential objectives is to ensure that the considerations and skills necessary for safety and productivity during operations can be imparted to novices. To achieve this, this chapter discusses the design mechanisms and corresponding framework for accident-enabled VR training scenarios, aiming to enhance the performance of tower crane operations. The proposed framework addresses safety implications by reflecting on potential accident cases and analysis results, further bridging these implications with existing knowledge and design for VR-based training protocols. The demonstrative implementation of such a training framework, for example, involved a system that embedded synthesized safety and productivity knowledge into a virtual, authentic public housing construction site. The tower crane lifting operation requirements for safety and productivity in such a building construction project were used to model the virtual training environment, along with the training agenda on their corresponding crane failures identified in previous accident investigations. The framework and design mechanism are expected to immediately benefit the safety of tower crane lifting operations while promoting the long-term development of emerging VR-based training scenarios in the construction industry.

1. Introduction

Occupational injuries and fatalities in the past ten years have been persistently occurring in the construction industry despite the current efforts of construction innovations. It showed no significant improvement as expected, especially in

[1-4] Department of Building and Real Estate, Faculty of Construction and Environment, The Hong Kong Polytechnic University, Hong Kong

* Corresponding author: hung-lin.chi@polyu.edu.hk

DOI: 10.1201/9781003514480-3

Hong Kong. Annual fatal cases that have occurred in the Hong Kong construction industry over the past two decades have remained at a consistent range. In the meantime, there has even been a slight rebound trend in non-fatal cases in recent years, following the beginning of the 21st century [8]. Hong Kong's fatal and non-fatal occupational injuries in the construction sector became the highest among all major sectors of the area, accounting for 20.5% and 9.4% in 2019 [13]. Safety concerns at construction sites have increased, and they are particularly essential to address, given the growing use of heavy machinery at sites, such as tower cranes. The operation of tower cranes is a highly specialized and hazardous profession characterized by significant skill requirements and a high level of safety awareness. Despite the implementation of stricter safety requirements in crane operations, a considerable number of accidents associated with tower crane operations have continued to occur over the past 20 years. 780 deaths among the 1125 tower crane accidents worldwide were reported over a decade, according to the data from the American National Standards Institute. An annual average of 27 fatal occupational injuries related to crane operations between 2003 and 2018 was also identified by the Census of Fatal Occupational Injuries in the U.S. In addition, the Ministry of Housing and Urban-Rural Development of the People's Republic of China reported nearly 125 tower crane accidents in 2020 [21]. Numerous statistics have shown that operators play a crucial role in most accidents [1], and 43% of crane accidents are attributed to the failures of the operators [7],[19].

Previous studies have not only developed systems to explore safety concerns on cranes but have also identified the essential elements for a successful crane operation. These elements include stakeholders, on-site crews, crane machinery, processes, and environments [20]. With the growing mechanized environment, advanced crane machinery, and the safety risk potentials coming behind, a few pilot studies in Hong Kong were conducted from both theoretical and empirical perspectives to investigate construction safety related to crane operation, which has been one major concern of the Labour Department of the HKSAR Government [17]. Based on serious accidents reported in various news media in Hong Kong, a pilot study conducted by the research team utilized 95 accident cases related to cranes to develop a structured accident database and propose a cluster approach and prioritize safety concerns in crane operations [2]. The preliminary results show that, combined with an insight into current regulations, legislation, and industrial codes of practice in Hong Kong, identifying the current lack of protections and generating applicable safety rules for construction crane operation are urgent needs for the future construction industry, solving the communication issues and enhancing training practice among crane operators and on-site crews. In fact, some research gaps remain, as presented by prior endeavors [14]. The risks to operators come from hazardous situations in or around the cabins. Although various solutions, including structure optimisation [6], safety pre-warning [16], and load monitoring [5], have been put forward to improve the safety of operators' working conditions, the potential solutions are generally considered to provide better training programmes or move the operators away from on-site, i.e., teleoperation [11],[15].

Recently, immersive training technologies, represented by Virtual Reality (VR), have been rapidly developed. It attempts to emulate a physical world through the means of a digital or simulated world, thereby creating a sense of immersion. Such a set of technologies involves multi-sensory experiences, including visual, auditory, and tactile, to create or recreate a virtual or remote environment. Many VR applications can be found across various fields, including gaming and entertainment, education, healthcare, and industrial design. Furthermore, it offers a new solution for enhancing tower crane operation training. The VR-based training programme completely alleviates trainees' safety concerns in the workspace while providing realistic simulations and immersive perceptual experiences.

This chapter describes a study to evaluate the potential of embedding crane training scenario design derived from operational accidents into VR training mechanisms. The study focuses on introducing direct safety implications derived from accident analysis outcomes and attempts to bridge these implications with existing safety knowledge as training objectives to enhance the trainees' performance. The proposed accident-enabled training framework details the related VR training scenario design mechanisms and utilizes a corresponding system development to validate its benefits comparatively. Synthesized safety knowledge is then embedded into a virtual construction site for demonstration, and a modeling process is discussed to determine safety and productivity requirements for tower crane lifts in such an environment. Additionally, the outcomes of accident reports from previous project cases identify potential crane failure modes, which have been integrated into the virtual construction sites as training elements. The corresponding accident causations were identified by linking trainees' safety violations to specific accident consequences in the virtual environment through a gaming mechanism. As long as the trainees fail to remove the safety constraints, the game requires them to restart the operational task until they identify hazards and take countermeasures to complete it within a single attempt, ensuring that no safety violations occur during the operation. To evaluate the proposed scenario design framework, a conventional VR training mode is employed, which provides trainees with stepwise safety guidance, accompanied by safety demonstrations, but without showing accident consequences or requiring a restart. A corresponding before-and-after experiment was conducted to evaluate the performance of trainees who experienced different training modes in terms of safety knowledge and operational skill acquisition. Trainees' perceptions of the VR training environment and different training modes were gathered through a post-test questionnaire survey to identify potential factors influencing their experiences. The evaluation outcomes, in general, indicated that accident-enabled VR training scenarios helped reduce safety violations and improved the operational productivity of trainees. It suggests the flexibility and effectiveness of accident-enabled training scenario design in helping to develop trainees' competence in navigating safety hazards and productivity pressures in real-world construction situations.

2. Crane Safety Training Practice

In the current crane safety training practice, the Hong Kong Labour Department, for example, has issued two codes of practice for safe crane operation [9],[10], with the inherent mission of ensuring that risks to occupational safety and health are managed adequately through enforcement, education, and promotion. Complied with the Occupational Safety and Health Ordinance (Cap. 509), the Labour Department also takes responsibility for investigating urgent occupational accidents. The Buildings Department is responsible for conducting site monitoring to ensure that relevant provisions of the Buildings Ordinance (Cap. 123) are complied with. The Housing Department also published and updated handbooks and booklets to enhance the "ZERO Incident" performance goal. The Construction Industry Council (CIC), authorized by the Hong Kong Government, is responsible for training, testing, and certifying crane operators (Cap. 583 and 587). Hong Kong Institute of Construction (HKIC) is a member organization of CIC and is committed to enhancing construction training. With a systematic accident analysis and insight into current legislation, training on tower cranes can further explore safety concerns neglected by academia or industry, and consider the human operator's factor in limited perception capacity during crane operations.

Two major challenges in today's Hong Kong construction industry have been identified that hinder improvements in crane safety. Firstly, accident analysis has been widely adopted in research on construction crane safety, encompassing qualitative surveys, statistical analysis, and data-driven modeling. However, despite the numerous accident reports that have garnered extensive societal attention in Hong Kong, there is a lack of a structured accident-driven safety training design that encompasses complete risks in scenarios, which can be used to systematically investigate operators' safety concerns. Second, the Hong Kong construction industry has faced challenges such as an aging workforce and a shortage of skilled crane operators [12]. To enhance effective crane safety management and implement more in-depth crane operator training practices, the HKIC committed to revamping the programme at its establishment in 2018. However, training systems to impart operational knowledge and hands-on experience to novices are still necessary. Additionally, proactive training data is not sufficiently utilized for research on innovative construction methods, nor is it used to tailor the training mode for crane operations.

3. Development of Vr Training Programme for Tower Cranes

In Hong Kong, the Labour Department has published the approval conditions for running mandatory safety training courses and issuing the Certificate of Crane Operator to trainees who complete the courses. The standardized part of the training courses must contain content relevant to legislation applicable to cranes, including handling and reporting procedures for accidents, as well as commonly used Personal Protective Equipment (PPE). In line with current training practices in Hong Kong, this study focuses on enhancing the design of training scenarios for VR-based training programmes that aid in crane operator training by supplementing the knowledge gained from a standardized training course. In this direction, we propose a VR-based training protocol for crane operators, focusing on safety and productivity. The authors design the VR training scenarios by embedding the knowledge of safety and productivity considerations during crane inspection and operation for both novice and experienced trainees. Afterward, a preliminary expert survey is conducted to evaluate and refine the training scenario design.

3.1 The Vr Training Scenario Design

The operation-level training tasks are set to be relevant to tower crane inspection and operation in this study. The authors consider a public housing project utilizing a hybrid building construction method, which incorporates cast-in-situ concrete for floor structures and precast concrete for wall structures. In Hong Kong, a key feature

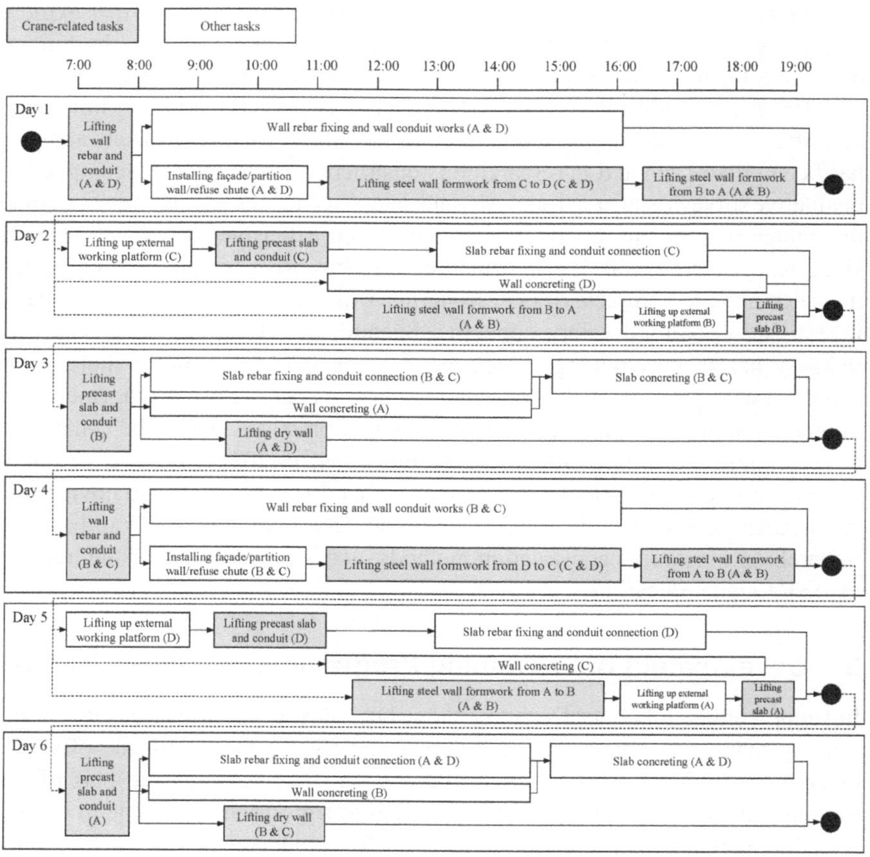

Fig. 1 Typical production activities in a six-day cycle of a public housing project.

of public housing projects is the adherence to a six-day cycle for a typical floor. According to the project information, each building block can have a symmetric design of four wings (Wing A, B, C, and D). The tower crane is responsible for lifting precast construction components to the intended position and transporting reusable temporary structures among different wings of a building block. The primary production activities in the studied building block during one six-day cycle are summarized in Figure 1. Notably, most crane-related activities are repetitive but essential to schedule control. As identified in the illustrations of the six-day cycle for this prefabrication housing project, the tower crane operational activities can be consolidated into five operational tasks: (1) Inspecting the tower crane before operation; (2) Rigging the precast construction components or reusable temporary structures for lifting; (3) Lifting the precast construction components or reusable temporary structures; (4) Unloading the precast construction components or reusable temporary structures from crane; and (5) Checking the tower crane before leaving.

In order to clarify the operational activities and their sequence during tower crane operation and inspection, the authors refer to the current legislation, regulations, industry codes of practice, and guidance notes in Hong Kong. Figure 2 illustrates the required procedures and precaution steps for the five operational tasks. Steps one to six are inspection duties of the tower crane operator before operation; steps seven to 14 are operational activities of the rigging process; steps 15 to 22 comprise the standard procedures for lifting the loads; steps 23 to 28 are operational activities of the unloading process, and steps 29 to 35 describe the operator's duties before leaving the tower crane.

As shown in Figure 3, the framework for embedding knowledge of tower crane safety into VR training scenarios is developed through three stages. Firstly, the content design of the VR training scenarios is derived from the five operational tasks and the relevant 35 precaution steps of tower crane inspection and operation. The trainees should comprehend the purpose of each safety precaution during the tutorial session and attempt to conduct them step by step to complete the training tasks. Secondly, the safety concerns and environmental hazards identified on construction sites through accident reports in Hong Kong [2] for tower crane operators are embedded as safety constraints that trainees should attempt to mitigate using each of the safety precautions. According to the results from the clustering and prioritization of operational safety concerns, environmental hazards, and crane failure modes, five training scenarios were developed to contain accident information from the five prioritized clusters, respectively. Thirdly, the trainees' performance from the safety perspective can be evaluated through their training outcomes. The indicators for identifying the trainees' performance are listed in Table 1. Under the circumstances that the trainees successfully identified all operational safety concerns and environmental hazards and removed all safety constraints through proper precautions, the outcome of the training exercise is expected to be the attainment of safe tower crane inspection and operation. On the other hand, safety constraints that are not properly identified and eliminated by the

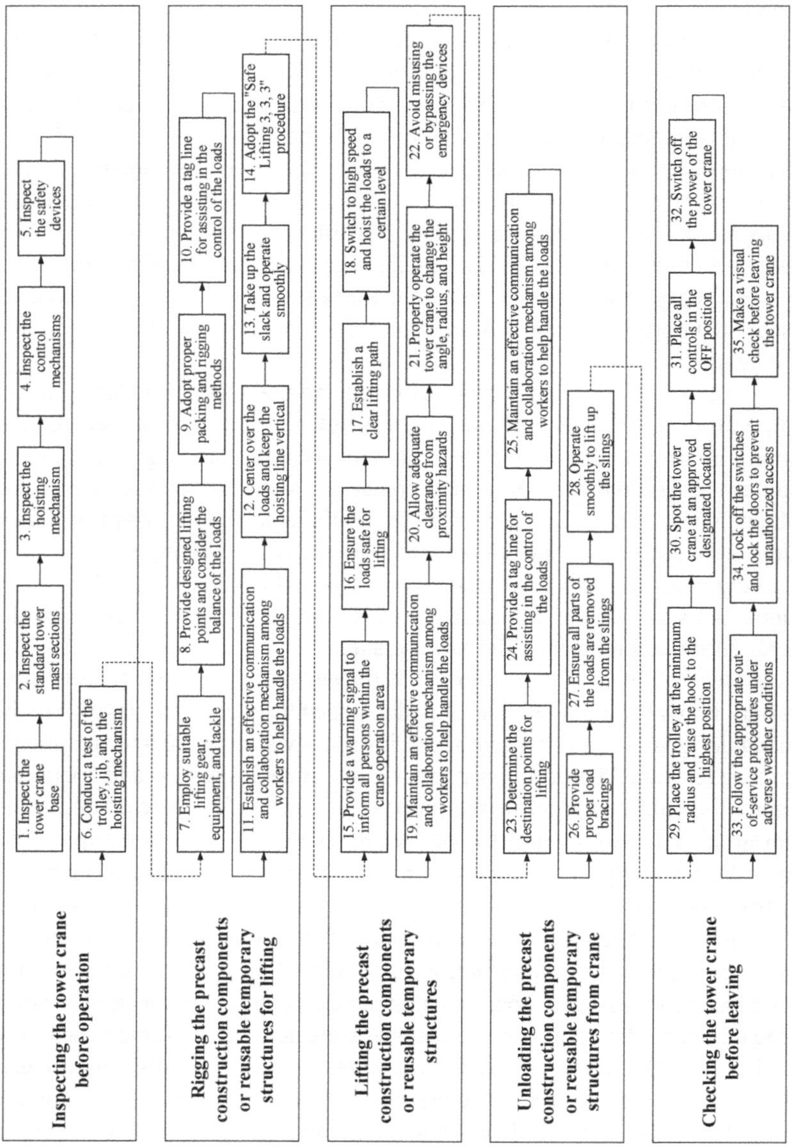

Fig. 2 The procedure and precautions of tower crane inspection and operation.

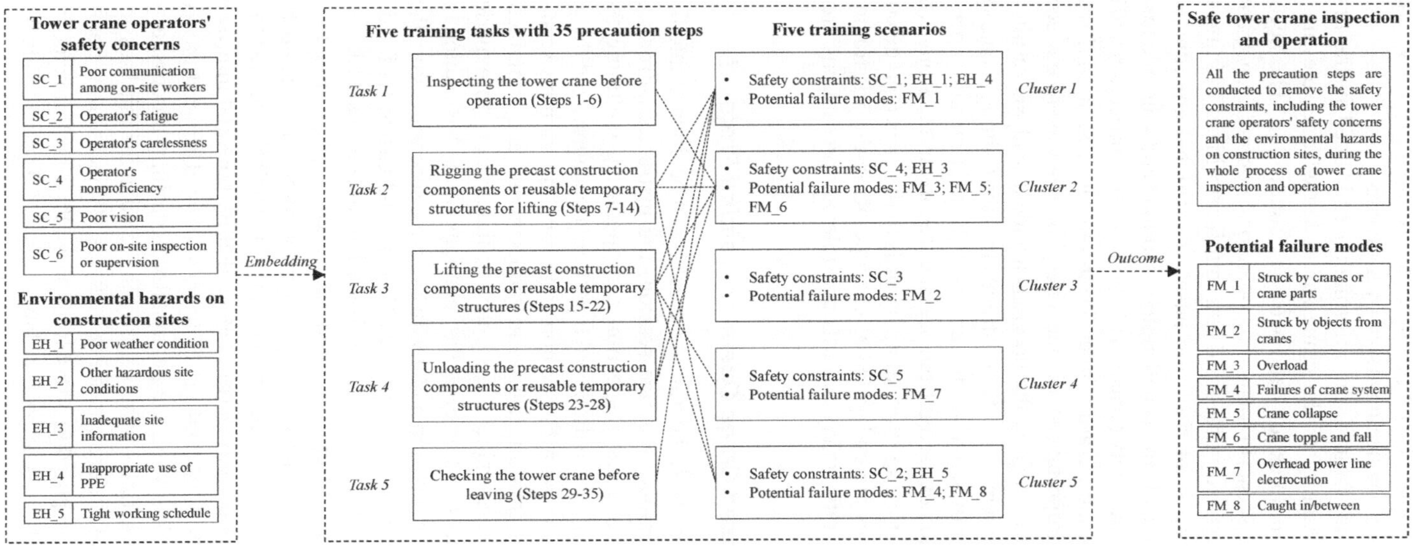

Fig. 3 The framework for embedding the knowledge of tower crane safety into the VR training scenario design.

trainees will be associated with potential failure modes of tower cranes, and the training outcome will be demonstrated by accidents.

As presented in Table 1, the information embedded in the first prioritized training scenario highlights the crane safety knowledge relevant to problems on PPE use, communication, and weather conditions. The second scenario prioritized training scenario focuses on operators' nonproficiency and inadequate site surveys. Underlined in the third and fourth prioritized training scenarios are unintentional human error risks and unsafe operational environments. The last scenario prioritized calls for a challenging workload and schedule for construction crane operations. The safety concerns of "Poor on-site inspection and supervision" and the environmental hazard "Other hazardous site conditions" were identified as being of great importance to crane safety, but they were not correlated with any potential failure mode in the previous accident analysis results. The current study also embeds them as safety constraints into the training scenarios to enhance trainees' safety knowledge of tower crane inspection and operation. Using accident information extracted from accident reports in the Hong Kong and U.S. construction industries, specific safety hazards of the eleven predetermined safety constraint categories are embedded as checkpoints in the VR training scenarios. The embedded specific safety hazards and relevant precautions are provided in Appendix 1.

3.2 The Vr -Based Crane Operator Training Protocol

As illustrated in Figure 4, the training process can be divided into three sessions. Firstly, in a tutorial-and-practice session, the instructor introduces the codes of practice of tower crane inspection and operation, including the applicable legislation, the accident handling and reporting procedures, and the required PPE. Then, in conjunction with the demonstration of the training scenario settings, the crane operator training tasks and requirements are explained. Under the instructor's guidance, trainees are given sufficient time to experience the VR learning environment and practice the functions of the VR equipment, including the VR visualization platform and joysticks. In the second session, the trainees are randomly assigned to two groups (i.e., Group I and Group II) for the first-round exercise. Trainees in the first group performed the operational tasks, and afterward, the instructor imparted crane-related safety knowledge after the operational exercise. For trainees in the second group, after completing the operational tasks, the instructor utilizes the Value Stream Mapping (VSM) tool [18] to help analyze their safety and productivity performance, identify potential improvements based on their training profiles, and generate personalized suggestions for achieving better operations. Following the personalized coaching process, a second-round exercise and a comparative analysis are conducted in the third session to validate the effectiveness of different training methods. Trainees from both the first and second groups are required to strive to improve their performance in the training tasks by applying the knowledge they have acquired. Their training performance is

Table 1 Indicators of the operational performance.

Indicators	*Description*
Processing time (PT)	PT refers to the time consumption between the beginning and the end of the crane inspection and operation process (from step 1 to step 8).
Lead time (LT)	LT refers to the time consumption of lifting the loads from the start location to the destination location (from step 2 to step 7).
Cycle time (CT)	CT is the frequency of completing the entire inspection and operation task by a process, which is set to 1.
Takt time (TT)	TT is the rate at which a lifting can be completed according to the project schedule, which is set at 5 minutes and 40 seconds.
Value-added time (VAT)	VAT is the part of LT indicating the value-adding lifting activities in each step.
Non-value-added time (NVAT)	Complementary to VAT, NVAT is the part of LT associated with non-value-adding activities in each step.
Waste time (WT)	WT is the portion of non-value-added time cost due to waste in each step.
Number of wastes (N_W)	N_W is the counted number of wastes in each step.
Number of violations (N_S)	N_S is the number of safety violations counted in each step.
Waste categories	Waste category refers to the type of waste in each step, including: (A) Waiting time: time consumed when the operation team is waiting for instructions from the operator and signaller;
	(B) Idle time: the time consumed when the operator is thinking without actively operating the crane;
	(C) Unnecessary transportation time: the time consumed when the loads are transported in an opposite direction from the destination.
	(D) Resting time: the time consumed when the operator suspends the operation to have a break or due to some physiological needs.
	(E) Reworking time: the time consumed when the operator conducts improper inspection or operation activities, and subsequently, rework is done.
Safety constraint categories	Safety constraint category refers to the type of tower crane operators' safety concerns (SC) or environmental hazards (EH) on construction sites identified through an investigation of 95 crane-related accidents, including: (SC_1) Poor communication among on-site workers;
	(SC_2) Operator's fatigue;
	(SC_3) Operator's carelessness;
	(SC_4) Operator's nonproficiency;
	(SC_5) Poor vision;
	(SC_6) Poor on-site supervision or inspection;
	(EH_1) Poor weather conditions;
	(EH_2) Other hazardous site conditions;
	(EH_3) Inadequate site information;
	(EH_4) Inappropriate use of PPE;
	(EH_5) Tight working schedule.

evaluated from the perspectives of time to complete tasks, value-added time, non-value-adding wastes, and violations of safety constraints. At the end of the training, participant feedback and expert opinions are collected through questionnaire surveys and interviews to inform the design of the training protocol.

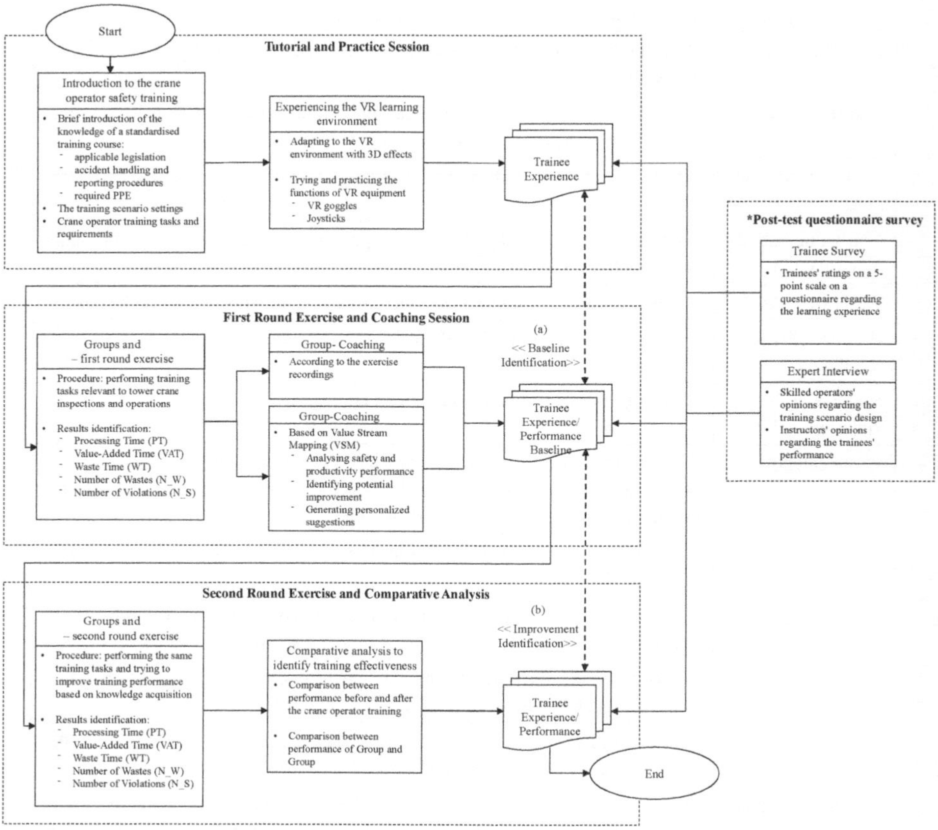

Fig. 4 The proposed VR crane operator training protocol.

3.3 An Expert Survey

To validate the VR-based crane operator training protocol design and the VR training scenario design, the authors conducted interviews with six experienced tower crane operators in Hong Kong. There are three sections in the preliminary expert survey: (i) evaluation of the clustered causal factors and crane accidents; (ii) suggestions for the training scenario design; (iii) discussion about the benefits and challenges of developing VR systems for crane operator training. In the first section, all six tower crane operators interviewed concur that the identified safety concerns and environmental hazards have the potential to cause crane accidents in the manner specified in the clustering results. For example, the interviewees considered it rational that the crane operator's misoperation, such as unintentionally bumping into the control, is likely to result in the loads falling. In the second section, valuable suggestions are collected to refine the training scenario design. For example, one experienced tower crane operator stated that daily inspections are essential for ensuring safe tower crane operation. Therefore, in this study, the procedure and precautions for tower crane inspection, both before and after operation, are included in the training tasks for crane operators. In the third section, the interviewees' opinions regarding the benefits and challenges are obtained. For example, one experienced tower crane operator mentioned that the immediate reaction to crane accidents is a missing part of traditional crane operator training, and it is meaningful for the trainees to learn how to properly respond to crane accidents in VR environments. On the other hand, another experienced tower crane operator expressed concern that problems, such as simulating the operators' view and their communication with signallers, should be addressed when developing the VR training scenarios.

3.4 Vr-Based Training Environment

Initially, the Cave Automatic Virtual Environment (CAVE), a semi-immersive VR technology, is employed to provide a simulated and interactive environment for trainees to develop their skills in addressing dynamic safety constraints on construction sites (as illustrated in Figure 5). Operational safety and productivity concerns identified through crane accident analysis are incorporated into VR training scenarios to impart the expertise of experienced crane operators to novices. Secondly, the proposed training protocol is devised to provide personalized coaching for trainees using the VSM tool. Finally, to evaluate the benefits of integrating lean thinking into the proposed training mechanism, the trainees' performance enhancement is measured from the perspectives of safety knowledge acquisition and operational skill development. The underlying mechanism by which the proposed system improves operational performance is also investigated by analyzing their responses in a post-test questionnaire survey.

To simplify the training process evaluation, the safety and productivity performance of trainees are further considered in the continuous flow of representative activities, as shown in Figure 6. From a safety perspective, representative hazards associated with each activity are embedded as safety constraints in the accident-enabled VR training scenarios. The virtual training scene is modeled after a public housing project with a hybrid construction method, incorporating cast-in-situ concrete for floor structures and precast concrete for wall structures. The operational task is devised as transporting a reusable steel formwork from the ground to the current working face. The Takt, as shown in Table 1 time for each lifting task is set at 5 minutes and 40 seconds, according to the adopted six-day schedule in this study (as shown in Figure 1). The accident-enabled VR training prototype is developed using the game engine Unity3D to incorporate safety

Fig. 5 The CAVE-enabled crane operator training environment (Adapted from [4]).

constraints from the first prioritized cluster, specifically inappropriate use of PPE, poor communication with on-site workers, and adverse weather conditions that can lead to struck-by accidents. The operators' safety performance can be observed and recorded in the VR training environment. If the operators successfully identify and remove all safety constraints, the training outcome will be the attainment of safe tower crane operation. Correspondingly, safety constraints that are not properly identified can lead to potential tower crane accidents, such as the assistant worker being struck by the crane loads.

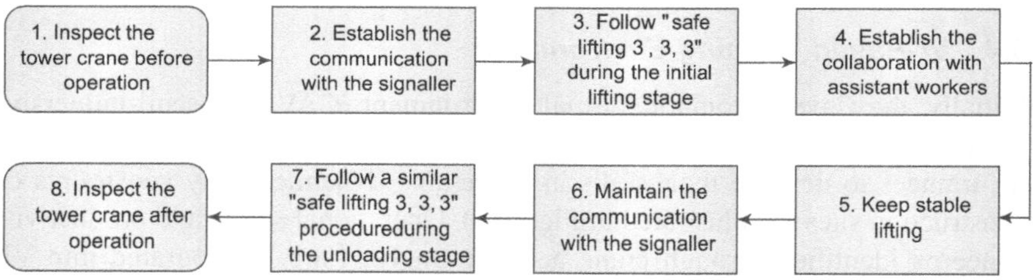

Fig. 6 The continuous flow of representative crane inspection and operation activities.

3.5 Indicators of the Evaluation

From the productivity perspective, as shown in Table 1, the trainees' operational performance is measured using time indicators from lean thinking. Time consumed by non-value-adding activities is visualized as non-value-added time (NVAT) in the performance. As part of the NVAT, waste time (WT) is associated with activities (also called wastes), including 1) waiting, 2) idling, 3) unnecessary transportation, 4) rest, and 5) rework. In addition, NVAT can also be linked to some non-value-adding but necessary activities, including 1) receiving instruction and 2) measuring and estimating. Complementary to the NVAT, the time consumed by value-adding activities, including 1) necessary transportation time and 2) inspection time, is visualized as value-added time (VAT).

The evaluation is conducted in a before-and-after experiment. VAT, WT, processing time (PT), the counted number of time wastes (N_W), the counted number of safety violations (N_S), waste categories, and safety constraint categories are recorded and calculated for operational safety and productivity assessment. Enlightened by previous studies that applied the VSM method to improve operators' performance [18], the VSM tools are utilized to identify the operational activities leading to time waste and safety violations, impart the fragmented safety and productivity knowledge through the standardized tower crane operation procedure, and generate personalized guidance for operators to achieve better performance. In the before-and-after experiment, the performance of trainees from Group I and Group II is compared to evaluate the benefits of adopting the proposed training mechanisms. A post-test questionnaire survey is devised based on the Slater-Usoh and Steed (SUS) presence questionnaire, technology acceptance model (TAM), and NASA task load index (NASA-TLX) to explore the factors influencing trainees' learning experience in terms of 1) ease of use, 2) effectiveness, 3) motivation, 4) efficiency, 5) intention to use, 6) cognitive workload, and 7) sense of presence.

3.6 Evaluation Outcomes

In summary, the accident-enabled VR training scenario aims to enhance trainees' safety knowledge acquisition by replicating hazardous operational conditions, whereas the proposed training protocol aims to assist trainees in developing their operational skills by identifying wasted time during their crane operation procedure. In the observation of the VSM tools' results, the process of keeping stable lifting contained most of the WT, related to work inactivity (e.g., idling) and ineffective work (e.g., unnecessary transportation and reworking). Additionally, most safety violations (e.g., failing to establish communication with the signaller) occurred during the initial lifting and unloading processes. After a discussion between the trainee and instructor, the trainee would strive to achieve an ideal productive performance without waste or safety violations through iterative practice.

The average performance of trainees from the two groups is provided in Table 2 and Figure 7. As indicated by the increased VAT values and decreased safety violations after training, it was found that trainees became more cautious about safety constraints during crane operation following the accident-enabled VR training scenario. The research findings also show that personalized VR training under a CAVE system reduced 96% of WT, 26% of PT, and 60% of counted waste, demonstrating its effectiveness in improving trainees' operational productivity as well as its advantages when compared with the conventional approach in VR training. Trainees' feedback on the CAVE-enabled training environment is shown in Figure 8. Trainees from the two groups expressed similar attitudes toward

Table 2 The average before-and-after training performance of the two groups of trainees.

Group No.	*Before training (min)*					*After training (min)*				
	VAT	*WT*	*PT*	*N_W*	*N_S*	*VAT*	*WT*	*PT*	*N_W*	*N_S*
Group I	2.70	0.70	3.57	2.67	4.33	4.00	0.20	4.40	1.33	3.00
Group II	2.83	1.60	4.70	1.67	4.00	3.27	0.07	3.47	0.67	2.00

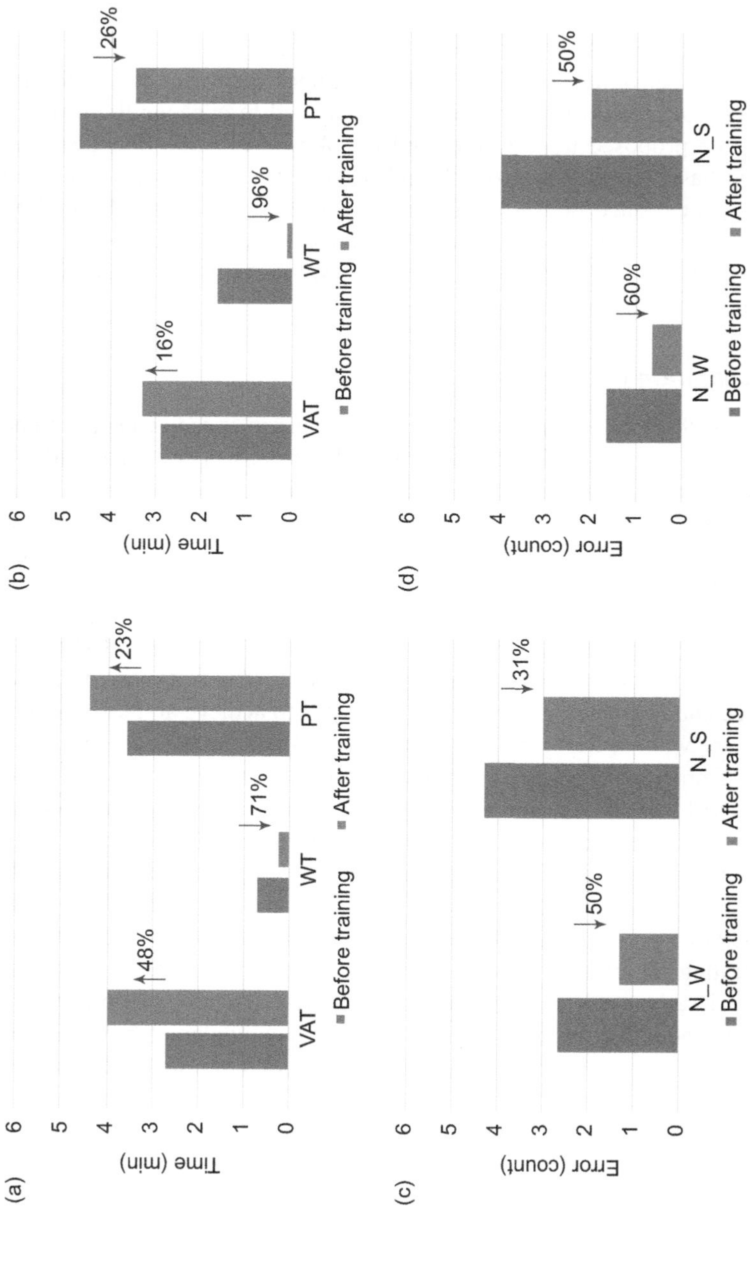

Fig. 7 The comparison of the conventional and accident-enabled VR training: (a) time indicators for Group II, (b) time indicators for Group I, (c) error counts for Group I, and (d) error counts for Group II.

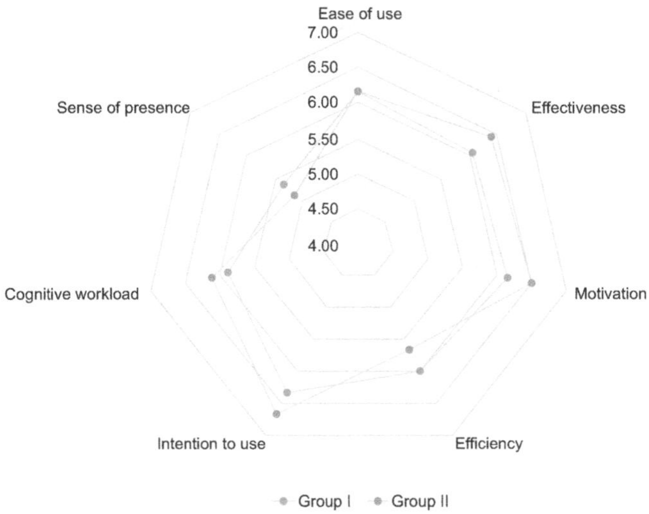

Fig. 8 Comparison of the feedback on training experience between the conventional and personalized VR training.

the CAVE system, especially in terms of ease of use and sense of presence. Additionally, trainees who received personalized training perceived the proposed training system as more effective and showed stronger motivation and a greater intention to use it in the future. However, they also reported a higher level of cognitive workload and a lower level of training efficiency, as the personalized training process required more time and effort from the trainees. Other limitations regarding the level of fidelity of the simulations (e.g., visual perception, control, environmental dynamics, and uncertainty in reality) should also be taken into consideration in future study efforts.

4. Summary and Conclusion

The work conducted in this chapter represents an attempt to design safety training scenarios for construction heavy machinery operations. The essential features include enhancements that consider existing accidents to systematically establish training checkpoints with corresponding operation failure modes that could occur in reality. Furthermore, a training programme should consider essential safety objectives and emphasize the importance of productivity, given that productivity pressure is a norm in the construction industry. This work, which develops training scenarios for tower crane lifting tasks based on accident analysis, is implemented and demonstrated in a VR training environment using a CAVE immersive visualization system. The corresponding training protocol further divided the participants into two groups: one experienced the proposed add-in features (i.e., accident-enabled training and personalized training that considered both safety and productivity), and the other followed a conventional training mechanism. The evaluations based on simulated lifting task training sessions showed significant improvements in the trainees' proficiency in efficient and effective operations, ensuring that safety and productivity concerns are taken rigorously. Such a VR-based training development mechanism can be further extended to the remote site scenarios to facilitate more machinery operation training programmes.

5. Acknowledgement

The project is supported by the Research Grants Council of Hong Kong through the General Research Scheme (PolyU 15223322). The authors hereby express their appreciation to the institute.

References

[1] Bedi, J.K., Rahman, R.A. and Din, Z. (2021). Heavy machinery operators: necessary competencies to reduce construction accidents. IOP Conf. Ser.: Earth Environ. Sci. 641, 012007. https://doi. org/10.1088/1755-1315/641/1/012007.

[2] Chen, J., Chi, H.L., Du, Q. and Wu, P. (2022). Investigation on operational concerns of construction crane operators: An approach integrating factor clustering and prioritization, *Journal of Management in Engineering, 38*(4), 04022020.

[3] Chen, Y., Zeng, Q., Zheng, X., Shao, B. and Jin, L. (2022). Safety supervision of tower crane operation on construction sites: An evolutionary game analysis. Safety Science 152, 105578. https://doi. org/10.1016/j.ssci.2021.105578.

[4] Department of Building and Real Estate, (2024). Immersive crane operator training environment. BRE Connected June 2024. https://www.polyu.edu.hk/bre/about-bre/publications/ (accessed July 2, 2025).

[5] Fang, Y., Chen, J., Cho, Y.K., Kim, K., Zhang, S. and Perez, E. (2018). Vision-based load sway monitoring to improve crane safety in blind lifts. *Journal of Structural Integrity and Maintenance 3*, 233–242. https://doi.org/10.1080/24705314.2018.1531348.

[6] Jin, L., Liu, H., Zheng, X. and Chen, S. (2020). Exploring the impact of wind loads on tower crane operation. *Mathematical Problems in Engineering* 2020, e2807438. https://doi. org/10.1155/ 2020/2807438.

[7] King, R.A. (2012). Analysis of crane and lifting accidents in North America from 2004 to 2010 (Thesis). Massachusetts Institute of Technology.

[8] Labour Department (2009). Occupational Safety and Health Statistics Bulletin. *Occup. Saf. Heal.* Branch, Labour Dep. 10, 1–8.

[9] Labour Department (2011). Code of Practice for Safe Use of Tower Cranes.

[10] Labour Department, (2017). Code of Practice for Safe Use of Mobile Cranes.

[11] Lee, J.S., Ham, Y., Park, H. and Kim, J. (2022). Challenges, tasks, and opportunities in teleoperation of excavator toward human-in-the-loop construction automation. Automation in Construction 135, 104119. https://doi.org/10.1016/j.autcon.2021.104119.

[12] Legislative Council (2018). Construction industry in Hong Kong.

[13] Legislative Council (2020). Occupational injuries in the construction sector 2019–2020.

[14] Sadeghi, S., Soltanmohammadlou, N. and Rahnamayiezekavat, P. (2021). A systematic review of scholarly works addressing crane safety requirements. Safety Science 133. https://doi.org/ 10.1016/j.ssci.2020.105002.

[15] Sitompul, T.A. (2022). Human–machine interface for remote crane operation: a review. Multimodal Technologies and Interaction 6, 45. https://doi.org/10.3390/mti6060045.

[16] Son, H., Seong, H., Choi, H. and Kim, C. (2019). Real-time vision-based warning system for prevention of collisions between workers and heavy equipment. *Journal of Computing in Civil Engineering 33*, 04019029. https://doi.org/10.1061/(ASCE)CP.1943-5487.0000845.

[17] Tam, V.W.Y. and Fung, I.W.H. (2011). Tower crane safety in the construction industry: A Hong Kong study. *Saf. Sci. 49*, 208–215. https://doi.org/10.1016/j.ssci.2010.08.001.

[18] Wang, P., Wu, P., Chi, H.L. and Li, X. (2020). Adopting lean thinking in virtual reality-based personalized operation training using value stream mapping, Autom. Constr. 119, 103355. https://doi.org/10.1016/j.autcon.2020.103355.

[19] Zaini, N.Z.M., Fikri Hasmori, M.F., Salleh, M.A.M., Yasin, M.N. and Ismail, R. (2020). Crane accidents at construction sites in Malaysia. IOP Conf. Ser.: Earth Environ. Sci. 498, 012105. https://doi. org/10.1088/1755-1315/498/1/012105.

[20] Zhou, W., Zhao, T., Liu, W. and Tang, J. (2018). Tower crane safety on construction sites: A complex sociotechnical system perspective. *Saf. Sci. 109*, 95–108. https://doi.org/10.1016/j. ssci.2018. 05.001.

[21] Zhong, H., Chen, L., Antwi-Afari, M. F., Bao, Z. and Chen, K. (2025). Dynamic risk assessment of tower crane operations by integrating functional resonance analysis method and Bayesian network. Developments in the Built Environment. 23, 100699. https://doi.org/10.1016/j. dibe.2025.100699

Appendix 1 The embedded specific safety hazards and relevant precautions.

Safety precautions during the tower crane inspection and operation procedure	*Embedded safety constraint category*	*Embedded specific safety hazard (checkpoint)*	*Potential failure mode*
1. Inspect the tower crane base	SC_6	Inappropriate welds of the base	–
2. Inspect the standard tower mast sections	SC_6	Loose jointing pins of the mast sections	–
		Inappropriate welds of the mast sections	
3. Inspect the hoisting mechanism	SC_6	The wear of the hoisting devices (chain slings, rope slings)	–
		The wear of connecting hardware (bolts, nuts, rings, hooks, clamps, shackles)	
4. Inspect the control mechanisms	SC_4	The malfunction of control switches, levers, and buttons in the operator's cabin	Crane topple and fall; Overload; Crane collapse
5. Inspect the safety devices	SC_6	The malfunction of the anemometer, the automatic safe load indicator, and the hook view camera	–
		A lack of the provision of PPE (gloves, ear protectors, reflective vests)	
6. Conduct a test of the trolley, jib, and the hoisting mechanism	SC_6	The malfunction of the hoisting devices (wire ropes, pulley blocks, end fastenings)	–
7. Employ suitable lifting gear, equipment, and tackle	EH_2	A lack of suitable lifting slings (e.g., fibre slings cannot be used for objects with sharp edges)	–
	EH_3	Inappropriate ground conditions for the crane standing	Crane topple and fall; Overload; Crane collapse
		Insufficient space available for crane access, erection, operation, and dismantling	
	EH_5	Some works are being conducted on or near the track when the tower crane is operating	Caught in/between; Failures of Cranes

Contd.

Contd.

Safety precautions during the tower crane inspection and operation procedure	*Embedded safety constraint category*	*Embedded specific safety hazard (checkpoint)*	*Potential failure mode*
8. Provide designed lifting points and consider the balance of the loads	EH_2	Loads are not being adequately secured due to the inappropriate selection of lifting points, which may lead to the loads swinging, becoming unstable, or sustaining damage when lifted.	–
9. Adopt proper packing and rigging methods	EH_2	Loads not being adequately secured due to selection of wrong rigging methods: (1) two single-leg slings used with direct attachment should be used for objects with lifting inserts; (2) choker hitch double wrapped method, straps and taglines should be used for loose materials such as steel bars; (3) lifting clamps, strops or slings should be used for some precast elements such as pre-stressed hollow-core floor slabs which may not have lifting inserts.	–
10. Provide a tag line for assisting in the control of the loads	EH_2	Loads are not being securely suspended due to a lack of provision of tag lines (functions of using tag lines: to keep the rigger at a safe distance and make the loads stable during lifting).	–
	EH_4	Improper use of PPE (gloves, ear protectors, reflective vests) by other practitioners.	Struck by cranes or crane parts

Contd.

Contd.

Safety precautions during the tower crane inspection and operation procedure	*Embedded safety constraint category*	*Embedded specific safety hazard (checkpoint)*	*Potential failure mode*
11. Establish an effective communication and collaboration mechanism among workers to help handle the loads	SC_1	A lack of effective communication between the tower crane operator and the signaller.	Struck by cranes or crane parts
		A lack of effective communication between the tower crane operator and other operators when there is an overlapping operation zone.	
	EH_2	A lack of effective communication between the signaller and other practitioners working nearby	–
		A lack of a competent signaller.	
12. Center over the loads and keep the hoisting line vertical	SC_4	Loads are not being securely suspended due to being inappropriately dragged along the ground.	Crane topple and fall; Overload; Crane collapse
13. Take up the slack and operate smoothly	SC_4	Loads are not being securely suspended due to inappropriately using high speed when initially lifting the loads.	Crane topple and fall; Overload; Crane collapse
14. Adopt the "Safe Lifting 3, 3, 3" procedure	EH_2	A lack of a separate operation zone (addressed by keeping 3 meters away from the loads being lifted).	-
	SC_4	Loads are not being securely suspended due to the initial unstable state of the loads (addressed by initially lifting up the loads 300mm from the ground and waiting for 3 seconds for stabilizing the loads before the lifting operation).	Crane topple and fall; Overload; Crane collapse

Contd.

Contd.

Safety precautions during the tower crane inspection and operation procedure	*Embedded safety constraint category*	*Embedded specific safety hazard (checkpoint)*	*Potential failure mode*
15. Provide a warning signal to inform people within the crane operation zone	EH_2	A lack of adequate supervision of the tower crane operation by trained and competent personnel.	–
16. Ensure the environmental conditions are safe for lifting the loads	SC_5	Insufficient lighting for the tower crane operation.	Overhead power line electrocution
17. Establish a clear lifting path	SC_4	Lifting the loads over people.	Crane topple and fall; Overload; Crane collapse
		Without keeping the minimum safety distance between the tower crane jib/the loads and the overhead power lines (less than 9m clearance from the overhead power lines).	
		Without keeping the minimum safety distance between the loads and the constructed structures during the vertical lifting process (less than 3m clearance from the structures).	
	EH_2	The risk of collision with another tower crane within the overlapping operation zone.	–
18. Switch to high speed and hoist the loads to a certain level	SC_2	Extensive working hours of tower crane operators	Caught in/between; Failures of Cranes
	EH_5	Pressured working schedule.	

Contd.

Contd.

Safety precautions during the tower crane inspection and operation procedure	*Embedded safety constraint category*	*Embedded specific safety hazard (checkpoint)*	*Potential failure mode*
19. Maintain an effective communication and collaboration mechanism among workers to help handle the loads	SC_1	A lack of effective communication between the tower crane operator and the signaller.	Struck by cranes or crane parts
		A lack of effective communication between the tower crane operator and other operators when there is an overlapping operation zone.	
	EH_2	A lack of effective communication between the signaller and other practitioners working nearby.	–
		A lack of a competent signaller.	
20. Allow adequate clearance from proximity hazards	SC_4	Lifting the loads over people.	Crane topple and fall; Overload; Crane collapse
		Without keeping the minimum safety distance between the tower crane jib/the loads and the overhead power lines (less than 9m clearance from the overhead power lines).	
		Without keeping the minimum safety distance between the loads and the constructed structures during the vertical lifting process (less than 3m clearance from the structures).	
21. Properly operate the tower crane to change the angle, radius, and height	SC_4	Abruptly swinging or stopping the tower crane.	Crane topple and fall; Overload; Crane collapse
		Reversing the tower crane jib too fast.	
	SC_3	Bumping into the control devices.	Struck by objects from cranes

Contd.

Contd.

Safety precautions during the tower crane inspection and operation procedure	*Embedded safety constraint category*	*Embedded specific safety hazard (checkpoint)*	*Potential failure mode*
22. Avoid misusing or bypassing the emergency devices	SC_4	Ignoring overload warnings when the weight of the loads exceeds the rated safe working load.	Crane topple and fall; Overload; Crane collapse
		Ignoring over-travel warnings when the load exceeds the maximum out or maximum in position.	
22. Avoid misusing or bypassing the emergency devices	SC_4	Ignoring collision warnings when there is another tower crane approaching the overlapping zone.	Crane topple and fall; Overload; Crane collapse
		Ignoring readings from the anemometer when there is a strong wind with a speed exceeding 62 km per hour.	
	EH_1	Extreme weather conditions (strong wind with a speed exceeding 62 km per hour; heavy rain causing poor soil condition, low visibility or with Black Rainstorm Warning Signal; thunderstorm).	Struck by cranes or crane parts
	EH_2	The risk of collision with another tower crane within the overlapping operation zone.	–
23. Determine the landing position of the loads	SC_4	The loads exceeding the predetermined maximum in or the maximum out position.	Crane topple and fall; Overload; Crane collapse
24. Provide a tag line for assisting in the control of the loads	EH_2	Loads are not being securely suspended due to a lack of provision of tag lines (functions of using tag lines: to keep the rigger at a safe distance and make the loads stable during lifting).	–
	EH_4	Improper use of PPE (gloves, ear protectors, reflective vests) by other practitioners.	Struck by cranes or crane parts

Contd.

Contd.

Safety precautions during the tower crane inspection and operation procedure	*Embedded safety constraint category*	*Embedded specific safety hazard (checkpoint)*	*Potential failure mode*
25. Maintain an effective communication and collaboration mechanism among workers to help handle the loads	SC_1	A lack of effective communication between the tower crane operator and the signaller.	Struck by cranes or crane parts
		A lack of effective communication between the tower crane operator and other operators when there is an overlapping operation zones.	
	EH_2	A lack of effective communication between the signaller and other practitioners working nearby.	–
		A lack of a competent signaller.	
26. Provide proper load bracings	SC_4	Loads are not being securely supported due to a lack of proper bracing.	Crane topple and fall; Overload; Crane collapse
27. Ensure all parts of the loads are removed from the slings	SC_4	Loads are not being securely supported due to the slings entangled with the loads when lifted.	Crane topple and fall; Overload; Crane collapse
28. Adopt a similar "Safe Lifting 3, 3, 3" procedure to lift up the slings	EH_2	A lack of a separate operation zone (addressed by keeping 3 meters away from the slings to be lifted up).	–
	SC_4	The slings are swinging or spinning due to inappropriately using high speed when initially lifting up the slings.	Crane topple and fall; Overload; Crane collapse
		The slings swinging or spinning due to the initial unstable state of the slings (addressed by initially lifting up the slings 300mm from the ground and waiting for 3 seconds for stabilizing the slings).	

Contd.

Contd.

Safety precautions during the tower crane inspection and operation procedure	*Embedded safety constraint category*	*Embedded specific safety hazard (checkpoint)*	*Potential failure mode*
29. Place the trolley at the minimum radius and raise the hook to the highest position	SC_6	Unsafe movement of the hoisting mechanism due to leaving the tower crane unattended.	–
30. Spot the tower crane at an anchorage point and attach rail clamps to the rails	SC_6	Insecurity of the tower crane due to leaving the tower crane unattended.	–
31. Place all controls in the OFF position	SC_6	Accidental operation of the tower crane due to leaving the tower crane unattended.	–
32. Switch off the power to the tower crane	SC_6	Unexpected electrical hazards occur when leaving the tower crane unattended.	–
33. Follow the appropriate out-of-service procedures under adverse weather conditions	EH_1	Extreme weather conditions (strong wind with a speed exceeding 62 km per hour; heavy rain causing poor soil condition, low visibility or with Black Rainstorm Warning Signal; thunderstorm).	Struck by cranes or crane parts
	SC_6	The risk of damage to the tower crane under adverse weather conditions due to leaving the tower crane unattended.	–
34. Lock off the switches and lock the doors to prevent unauthorized access	SC_6	Unauthorized access due to leaving the tower crane unattended.	–
35. Make a visual check before leaving the crane	SC_6	Inadvertent operation before leaving the tower crane unattended.	–

4 VR-Enabled Digital Twin Systems for Tunnel Construction Management

Yishuo Jiang[1], Hongda Lin[2*] and Tianhang Zhang[3]

Tunnel construction is essential for supporting urban infrastructure and facilitating efficient transportation. Traditional tunnel construction methods, such as cut-and-cover, conventional mining, drill-and-blast, shield tunneling (STM), tunnel boring machine (TBM), immersed tube, and the New Austrian Tunneling Method (NATM), each offer distinct advantages and disadvantages. While these methods have been continuously optimized to better adapt to complex environments, challenges such as accurate geological forecasting, ground deformation control, and risk management persist, highlighting the need for continued innovation in tunnel engineering practices. This chapter discusses the integration of Virtual Reality (VR) and Digital Twin (DT) technologies in tunnel construction, which introduces an integrated VR-DT framework, modular resource classification, and the services these technologies offer. Then, a case study is presented to demonstrate its effectiveness in enhancing both operational efficiency and overall safety. The use of VR and DT technologies enables real-time monitoring and immersive visualization, improving tracking of excavation, reinforcement, and logistics. These systems help engineers and managers identify risks, test strategies, and make adjustments during construction to reduce delays and cost overruns. As tunnel projects become increasingly complex, the continued integration of DT, VR, AI, and IoT technologies will be crucial for creating smarter, safer, and more efficient construction processes.

1. Introduction

With the acceleration of urbanization and the continuous improvement of infrastructure, the demand for transportation is steadily increasing. Tunnel construction has become a crucial means of alleviating traffic congestion. Particularly in large cities and densely trafficked areas, tunnels can effectively reduce travel time and enhance transportation efficiency. For example, the Zizhi tunnel is 14.4 kilometers long which greatly shortens the traffic time from the northwest to the south of the city, effectively solving the traffic network problem that has plagued the southwest of Hangzhou for a long time, and ending the history

[1] Department of Systems Engineering, Cornell University, Ithaca, USA.
[2,3] Research Center for Urban Fire Safety Engineering, Zhejiang University, Hangzhou, Chin.
* Corresponding author: linhongda@zju.edu.cn

DOI: 10.1201/9781003514480-4

of no urban main road in this area (as shown in Figure 1 (a)); The Yellow Crane Tower tunnel in Wuhan is 2,190 meters long, which has opened up the "dead road" for many years and greatly improved the travel efficiency. In the past, from the direction of Jiyu Bridge to Zhangzhidong Road, it took at least 20 minutes to drive, and now it takes only 3 minutes (as shown in Figure 1 (b)).

Fig. 1 (a) Zizhi Tunnel, (b) Yellow Crane Tower Tunnel.

Tunnels can also protect transportation infrastructure from adverse weather conditions and natural disasters. Moreover, relocating roads, railways, or other transportation systems underground can alleviate surface traffic congestion and reduce environmental pollution. This is particularly crucial for large cities, where traffic congestion and air pollution have become global challenges. In addition, tunnels serve as conduits for energy supply and transmission. For instance, underground cable tunnels provide a secure means of transmitting electricity, and can also be used for the transportation of natural gas and oil pipelines. Furthermore, they play a vital role in the construction of water supply and drainage systems, ensuring urban water security and efficient wastewater management. As urban development demands continue to grow, the role of tunnels will further expand and deepen. Consequently, advancements in tunnel construction technologies must evolve to meet the requirements of modern society. Tunnel construction technology is constantly evolving, including the Cut and Cover Method, Conventional Mining Method, Drill and Blast Method, Shield Tunneling Method (STM), Tunnel Boring Machine Method (TBM), Immersed Tube Method, and New Austrian Tunneling Method (NATM) (as shown in Figure 2).

The cut-and-cover method, dating back to ancient times (e.g., Babylonian irrigation channels), is the oldest tunnel construction technique. It involves excavating the ground surface, building the tunnel in an open trench, and then restoring the surface. This method is widely used for shallow-buried tunnels and is one of the most basic and commonly employed methods in soft-soil underground engineering. This method is commonly used to construct short tunnels [1, 2]. This method is the oldest in the history of tunnel construction design, which involves cutting and deepening the soil, forming a large ditch space. The tunnel box element is installed after it is closed to restore the surface by a support method capable of bearing the load on the surface [3]. This method is widely used in the construction of subway stations in fast-paced urban areas. It uses piles and wall systems to construct tunnel walls that have been cut and closed, and is the most popular technique because excavation can be done economically [4]. However, the cut and cover method has a clear disadvantage if implemented in urban areas with noise and vibration [5], which have a significant impact on city traffic and the lives of residents. Furthermore, a major environmental risk of the cut-and-cover method is the possibility of soil movement in soft soil areas. Thus, the application of the cut-and-cover method needs to consider several aspects, including environmental conditions, geology, hydrology, depth of work, and economic indicators [6].

As tunneling advanced into the Middle Ages, the conventional mining method emerged to minimize surface disruption caused by cut-and-cover. This technique, inspired by traditional mining, involves sequential excavation and immediate support using blasting, steel arches, and shotcrete [7]. It is more flexible in confined spaces and avoids major surface impacts. During construction, steel arches and shotcrete are used for initial support, followed by a secondary concrete lining that closely fits the surrounding rock or soil to form a thin-wall structure [8]. However, it requires complex support systems, consumes a large amount of timber, and progresses slowly with high labor demand [9]. In the 17th to 18th centuries, the drill and blast method evolved to address the inefficiency and low mechanization of conventional mining. It introduced a repeatable cycle of drilling, charging, blasting, ventilation, mucking, and support installation [10], enabling better rock control and adaptability to various tunnel shapes. While still labor-intensive, it offered improved excavation efficiency in hard rock environments and allowed flexible use of temporary support [11]. A major safety concern arises during mucking and support installation, as the freshly blasted area remains unsupported and poses a high risk of rock falls [12].

By 1843, the development of mechanized tunneling led to the shield tunneling method (STM), first used in the Thames Tunnel. This method overcame the safety and manual labor risks of drill and blast by using a mechanized shield machine to excavate and support the tunnel simultaneously [13]. It offered high safety and minimal surface disruption, particularly in soft soil and urban areas. However, its performance can be affected by complex geology, leading to risks such as face instability, surface settlement, and tool wear. In addition, the high equipment cost, segment production, and maintenance requirements make it less economical for small-scale or variable ground conditions, resulting in significantly higher overall project expenses [14]. To meet the demands of deep and long tunnels across rivers and straits, the tunnel boring machine (TBM) method emerged around 1910. TBMs mechanized rock excavation with rotating cutters and advanced systems, reducing surface disturbance and enhancing safety. Compared to shield tunneling, TBMs offered higher efficiency and were ideal for hard rock [15]. Key components include the cutterhead, drive motors, support frame, and propulsion system [16]. Still, they required high upfront investment and performed poorly in highly heterogeneous geological conditions [17]. Despite its technical advantages, TBM use involves high initial investment and equipment costs. In the 1950s, the immersed tube method was introduced for underwater crossings, solving limitations of TBM and shield methods in wide or deep water [18]. It involves prefabricating tunnel segments, sinking them into dredged trenches, and joining them underwater. This method simplified equipment needs and construction speed, especially for large-scale underwater tunnels. However, the method also presents significant challenges. One of the most critical issues is settlement control, which is influenced by numerous factors such as subsoil variability, foundation treatment, dredging techniques, siltation, and external loads [19]. Finally, in the 1960s, the New Austrian Tunneling Method (NATM) was developed to address the rigidity and high costs of mechanized methods. NATM leverages the self-supporting capacity of surrounding rock, applying flexible support like shotcrete and bolts based on real-time monitoring [20]. It adapts well to changing ground conditions, especially in mountainous regions, and reduces equipment demands. However, its success depends on accurate geological assessment and timely interpretation of monitoring data. Delays or errors in these processes can lead to excessive deformation or collapse. Additionally, NATM is labor-intensive and slower than mechanized methods like shield tunneling and may involve greater management complexity in unstable ground conditions [21].

Fig. 2 Development of tunnel construction methods.

Over time, many tunnel construction methods have been developed to solve problems like ground disturbance, safety risks, or difficult geological conditions. Although methods like shield tunneling and TBM offer high automation and safety, their high cost and unsuitability for shallow tunnels make the cut-and-cover method, with its low cost, simple process, and effectiveness in urban shallow-depth projects, still the most widely used in current practice. Despite technological progress, several key research challenges still hinder the optimization of tunnel construction. One major issue is the lack of accurate predictive models for geological conditions ahead of the tunnel face. Although large amounts of geotechnical data are collected, current models fall short of forecasting with precision, which affects safety, cost control, and project timelines. Another gap lies in the limited automation of construction processes. While tunnel boring machines and some equipment are partly automated, many tasks such as lining installation, quality checks, and site coordination still rely heavily on manual labor. In addition, real-time monitoring and integration of data from sensors, machinery, and environmental systems remain underdeveloped. Better integration could greatly enhance decision-making and reduce risks and delays. Sustainable construction practices also need more attention, particularly in finding alternative materials, reducing energy use, and improving recycling. Lastly, while safety technologies have improved, more research is needed to design automated systems that protect workers, especially in unstable or flooded conditions.

Addressing these research gaps above could significantly improve the efficiency and safety of tunnel construction. In this regard, emerging technologies, such as Virtual Reality (VR) and Digital Twins (DT) offer promising solutions to optimize planning, design, execution, and post-construction phases. Virtual Reality (VR) and Digital Twins (DT) are emerging as potentially transformative technologies for the construction industry. These tools offer several advantages that could optimize the tunnel construction process by enhancing the planning, design, execution, and post-construction phases [22]. VR is a technology that uses computer-generated simulations to create interactive, immersive 3D environments, allowing users to visualize and experience tunnel designs or construction processes before physical work begins. DT involves creating dynamic, real-time virtual models of physical infrastructure, including tunnels, that continuously receive data from sensors and other sources to monitor performance and predict future conditions. By offering realistic simulations for planning and training, VR helps improve predictive modeling and safety training, while DT provides continuous real-time data integration, enabling better monitoring, decision-making, and adaptation to changing conditions during tunnel construction. This can ultimately optimize project timelines, reduce costs, and enhance worker safety.

Currently, VR and DT technologies have been widely applied in construction engineering. Li et al. [23] identified key areas for mucking task lighting by eye-tracking technology as the upper and lower edges of the bucket, debris being moved, and expected stacking area, which aims to address the issue of uneven illumination distribution at construction sites. Ren et al. [24] developed an immersive educational tool to visualize computational fluid dynamics (CFD) modeling datasets of ventilation and respirable dust flow characteristics in the tunnel face in a virtual reality (VR) environment. An algorithm was developed for processing the large and complex CFD datasets into a form that can be processed and visualized using standalone VR headsets with limited processing power. The VR-CFD system was assessed by an industry expert and via many industries showcases, where regular feedback was received for making significant improvements in this education tool. This tool has been developed as a training platform to allow frontline workers to better understand the results of decisions made during tunneling operations and the best practices for dust control. Li et al. [22] present a DT framework consisting of three key parts with specific components (as shown in Figure 3): real object corresponds to a wide range of natural and artificial components. Interaction is a virtual-real mapping process, composed of real-time sensing and management. Digital object includes spatial model, contextual model, real-time update, and evaluation model.

2. Framework for VR-Enabled Digital Twin Systems in Tunnel Construction

Tunnel construction is one of the most challenging forms of infrastructure development due to its underground environment, geological uncertainties, and the necessity for precise excavation and reinforcement strategies[25]. Unlike traditional surface-level construction, where conditions are generally predictable, tunnel construction requires continuous adaptation to dynamic underground conditions, material logistics in confined spaces, and the management of specialized machinery such as Tunnel Boring Machines (TBMs), drilling rigs, and ventilation systems[26, 27]. Additionally, tunnel excavation often involves high-risk working environments, requiring real-time monitoring and proactive decision-making to prevent structural failures, hazardous gas accumulations, or unexpected ground collapses[28].

Traditional construction monitoring and project management approaches lack the ability to synchronize excavation activities with real-time geotechnical analysis, material tracking, and safety assessments[27, 29, 30]. In response to these challenges, the integration of Digital Twin and VR technologies introduces a real-time, data-driven, and immersive approach to tunnel construction management. The Digital Twin functions as an advanced virtual counterpart of the tunnel site, providing continuous, AI-enhanced updates on excavation progress, structural stability, and material utilization[31, 32]. VR extends this capability by enabling immersive interaction with real-time construction data, allowing engineers and workers to visualize excavation sequences, predict potential geotechnical risks, and refine reinforcement strategies in a fully interactive 3D environment[22, 26] By combining real-time monitoring, AI-driven analytics, and VR-based visualization, the proposed framework ensures that tunnel construction projects achieve greater efficiency, accuracy, and adaptability to underground conditions.

2.1 Integrated VR-DT Framework for Tunnel Construction

The VR-DT framework is structured around three key components: the Physical Space, the Digital Twin Space, and the Smart Mobile Gateway, which together ensure seamless synchronization between the underground tunnel site and its real-time virtual representation[33].

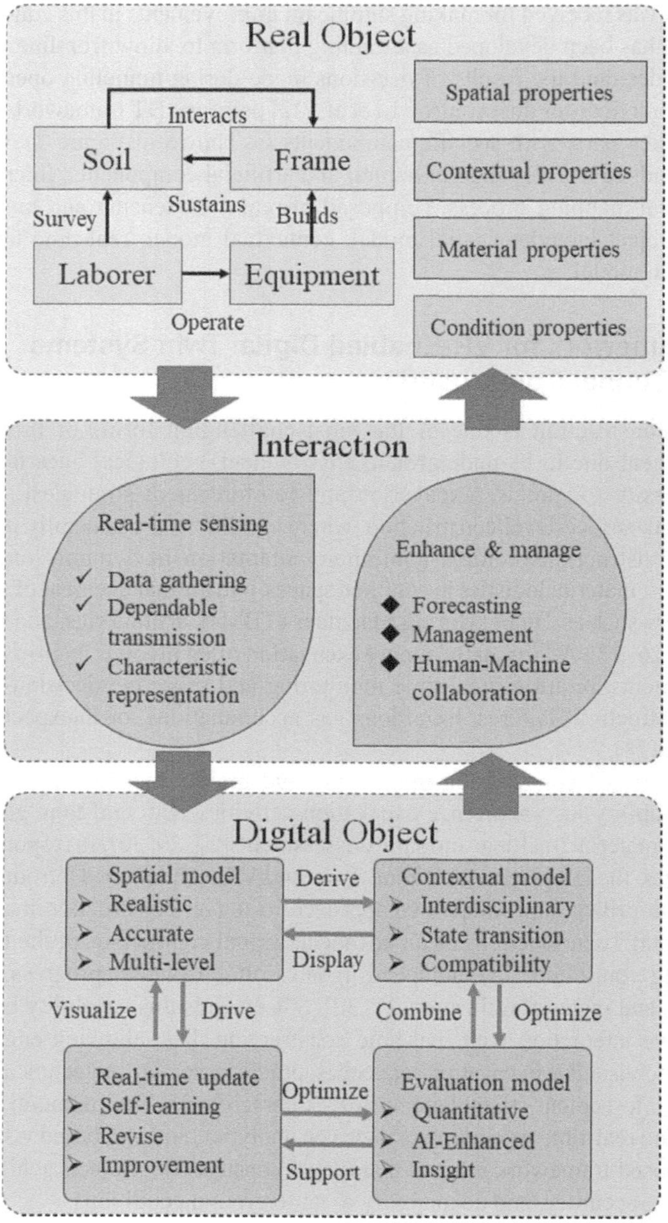

Fig. 3 A conceptual framework of DT [22].

The Physical Space represents the actual tunnel excavation environment, where workers, specialized tunnel construction machinery, and materials interact within a confined and dynamically changing underground setting. Unlike surface-level construction sites, tunnel projects involve limited visibility, complex ventilation requirements, and continuous exposure to geological uncertainties[25]. Workers operate in restricted spaces, necessitating real-time location tracking and safety monitoring[34]. TBMs, drilling rigs, and support machinery must be continuously monitored for wear, overheating, and alignment precision to prevent costly breakdowns. Construction materials, including precast tunnel segments, shotcrete, reinforcement bars, and waterproofing membranes, must be efficiently transported and positioned within the confined underground workspace, where storage and logistics require precise coordination[35].

The Digital Twin Space serves as the real-time, AI-enhanced virtual representation of the tunnel construction process, ensuring real-time updates on excavation progress, equipment performance, and structural stability[22, 36]. Unlike traditional project planning tools that rely on static models, the Digital Twin is a fully dynamic system, continuously fed with sensor data from the tunnel site. This virtual environment integrates geotechnical monitoring, AI-driven predictive modeling, and real-time simulation tools to provide engineers with an interactive, high-fidelity representation of tunnel conditions. Within this space, machine learning algorithms analyze excavation patterns, predict geological instabilities, and generate automated recommendations for optimizing excavation speeds and reinforcement schedules.

The Smart Mobile Gateway acts as the central data processing and communication interface, ensuring that real-time data flows seamlessly between the physical tunnel site and its digital twin. The gateway operates through a three-phase process: data acquisition, real-time AI processing, and automated decision execution. In the data acquisition phase, sensors embedded in TBMs, drilling rigs, conveyor belts, and ventilation systems continuously monitor machine health, ground stability, and environmental conditions such as air quality and gas levels[37]. These data streams are then processed within the Digital Twin's AI-driven analytics engine, where risk factors, structural behaviors, and excavation progress trends are analyzed. The final phase involves sending optimized decisions back to the tunnel site, where AI-guided automation adjusts machine parameters, optimizes material transport logistics, and alerts workers to safety risks in real time.

The integration of VR-based visualization within the Digital Twin provides tunnel engineers with a completely immersive decision-support system[38]. Unlike traditional 2D models, which lack interactivity, VR enables project teams to step inside a real-time virtual representation of the tunnel, navigate through excavation areas, and interact with AI-generated risk simulations[39]. Through VR headsets and motion-tracking controllers, engineers can test reinforcement strategies, visualize subsurface geological formations, and practice emergency response drills in a fully interactive 3D environment[40].

The effectiveness of this framework is illustrated in Figure 4, which presents a Digital Twin-enabled tunnel construction site, emphasizing how real-time excavation progress, material movement, and worker activities are continuously updated within the virtual model. The Smart Mobile Gateway facilitates instant

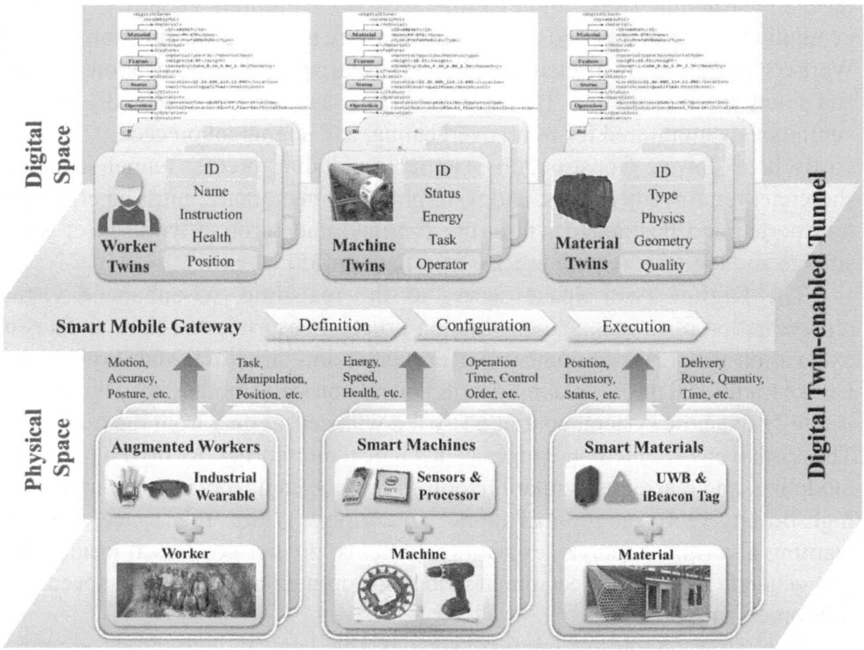

Fig. 4 Digital twin framework for tunnel.

data synchronization, ensuring that any unexpected geological shifts, equipment malfunctions, or safety hazards detected in the tunnel are immediately reflected in the Digital Twin. The VR component enhances situational awareness, allowing project managers to engage in immersive planning sessions, test alternative excavation methodologies, and simulate emergency scenarios before implementing real-world actions.

Beyond monitoring and simulation, the VR-DT framework enhances collaborative decision-making by allowing multiple stakeholders to remotely interact with the tunnel construction model in a shared virtual environment[41, 42]. Instead of relying on static reports, project teams can enter a virtual replica of the tunnel, discuss AI-generated predictive models, and collectively refine excavation plans. This VR-enabled collaboration improves communication, accelerates decision-making, and ensures that all stakeholders remain aligned on risk mitigation strategies and operational adjustments. To further optimize resource tracking, Figure 5 presents a structured classification of tunnel construction components, categorizing materials, machinery, and workforce elements to facilitate automated inventory tracking, AI-driven logistics optimization, and real-time equipment utilization monitoring[43, 44]. By leveraging predictive analytics and real-time material flow simulations, this system ensures that construction materials are precisely allocated, preventing delays and reducing waste.

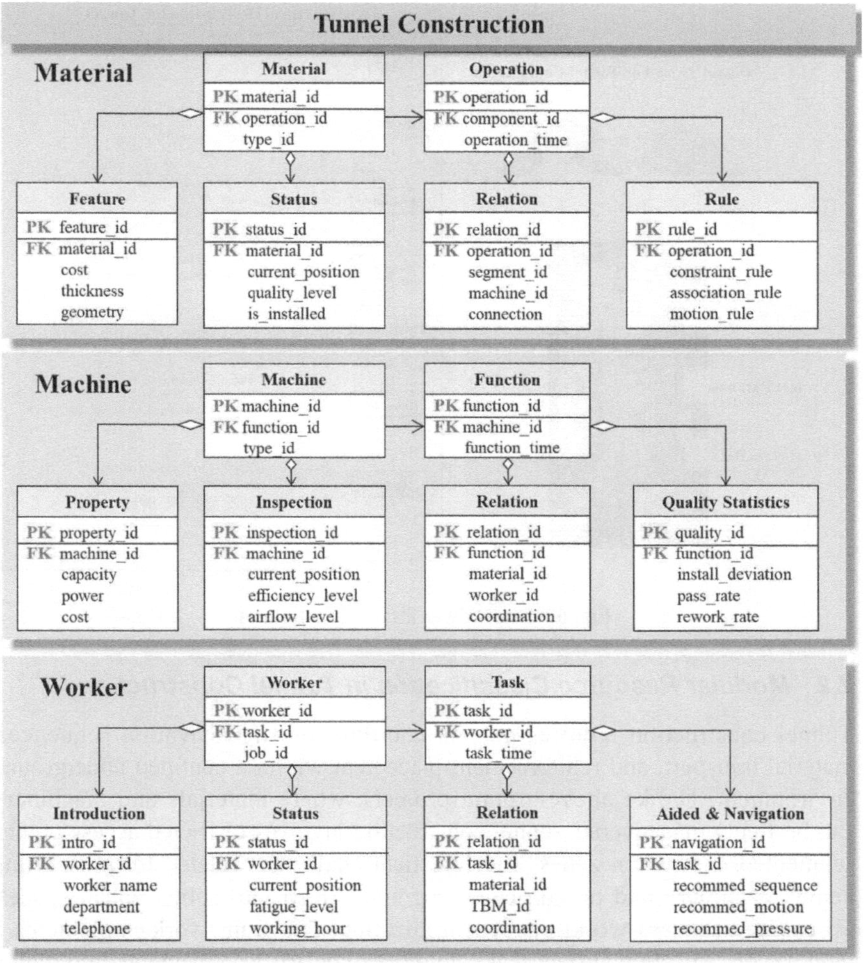

Fig. 5 Digital twin template for tunnels.

By combining sensor-equipped excavation machinery, AI-driven risk modeling, and VR-based interactive visualization, this framework transforms tunnel construction management into a data-driven, predictive, and highly efficient process. A digital twin instance is shown in Figure 6. The integration of real-time monitoring, automated excavation optimization, and immersive digital simulations enables tunnel construction projects to dynamically adapt to underground conditions, mitigate risks, and improve operational efficiency at every stage of the project lifecycle.

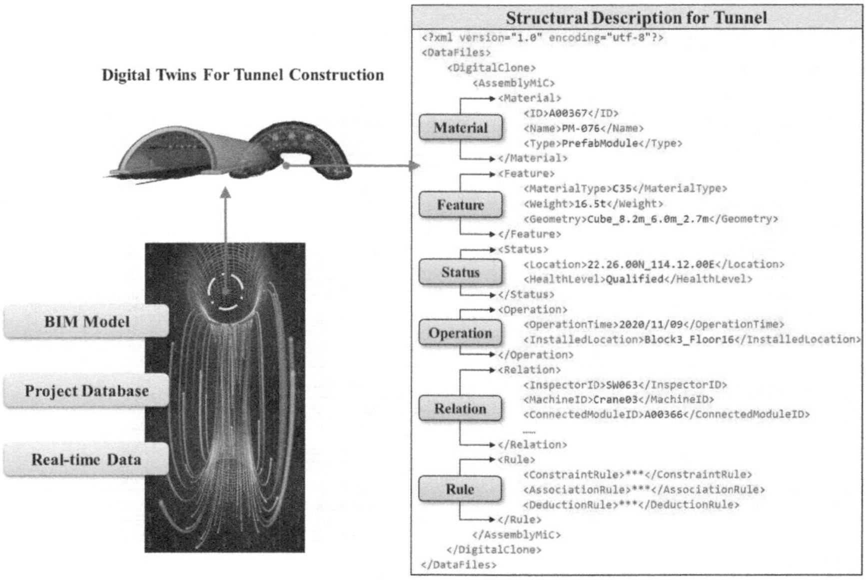

Fig. 6 Digital twin instance for tunnels.

2.2 *Modular Resource Classification in Tunnel Construction*

Tunnel construction requires precise coordination of excavation sequences, material transport, and reinforcement placement within a confined underground environment. Unlike above-ground projects, where materials and machinery can be freely maneuvered, tunnel construction involves restricted access points, segmented excavation zones, and logistical challenges related to transporting heavy machinery and prefabricated segments deep into subterranean spaces. To ensure seamless workflow synchronization, this framework implements a modular classification system that categorizes construction elements into digital representations, enabling real-time monitoring, predictive logistics, and automated resource allocation.

Within this system, excavation machinery such as TBMs, drilling rigs, and automated muck disposal systems are equipped with real-time sensor networks that continuously track machine performance, maintenance requirements, and operational efficiency. AI algorithms analyze these data streams to predict TBM cutter head wear rates, detect deviations in boring trajectories, and adjust excavation speeds based on real-time geological feedback. Unlike surface-level construction, where machinery can be easily accessed for maintenance, tunnel excavation demands predictive failure detection and automated scheduling of repair interventions, minimizing downtime and avoiding costly project delays[45].

Reinforcement materials such as shotcrete, rock bolts, and prefabricated tunnel segments are digitally cataloged within the Digital Twin, allowing for automated material tracking, predictive supply chain coordination, and just-in-time delivery scheduling[46]. The Smart Mobile Gateway ensures that material inventories are continuously updated, preventing stock shortages and optimizing tunnel lining placement in accordance with excavation progress. This approach is particularly critical in long tunnel drives, where delays in material transportation can halt excavation operations, leading to costly inefficiencies. Geotechnical monitoring tools, including ground-penetrating radar (GPR), Lidar scanners, and stress-strain sensors embedded in tunnel walls, provide continuous updates on ground stability, fault line movement, and hydrogeological risks. These tools generate real-time alerts when settlements, rock fractures, or sudden shifts in tunnel lining pressures are detected. The Digital Twin integrates these data streams to create predictive risk models, allowing engineers to visualize underground conditions in a VR-enhanced simulation, where reinforcement strategies can be tested before implementation. Workforce tracking systems embedded within wearable IoT devices, biometric sensors, and smart helmets ensure continuous safety monitoring, fatigue detection, and automated hazard alerts. In underground environments, where oxygen levels, toxic gas concentrations, and temperature fluctuations pose serious risks, the Digital Twin synchronizes real-time worker vitals with environmental monitoring systems, triggering automated evacuation alerts and VR-assisted emergency simulations when hazardous conditions arise.

This modular classification system, when fully integrated with the VR-DT framework, transforms tunnel construction into a predictive, data-driven process, ensuring that excavation sequences, reinforcement schedules, and safety procedures remain dynamically optimized in response to evolving underground conditions.

2.3 Services of the VR-DT Framework in Tunnel Construction

The VR-DT framework offers a groundbreaking solution for addressing the unique challenges of underground excavation, risk mitigation, and workforce management[47]. Unlike traditional construction monitoring systems that rely on periodic inspections and static project schedules, this framework integrates real-time sensor networks, AI-powered automation, and immersive VR-based visualization, ensuring that tunnel construction remains dynamically optimized at every phase. The core services provided by this framework include real-time geotechnical monitoring and risk assessment, AI-driven excavation workflow optimization, VR-based safety training, and predictive maintenance for tunnel machinery and materials.

(1) Real-Time Geotechnical Monitoring and Risk Assessment

One of the most critical challenges in tunnel construction is the continuous assessment of geological stability, stress distribution, and structural reinforcement requirements[48]. Unlike above-ground projects, where construction sites are exposed to visible environmental conditions, tunnels operate in high-risk, concealed environments, requiring advanced monitoring tools to detect potential failures before they occur. The Digital Twin continuously processes data from Lidar scanners, GPR imaging systems, and strain sensors embedded within tunnel walls, generating high-fidelity 3D models that dynamically update as excavation progresses. AI-driven risk analysis tools assess rock mass behavior, hydrogeological variations, and potential fault line disturbances, identifying areas that may require additional reinforcement or adjusted excavation techniques. These findings are immediately visualized in the VR interface, where engineers can immerse themselves in the tunnel's digital twin, interact with live geotechnical data, and test alternative reinforcement strategies before implementation. Through VR-enhanced safety simulations, workers can also receive real-time hazard alerts when ground deformations or unexpected subsurface pressures are detected. Instead of relying on static 2D schematics, project teams can physically navigate a fully interactive virtual tunnel model, where they can visualize stress points, identify hazardous excavation zones, and collaboratively decide on adaptive reinforcement solutions.

(2) AI-Driven Workflow Optimization and Process Automation

Tunnel construction requires precise coordination of excavation speeds, material removal logistics, and reinforcement placement, all of which must adapt to constantly changing underground conditions. The VR-DT framework incorporates AI-driven workflow optimization, ensuring that excavation sequences remain dynamically adjusted to evolving ground conditions. The Digital Twin continuously receives sensor data from TBMs, drilling rigs, ventilation systems, and conveyor belts, analyzing these inputs to predict equipment wear rates, optimize muck removal strategies, and prevent bottlenecks in tunnel segment installation. Unlike traditional construction sites, where machinery can be moved or repositioned freely, tunnel excavation requires precise logistical sequencing, with each phase influencing the next. The VR interface allows engineers to explore multiple excavation strategies, test different material flow optimizations, and simulate alternative TBM speed adjustments in real time. AI-driven excavation planning also enhances ventilation system efficiency, ensuring that airflow dynamics are optimized to prevent the accumulation of toxic gases and dust particles within confined tunnel spaces. The VR simulation environment allows engineers to visualize air circulation patterns, adjusting ventilation duct placements and fan speeds within the virtual model before implementing changes on-site.

(3) VR-Based Training and Safety Simulations for Workers

Tunnel construction poses high-risk safety challenges, requiring workers to operate within dark, confined spaces, where structural collapses, ventilation failures, and equipment malfunctions pose life-threatening dangers[49]. Unlike traditional training methods that rely on classroom instruction and theoretical manuals, the VR-DT framework provides a fully immersive, scenario-based safety training platform. Workers can enter VR-simulated emergency drills, where they practice evacuation procedures for rock collapses, underground flooding, or toxic gas leaks. These simulations include haptic feedback and AI-driven scenario randomization, ensuring that each training session adapts dynamically to evolving risk conditions. Instead of reacting to emergencies for the first time in real-world situations, tunnel workers are already familiar with critical response protocols, having undergone multiple VR-assisted emergency simulations. For equipment operators, VR-based training modules provide interactive digital replicas of TBM control systems, drilling rig operations, and material handling equipment, allowing them to practice operating machinery in a risk-free virtual environment before using real-world equipment. This approach significantly reduces on-site errors, machinery accidents, and costly delays due to operator inexperience.

(4) Automated Predictive Maintenance and Material Management

Unlike surface construction, where equipment can be readily accessed for repairs, tunnel machinery is often deep underground, requiring highly predictive maintenance scheduling to prevent costly failures[50]. The VR-DT framework integrates AI-powered predictive maintenance algorithms, which continuously monitor TBM cutter headwear, drilling rig vibration frequencies, and conveyor belt motor efficiency. The Digital Twin preemptively identifies maintenance needs before breakdowns occur, preventing costly excavation stoppages and ensuring uninterrupted tunnel boring operations. Material logistics are also optimized through VR-enhanced inventory visualization, where engineers can enter a virtual material tracking interface, view live tunnel segment inventory, and adjust delivery schedules based on real-time excavation progress.

Tunnel construction presents unique challenges due to its complex geological conditions, confined workspaces, and continuous excavation requirements. Unlike traditional surface-level construction, tunnel projects demand precise real-time monitoring, adaptive risk management, and seamless coordination of excavation machinery, materials, and workforce. These constraints require a data-driven, predictive approach that ensures construction remains efficient, safe, and adaptable to underground conditions. The VR-DT framework provides an integrated solution by combining real-time Digital Twin modeling, AI-powered predictive analytics, and immersive VR-based visualization. The Digital Twin continuously synchronizes excavation progress, structural stability, and environmental factors, ensuring that all operations remain dynamically optimized. The Smart Mobile Gateway facilitates real-time data exchange, allowing engineers to adjust excavation workflows, predict geological risks, and automate reinforcement strategies. The incorporation of VR-based simulation and visualization further enhances decision-making and safety training, enabling engineers and workers to immerse themselves in real-time tunnel models, interact with AI-driven risk assessments, and test excavation strategies before implementation. This capability transforms tunnel construction into a proactive, simulation-driven process, where risks are identified and mitigated before they impact real-world operations.

By integrating sensor-driven monitoring, AI-guided process automation, and VR-assisted planning, the VR-DT framework redefines tunnel construction management, ensuring greater efficiency, enhanced safety, and cost-effective resource utilization. As advancements in AI, VR collaboration, and real-time geotechnical modeling continue to evolve, this system will further improve automation, predictive decision-making, and intelligent underground infrastructure development.

3. Case Study: Digital Twin-Enabled VR System for Tunnel Construction

3.1 Project Overview

Tunnel construction presents a unique set of engineering and operational challenges, particularly in complex underground environments where space is confined, geological conditions are uncertain, and construction sequences must be precisely coordinated. Unlike surface construction, tunnel projects operate within restricted access zones and require continuous adaptation to dynamic subsurface conditions. These complexities demand advanced tools for real-time monitoring, data-driven decision-making, and seamless coordination across excavation, reinforcement, and logistics workflows.

This case study focuses on the application of a Digital Twin-enabled VR system to the Wangzhai Tunnel project in Wencheng County, which began construction in 2013. The tunnel is located along the G322 Ruiyou highway, configured as a two-lane, bi-directional single-bore tunnel with a total length of 2,240 meters, including 28 meters of cut-and-cover and 2,212 meters of mined tunnel. The tunnel has a clear width of 9 meters and features both straight and curved alignments. Specifically, the segment from K64+635 to K66+730.5 lies on a straight path, while K66+730.5 to K66+875 follows a 650-meter radius curve, with a transverse slope of 2% and a longitudinal slope of 2.82% in an upward direction. The mined section of the tunnel was constructed using the New Austrian Tunneling Method (NATM), while the cut-and-cover section was built using open excavation.

In this context, the Digital Twin-enabled VR system was introduced to provide a comprehensive, real-time digital representation of the tunnel construction process, enabling precise control over excavation activities, segment assembly, structural monitoring, and material logistics. The system integrates IoT sensors, predictive AI algorithms, and immersive VR interfaces, allowing stakeholders—including clients, project managers, construction workers, and crane operators—to coordinate operations through a shared, data-rich digital environment.

As shown in Figure 7, the system is structured into three primary service modules: Asset Management Service, which tracks materials and monitors segment assembly and position in real time; Real-Time Supervision Service, which supports excavation progress monitoring, analytic reporting, and structural quality assessments; and Construction Execution Service, which verifies tunnel

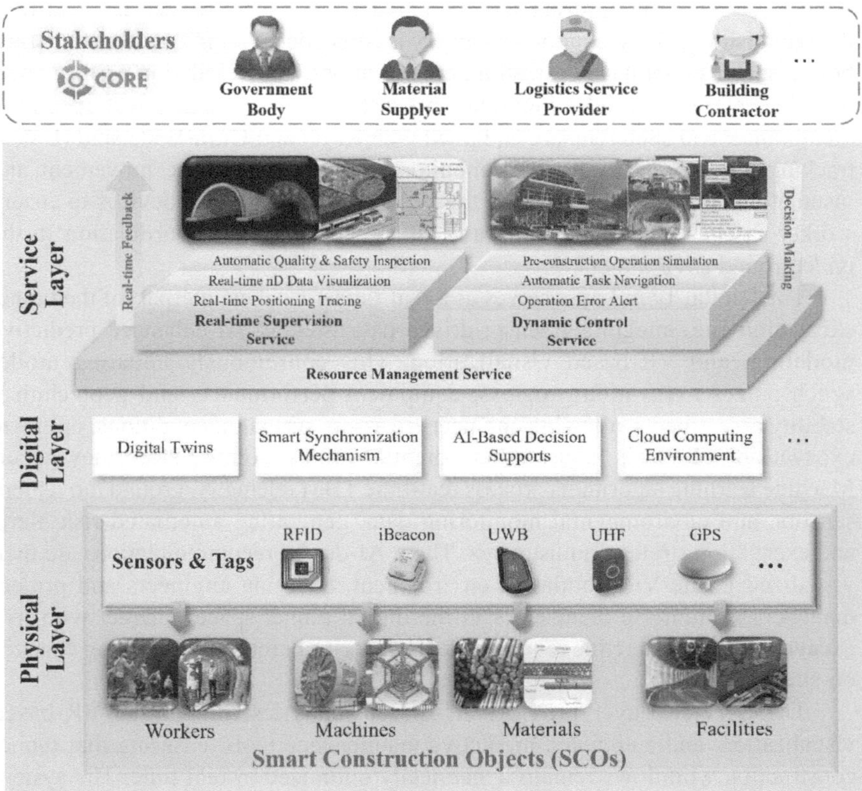

Fig. 7 Multi-Layer Architecture of Smart Construction System.

tasks, provides operational guidance, and issues error alerts when deviations are detected. Together, these services form the backbone of a smart, data-driven tunnel management approach, supporting safe, efficient, and coordinated construction across the full length of the Wangzhai Tunnel.

3.2 Digital Twin-Enabled VR System for Tunnel Construction

The implementation of the Digital Twin-enabled VR system in this tunnel construction project follows a three-layered architecture: the Physical Layer, the Digital Twin Layer, and the Service Layer, each of which plays a critical role in ensuring real-time cyber-physical interoperation.

The Physical Layer consists of workers, excavation machinery, reinforcement materials, and geotechnical monitoring tools operating within the confined underground tunnel space, as shown in Figure 8. Unlike surface construction sites, tunnel projects rely on highly specialized excavation equipment such as TBMs, drilling rigs, automated shotcrete sprayers, and ventilation systems, all of which require continuous monitoring and predictive maintenance. To facilitate real-time data acquisition, all excavation machinery is embedded with IoT sensors that track boring speed, cutter headwear, structural vibrations, and ventilation performance. Reinforcement materials, including precast tunnel segments, waterproofing membranes, and shotcrete, are equipped with RFID and Ultra-Wideband (UWB) tracking systems, ensuring precise monitoring of material storage, movement, and installation. The workforce is also tracked using wearable IoT devices to ensure worker safety, fatigue detection, and emergency response coordination in the underground environment.

The Digital Twin Layer serves as a real-time virtual counterpart of the tunnel excavation site, integrating sensor-driven data streams, AI-enhanced predictive modeling, and VR-based visualization. This continuously updating model synchronizes excavation progress, equipment performance, and geotechnical stability assessments, allowing engineers to assess structural conditions, optimize excavation sequences, and predict potential failure points before they occur. Machine learning algorithms process real-time TBM telemetry, ground movement sensors, and environmental monitoring data, generating automated risk alerts and excavation strategy adjustments. These AI-driven recommendations are then visualized in the VR simulation environment, allowing engineers and project managers to immerse themselves in the digital tunnel space, interact with live excavation data, and refine reinforcement strategies before implementing changes on-site.

The Service Layer provides AI-driven analytics, immersive VR-based visualization, and automated predictive maintenance tools, ensuring that tunnel construction workflows remain dynamically optimized in real time. The system delivers three key services: Asset Management Service, Real-Time Supervision Service, and Construction Execution Service. The Asset Management Service enables precise tracking of tunnel lining segments, material staging, and real-time position tracking of key construction elements. The Real-Time Supervision Service provides live excavation progress monitoring, analytical chart reviewing, and structural quality tests, ensuring that excavation sequences remain optimized. The Construction Execution Service includes task verification, aided operation guidance, and real-time error alerts, allowing workers and managers to proactively address construction deviations. By integrating these services, the Digital Twin-enabled VR system ensures continuous automation, real-time decision support, and enhanced worker safety throughout tunnel construction operations.

3.3 Case Study Implementation: Real-Time Cyber-Physical Interoperation

The Digital Twin-enabled VR system is implemented in a high-capacity transportation tunnel project, where precise excavation control and structural reinforcement are critical to project success. The integration of IoT-enabled excavation machinery, AI-driven predictive modeling, and immersive VR-assisted decision support enables real-time cyber-physical interoperation, ensuring optimized excavation efficiency, enhanced worker safety, and automated risk mitigation. This system provides continuous monitoring of TBM operations, ground stability, ventilation performance, and material logistics, allowing engineers to adjust excavation sequences, predict failures, and enhance construction safety through VR-based simulations. The service architecture of this system is visualized in Figure 8, which categorizes the functionality into three core modules: Asset Management Service, Real-time Supervision Service, and Construction Execution Service.

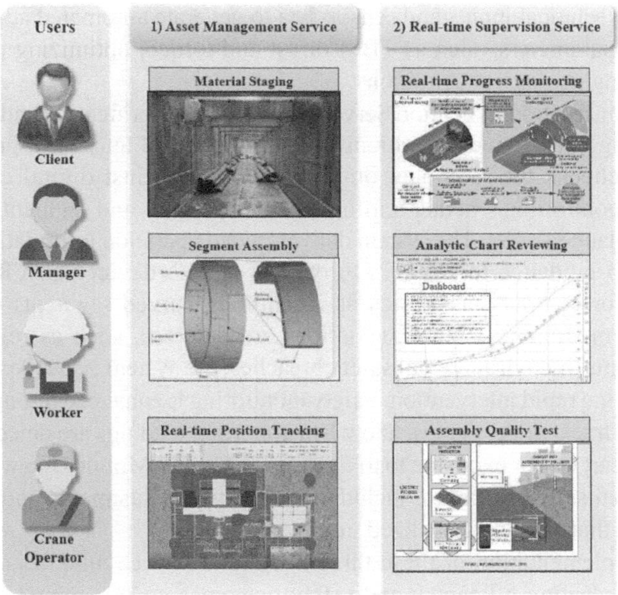

Fig. 8 Digital twin services for tunnels.

The Asset Management Service focuses on organizing and tracking physical resources in the tunnel environment. It enables real-time management of materials, particularly in areas such as material staging and segment assembly. In tunnel construction, these elements are logistically complex due to constrained underground access and limited storage space. The system ensures that all materials—such as precast tunnel segments, reinforcement bars, and sealing elements—are digitally cataloged and tracked throughout the construction cycle. With real-time position tracking, workers and equipment are monitored as they move through excavation zones, ensuring safe operations and efficient task coordination. This module prevents misplacement of resources and supports automated alerts when materials are not staged or assembled in accordance with excavation progress.

The Real-time Supervision Service enables dynamic excavation monitoring, integrating real-time data from TBMs, ventilation systems, and environmental monitoring sensors. Engineers can review excavation performance through visual dashboards and analytic chart reviews, which visualize progress trends, equipment utilization, and productivity metrics. This module also includes assembly quality tests, which compare actual segment installation with predefined standards, helping project managers quickly detect deviations or installation errors. All of this information is continuously synchronized with the Digital Twin and projected into the VR interface, where stakeholders can immerse themselves in the evolving tunnel model to assess progress and identify risks. The system's AI components process geotechnical inputs and sensor data to generate automated adjustments to excavation parameters, such as TBM thrust and torque, optimizing productivity under varying geological conditions.

The Construction Execution Service supports task verification and operational safety during daily tunnel operations. As illustrated in Figure 8, it provides task validation checks that compare on-site activities against digital construction sequences, confirming whether reinforcement operations and segment placements follow the planned order. The system delivers aided operation information, offering step-by-step guidance to workers through wearable devices or VR terminals, especially for complex tasks such as assembling curved segments or applying waterproof membranes. When deviations are detected, such as reinforcement misalignments or environmental safety breaches, the system issues real-time error alerts, allowing rapid intervention. Safety monitoring is enhanced through wearable sensors that track worker vitals, and when abnormal readings are detected-such as increased heart rate or exposure to toxic gases-the Smart Mobile Gateway activates a coordinated safety response, including ventilation adjustments and evacuation instructions that are also visualized in VR.

The implementation of these three integrated service modules significantly improved execution efficiency and safety outcomes on this tunnel project. The Asset Management Service maintained a high level of material flow control and segment traceability, minimizing construction delays due to material shortages or misplacement. The Real-time Supervision Service allowed engineers to evaluate progress and quality with a high degree of granularity, while AI-driven adjustments ensured optimal TBM operation across varying strata. The Construction Execution Service supported both operation guidance and safety enforcement, with real-time alerts and VR-based collaboration enabling fast responses to on-site anomalies. By connecting these layers of digital support into a unified cyber-physical system, the project team transformed traditional tunnel construction practices into a proactive, simulation-enhanced process with data-driven planning and intelligent control at its core.

4. Conclusions

Tunnel construction methods have experienced continuous development to address the increasing demands of urban expansion and complex geological conditions.

Early techniques, such as the cut-and-cover method, which featured simple procedures and low construction costs, established the foundation for urban tunnel building. As cities grew in size and density, the disadvantages of early methods, including environmental disturbance and traffic obstruction, became increasingly evident. To overcome these issues, shield tunneling methods were introduced, improving ground stability and minimizing impacts on the surface environment. The adoption of TBM further enhanced construction speed and safety, particularly for long-distance and deep tunnels constructed under difficult conditions. Later, the New Austrian Tunneling Method was developed, emphasizing the use of surrounding rock strength and the timely installation of support structures, which improved adaptability to varied geological environments. Overall, each method represents a trade-off among construction efficiency, safety, environmental protection, and cost control. Despite these technological improvements, challenges such as accurate geological forecasting, control of ground deformation, and management of construction risks remain. These challenges continue to affect the selection and application of construction techniques and highlight the need for further refinement of tunnel engineering practices.

The implementation of a Digital Twin-enabled VR system in tunnel construction significantly improves excavation management, safety, and decision-making. By integrating sensor-driven monitoring, AI-powered modeling and immersive VR-based visualization synchronizes excavation progress, structural reinforcement, and logistics. Unlike conventional methods relying on delayed feedback, this framework supports real-time cyber-physical integration, allowing project teams to continuously monitor TBM performance, ground stability, and worker safety within an intelligent ecosystem.

Real-time excavation visualization, immersive VR-based ground assessment, and scenario simulation have significantly advanced the quality and efficiency of underground project execution. Engineers can immerse in a virtual tunnel to evaluate potential risks, validate reinforcement strategies, and test excavation sequences before field implementation. AI-driven analytics provide automated recommendations for adjusting TBM thrust, ventilation, and material transport, transforming construction workflows adaptive and predictive. This integration substantially reduces delays, prevents cost overruns, and minimizes safety hazards, resulting in a more streamlined and intelligent tunnel construction process.

As tunnel construction projects grow in scale, depth, and complexity, especially in urban, hydropower, and underground utility sectors, Digital Twin and VR technologies will be increasingly indispensable. Integrating IoT data, AI optimization, and VR planning offers a powerful foundation for next-generation tunnel infrastructure. Future advances in multi-user VR collaboration, geotechnical simulation, and automated excavation will further enhance system capabilities, making tunnel construction safer, more efficient, intelligent, adaptive, and simulation-driven.

References

[1] Liu, X.-Y., Fan, Z.-J. and B.-F. Yang. *Research on Safety Measures of Shallow-buried and Undercut Tunnel*. in *E3S Web of Conferences*. 2021. EDP Sciences.

[2] Li, M.-G., et al., *Comparative study of construction methods for deep excavations above shield tunnels.* Tunnelling and Underground Space Technology, 2018. 71: pp. 329–339.

[3] ILANCHELVAN, A., 2020. *Tunnel-Pile-Soil Interaction.*, Universiti Teknologi Malaysia.

[4] Ardeshana, A., Pitroda, J. and Bhavsar, J. *Tunnels and Tunneling Operations: Introduction To Old and New Era.* in *Conference: International Conference on: "Engineering: Issues, Opportunities and Challenges for Development" ISBN: 978-81-929339-1-7At: SN Patel Institute of Technology & Research Centre, Umrakh, Bardoli.* 2015.

[5] Abdullah, R., A review on selection of tunneling method and parameters effecting ground settlements. *Electronic Journal of Geotechnical Engineering*, 2016. 21(14): pp. 4459–4475.

[6] Khan, M., Sadique, M. and Alam, M. Stress analysis of tunnel in soft soil: a state of art report. *International Journal of Advance Research in Science and Technology*, 2017. 6(03): pp. 187–195.

[7] Li, P., Yan, L. and Yao, D. *Study of tunnel damage caused by underground mining deformation: calculation, analysis, and reinforcement.* Advances in Civil Engineering, 2019. 2019(1): pp. 4865161.

[8] Wang, Q., et al., *The safety risk assessment of mine metro tunnel construction based on fuzzy Bayesian network.* Buildings, 2023. 13(7): pp. 1605.

[9] He, M., et al., Compensation excavation method control for large deformation disaster of mountain soft rock tunnel. *International Journal of Mining Science and Technology*, 2022. 32(5): pp. 951–963.

[10] Sun, Z., Zhang, D. and Fang, Q. *Technologies for large cross-section subsea tunnel construction using drilling and blasting method.* Tunnelling and Underground Space Technology, 2023. 141: p. 105161.

[11] Mottahedi, A., Sereshki, F. and Ataei, M. *Development of overbreak prediction models in drill and blast tunneling using soft computing methods.* Engineering with computers, 2018. 34: pp. 45–58.

[12] Girmscheid, G. and Schexnayder, C. *Drill and blast tunneling practices.* Practice periodical on structural design and construction, 2002. 7(3): pp. 125–133.

[13] Chen, X., et al., *A review of seismic resilience of shield tunnels.* Tunnelling and Underground Space Technology, 2023. 136: pp. 105075.

[14] Zhou, C., et al., *Deep learning technologies for shield tunneling: Challenges and opportunities.* Automation in Construction, 2023. 154: p. 104982.

[15] Chen, E. and Leung, C.K. *Mechanical aspects of simulating crack propagation in concrete under steel corrosion.* Construction and Building Materials, 2018. 191: pp. 165–175.

[16] Zhang, Y., et al., *A visual survey of tunnel boring machine (TBM) performance in tunneling excavation: Mainstream direction, brief review and future prospects. Applied Sciences*, 2024. 14(11): p. 4512.

[17] Rostami, J., *Performance prediction of hard rock Tunnel Boring Machines (TBMs) in difficult ground.* Tunnelling and Underground Space Technology, 2016. 57: pp. 173–182.

[18] Li, W., et al., *Model test of immersed tube tunnel foundation treated by sand-flow method.* Tunnelling and underground space technology, 2014. 40: pp. 102–108.

[19] Gursoy, A., *Immersed steel tube tunnels: An American experience.* Tunnelling and Underground Space Technology, 1995. 10(4): pp. 439–453.

[20] Golser, J., Keuschnig, M. and Weichenberger, F.P. *NATM–Review and Outlook.* Geomechanics and Tunnelling, 2020. 13(5): pp. 466–474.

[21] Du, B., et al., *Conception and Exploration of Using Data as a Service in Tunnel Construction with the NATM.* Engineering, 2018. 4(1): pp. 123–130.

[22] Li, T., et al., *Digital twin for intelligent tunnel construction.* Automation in Construction, 2024. 158: p. 105210.

[23] Li, J., Zhu, J. and Guan, C. *Mitigating Fatigue in Tunnel Construction through Illumination: Evidence from Virtual Reality.* Journal of Construction Engineering and Management, 2024. 150(11): p. 04024165.

[24] Ren, T., et al., *Development and evaluation of an immersive VR-CFD-Based tool for dust exposure mitigation in underground tunnelling operations.* Tunnelling and Underground Space Technology, 2024. 143: p. 105496.

[25] Chapman, D.N., Metje, N. and Stark, A. *Introduction to tunnel construction.* 2017: Crc Press.

[26] Mroueh, H. and Shahrour, I., *A simplified 3D model for tunnel construction using tunnel boring machines.* Tunnelling and Underground Space Technology, 2008. 23(1): pp. 38–45.

[27] Liu, H., Li, P. and Liu, J., *Numerical investigation of underlying tunnel heave during a new tunnel construction.* Tunnelling and Underground Space Technology, 2011. 26(2): pp. 276–283.

[28] Sousa, R.L. and Einstein, H.H., *Lessons from accidents during tunnel construction.* Tunnelling and Underground Space Technology, 2021. 113: p. 103916.

[29] Sousa, R.L. and Einstein, H.H., *Risk analysis during tunnel construction using Bayesian Networks: Porto Metro case study.* Tunnelling and Underground Space Technology, 2012. 27(1): pp. 86–100.

[30] Song, Z., et al., *Study of the stability of tunnel construction based on double-heading advance construction method.* Advances in Mechanical Engineering, 2020. 12(1): p. 1687814019896964.

[31] Liu, L., Song, Z. and Li, X. *Artificial intelligence in tunnel construction: A comprehensive review of hotspots and frontier topics.* Geohazard Mechanics, 2024. 2(1): pp. 1–12.

[32] Lin, P., Zhang, L. and Tiong, R.L., *Multi-objective robust optimization for enhanced safety in large-diameter tunnel construction with interactive and explainable AI.* Reliability Engineering & System Safety, 2023. 234: p. 109172.

[33] Al-Bataineh, M., AbouRizk, S. and Parkis, S., *Using simulation to plan tunnel construction.* Journal of Construction Engineering and Management, 2013. 139(5): pp. 564–571.

[34] Sommovigo, V., et al., Tunnel construction workers' well-being: the role of job control and supervisor support. *International Journal of Construction Management*, 2021. 21(9): pp. 945–957.

[35] Zhang, C., et al., *Formulation and performance of grouting materials for underwater shield tunnel construction in karst ground.* Construction and Building Materials, 2018. 187: pp. 327–338.

[36] Ye, Z., et al., *A digital twin approach for tunnel construction safety early warning and management.* Computers in Industry, 2023. 144: p. 103783.

[37] Buca, R. and Mitrosz, O., *Complex geotechnical engineering for Port of Gdansk development–gateway to Central-Eastern Europe.* in *Proceedings of 13th Baltic Sea Geotechnical Conference, Vilnius.* 2016.

[38] Fan, Y., et al., A digital-twin visualized architecture for Flexible Manufacturing System. *Journal of Manufacturing Systems*, 2021. 60: pp. 176–201.

[39] Zhao, K., Janutolo, M. and Barla, G., *A completely 3D model for the simulation of mechanized tunnel excavation.* Rock mechanics and rock engineering, 2012. 45: pp. 475–497.

[40] Kinateder, M., et al., *Human behaviour in severe tunnel accidents: Effects of information and behavioural training.* Transportation research part F: traffic psychology and behaviour, 2013. 17: pp. 20–32.

[41] Li, X., et al., *Digital-twin-enabled JIT design of rock tunnel: Methodology and application.* Tunnelling and Underground Space Technology, 2023. 140: p. 105307.

[42] Wu, Z., et al., *A novel method for tunnel digital twin construction and virtual-real fusion application.* Electronics, 2022. 11(9): p. 1413.

[43] Zhao, J., et al., *Real-time resource tracking for analyzing value-adding time in construction.* Automation in Construction, 2019. 104: pp. 52–65.

[44] Haas, M., et al., *Waste or valuable resource–a critical European review on re-using and managing tunnel excavation material.* Resources, Conservation and Recycling, 2020. 162: p. 105048.

[45] Galli, G., Grimaldi, A. and Leonardi, A. *Three-dimensional modelling of tunnel excavation and lining.* Computers and Geotechnics, 2004. 31(3): pp. 171–183.

[46] Bobet, A. and Einstein, H. *Tunnel reinforcement with rockbolts.* Tunnelling and Underground Space Technology, 2011. 26(1): pp. 100–123.

[47] Cao, E., Guo, F. and Wang, K. *A digital twin (DT) framework at design and construction phases.* in *International Conference On Sustainable Buildings And Structures Towards A Carbon Neutral Future*. 2023. Springer.

[48] Li, S., et al., *Predicting geological hazards during tunnel construction. Journal of Rock Mechanics and Geotechnical Engineering,* 2010. 2(3): pp. 232–242.

[49] Ulvestad, B., et al., *Increased risk of obstructive pulmonary disease in tunnel workers.* Thorax, 2000. 55(4): pp. 277–282.

[50] Zheng, Y., Zhang, Q. and Zhao, J. *Challenges and opportunities of using tunnel boring machines in mining.* Tunnelling and Underground Space Technology, 2016. 57: pp. 287–299.

[51] Li, J., et al., *Assessment of underground tunnel stability to adjacent tunnel explosion.* Tunnelling and Underground Space Technology, 2013. 35: pp. 227–234.

5 A Virtual Reality-Based Integrated System for Fire Protection Design, Monitoring, and Emergency Evacuation in Tunnels

Xinyu Kou[1], Xiaojun Li[2*] and Chao Chen[3]

Tunnels are characterized by an enclosed environment and dense pedestrian traffic, necessitating special attention to fire safety issues. The application of virtual reality (VR) technology in tunnel fire safety simulation provides an immersive experience that can reduce unexpected human behaviors in fire scenarios. This chapter introduces an integrated system for tunnel fire safety design, monitoring, and emergency evacuation based on virtual reality. First, the system allows engineers, designers, and the general public to optimize fire safety designs for tunnel structures, ventilation systems, and firefighting facilities through an interactive module. Users can verify the effectiveness of design solutions by selecting various fire scenarios. Second, the system integrates real-time monitoring and early warning features, providing dynamic data on smoke dispersion and temperature distribution to assist professionals in quickly locating the fire source and issuing warnings, while also offering emergency response guidance to the public. Finally, the system employs a collaborative hardware-software approach (including control, simulation, data acquisition, and electronic supply systems) to create a highly immersive virtual environment that simulates the fire evacuation process. This enhances the emergency preparedness of both test participants and the general public in real fire situations. By integrating 3D modeling, real-time data visualization, and interactive simulation, the system not only improves the scientific and safety aspects of fire safety design but also provides a comprehensive solution for the prevention, monitoring, and evacuation training in tunnel fire emergencies.

1. Introduction

Tunnels are a crucial part of modern infrastructure, serving essential roles in transportation, water management, resource extraction, and military defense [1]. However, because they are often located underground or within mountains, tunnels face significant fire safety challenges. A fire inside a tunnel can have devastating consequences, including loss of life, structural damage, and long-term disruptions to transportation and other services [2–4].

[1-3] Department of Geotechnical Engineering, Tongji University, Shanghai, 200092, China.
* Corresponding author: lixiaojun@tongji.edu.cn

DOI: 10.1201/9781003514480-5

One of the major challenges in tunnel fire safety is the limited number of evacuation routes. Unlike buildings above ground, tunnels usually have only a few exits available during emergencies. This can lead to congestion and delays, especially when large numbers of people need to be evacuated quickly. Additionally, the confined nature of tunnels makes fire and smoke spread much faster, lowering visibility, increasing temperatures, and depleting breathable air. These factors make it extremely difficult for people to escape safely during a fire [5–6]. In addition to these structural challenges, emergency responders also face difficulties in accessing fire-affected areas within tunnels. Due to the narrow and often complex underground spaces, rescue operations tend to be slower than in above-ground buildings. Moreover, human behavior during emergencies is often unpredictable, with panic and poor decision-making potentially blocking exits and hindering evacuation efforts, which can even lead to secondary disasters [7].

Traditional fire safety training for tunnels often focuses on theoretical knowledge and static, non-interactive drills. QIU et al. [8] used Legion software to simulate and optimize how factors, such as train corridor width, door width, and seat spacing affect evacuation efficiency. NAJ MANOVA et al. [9] conducted emergency evacuation experiments on trains, analyzing how different passenger groups, exit widths, and exit types influence movement patterns and path choices. While these methods are useful for laying the groundwork, they aren't enough when it comes to preparing professionals and drivers to handle the real-life complexities and urgency of tunnel fires [10–11]. These methods can't replicate the fast-changing dynamics of an actual fire or the stress people experience during emergencies. The limitations of these approaches are particularly clear when dealing with the specific challenges posed by fires in confined spaces like tunnels.

Due to the high immersion and strong interactivity of Virtual Reality (VR) technology, it has become an important tool for conducting emergency drills in high-risk industries. Compared to traditional tunnel fire simulation techniques, VR allows the creation of detailed interactive simulations to replicate fire, smoke, and other hazards in confined spaces such as tunnels [12]. These simulations offer an immersive experience, enabling relevant personnel to practice and improve their skills in a safe and controlled environment.

Currently, existing research primarily focuses on system architecture design, industry-specific scenario implementation, and personnel capability training. System architecture design typically employs C/S (Client/Server) architecture and the Unity3D game engine, facilitating modular construction of easy-to-operate, feature-rich platforms. For example, a subway station fire evacuation drill platform has been developed based on the Unity engine and modular functional analysis [13]. Additionally, open-source graphics [14], Unreal Engine [15], HMD [16], and Cyber Session [17] are often used for simulating fire scenarios.

The application of industry-specific scenarios mainly focuses on typical disaster situations and the development of emergency equipment in high-risk industries, with an emphasis on emergency drills tailored to responding to these scenarios. First, in the firefighting sector, training is conducted on the human-machine interaction and simulation of firefighting equipment. VR fire simulations are used to train operators in handling different firefighting tools, improving their proficiency. For instance, immersive VR training simulations have been designed to teach the proper use of fire extinguishers. In addition, emergency response drills for unexpected incidents are also conducted in other high-risk industries. VR-based evacuation exercises have shown that spatial awareness and crowd movement patterns significantly impact evacuation performance. ARTEM D. O. et al. [21] used VR to train personnel in the coal and mining industries, helping them develop critical emergency response skills and knowledge. Zhou et al. [22] applied a social force model to categorize crowds into three groups, finding that the distribution of these groups affects evacuation time. Moreover, placing warning signs at bottleneck points during evacuations can improve efficiency. Zhang et al. [23] employed VR technology to explore how crowd movement influences individual decision-making during emergency evacuations, revealing the impact of crowd flow on path selection and evacuation speed.

Training for personnel focuses primarily on core job skills, with a strong emphasis on genuinely enhancing their emergency response capabilities. KOUTITAS G. et al. [24] developed a virtual interactive training system aimed at improving the speed and accuracy of personnel during rescue tasks. DANG P. et al. [25] created a multi-participant VR fire evacuation framework, ensuring that all participants share the same visual perspective, which increases evacuation safety and the likelihood of successful evacuations by multiple people.

Current tunnel fire simulation technologies make it possible to generate and control multiple scenarios, allowing users to view these situations from various locations and perspectives. This creates a nearly realistic fire experience while ensuring safety, cost-effectiveness, and minimal environmental impact. However, current VR fire simulations typically focus on engineers, operators, and the public, often overlooking the impact of multi-role collaboration on tunnel fire evacuation. Engineers, for instance, need to test and observe firefighting strategies within tunnel fire scenarios, which helps improve firefighting protocols [26-27]. Experts, on the other hand, must issue evacuation warnings quickly and accurately during a fire, thereby improving evacuation success rates. However, firefighting strategies and evacuation warnings are not solely the concern of engineers and experts. For the general public, familiarity with the building's structure plays a decisive role in choosing the safest evacuation routes during a fire [28-29]. The subway, with its closed environment and large crowds, can easily lead to panic among the public during a fire [30-31]. The ability to quickly identify the source of the fire in its early stages is crucial for determining the direction of escape and reducing the level of panic. Therefore, the public needs to understand fire monitoring and early warning systems.

This chapter explains how to develop a VR system for simulating fire evacuations in tunnels. The system serves four primary functions. Engineers can use it to improve fire safety designs for the structure, buildings, ventilation, and firefighting systems. Experts can monitor real-time data to identify the fire's location and issue early warnings. The system's various functions are tested through these fire evacuation simulations.

2. Virtual Reality-Based Integrated System

The study of tunnel fires focuses on key areas such as fire design, structural protection, smoke control, the use of fire suppression systems, and understanding fire behavior [32]. Building on this, Figure 1 shows that the VR-based fire simulation system includes four main functions. A basic knowledge base for anyone who may enter the tunnel. A fire safety design module for professionals. A fire monitoring and early warning system for both professionals and passengers. An immersive

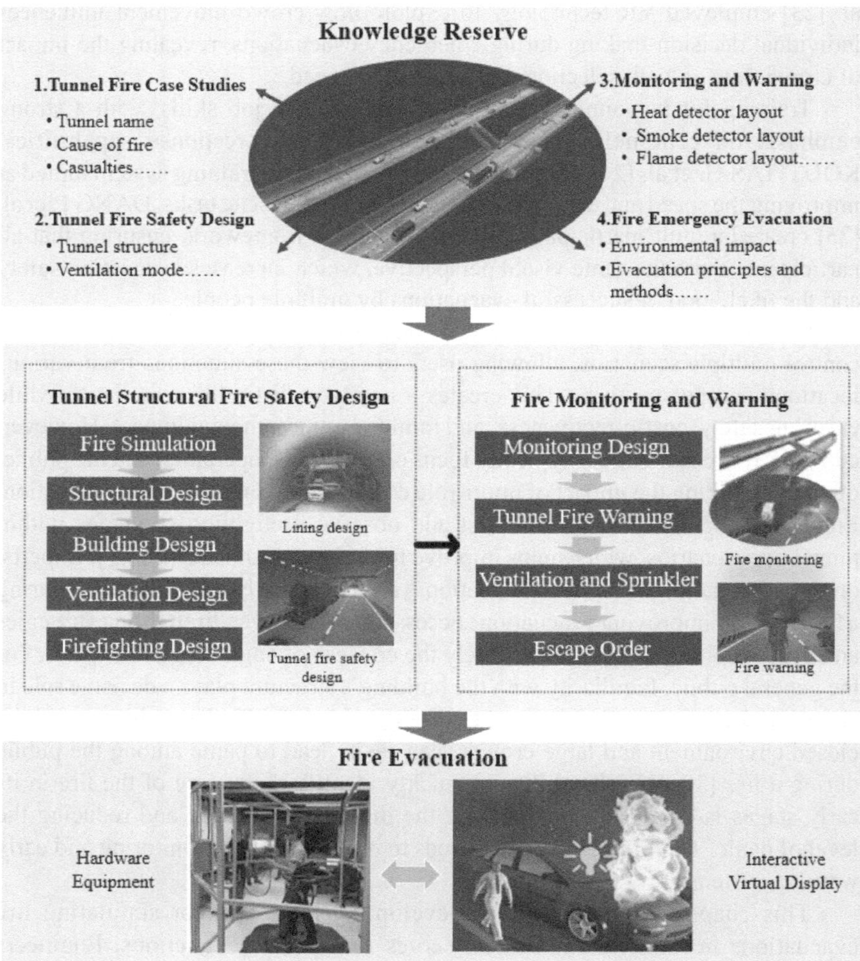

Fig. 1 System Architecture Diagram.

fire evacuation simulation for anyone who may enter the tunnel. The knowledge reserve module includes case studies, tunnel design methodologies, types of fire monitoring and early warning systems, and fundamental evacuation principles. It is designed to strengthen the professional expertise of participants. The fire design feature allows users to interact with tunnel construction, ventilation, and fire suppression plans, enabling them to create safe and effective fire protection designs by selecting different fire scenarios. The fire monitoring and early warning system provides real-time data on smoke and temperature distribution during a fire, helping operators and passengers quickly locate the source of the fire and respond appropriately. The fire evacuation simulation creates a VR model based on the fire protection design, including the tunnel's layout and safety features. This model is then imported into a virtual reality engine for visualization. By linking hardware equipment with software, the system maximizes the realism and immersion of fire evacuation simulations, enhancing the experience for everyone involved.

3. Knowledge Reserve

3.1 Knowledge Base Module

The knowledge base module of the tunnel fire cognition system is built upon real-world cases and engineering practices. It provides a comprehensive framework that integrates accident analysis, fire protection design, fire detection and warning technologies, and emergency evacuation strategies. This module serves as a theoretical foundation to support systematic tunnel fire prevention and control.

Case studies enhance participants' understanding of tunnel fire hazards by presenting key information such as fire duration and casualties. For example, in 2019, a heavy semi-trailer caught fire in the Maoliling Tunnel in China due to tire combustion. The resulting blaze lasted three hours, caused five deaths and 31 injuries, and led to severe damage to the tunnel structure and vehicles. This case illustrates the considerable risks posed by heavy vehicle fires, characterized by high thermal loads and rapid-fire spread. Early-stage emergency response and well-designed evacuation pathways are critical factors in minimizing casualties.

3.2 Fire Protection Design Principles

Tunnel fire protection design encompasses four key areas: control of ignition sources, fire-resistant structural design, smoke control and ventilation, and the configuration of firefighting facilities. As shown in Figure 2, vehicles are the primary fire sources and can be categorized into cars, buses, heavy coaches, and tankers, each exhibiting distinct fire behaviors and thermal loads.

Structural protection focuses on mitigating spalling in concrete linings under high temperatures. This phenomenon, explained by thermophysical mismatch theory, arises from differences in thermal expansion between reinforcement and concrete or aggregate and paste. The use of fiber-reinforced concrete (e.g., steel

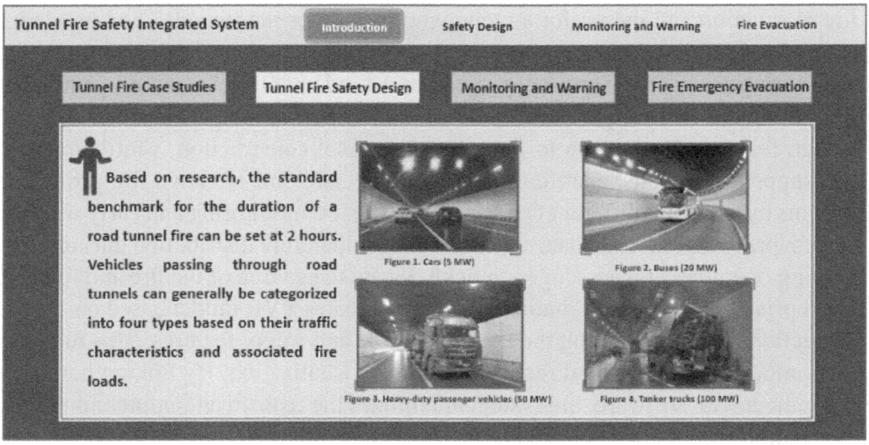

Fig. 2 Different sources of fire.

or polypropylene fibers) enhances thermal resistance by forming stress-relieving microstructures. Evacuation systems are designed to suit tunnel geometry and are generally classified into three types: (1) lateral evacuation, using cross-passages to neighboring tunnels; (2) longitudinal evacuation, using escape routes beneath the roadway; and (3) hybrid systems, which combine both to provide redundancy. Ventilation primarily adopts longitudinal airflow, achieved through jet fans that control smoke movement in a single direction. For localized smoke removal, as shown in Figure 3, zoned smoke extraction systems employ remote-controlled vents and ducts near the fire source to extract smoke efficiently. Firefighting facilities must comply with classification standards: hydrants and extinguishers

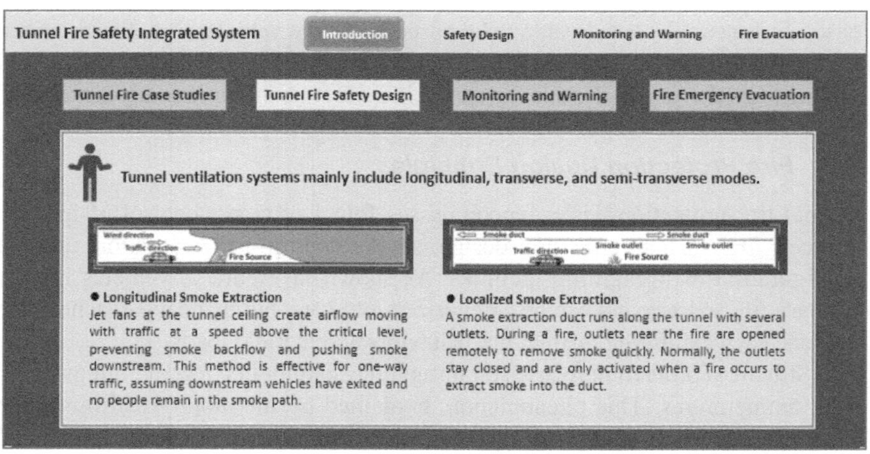

Fig. 3 Knowledge of Ventilation Design in Tunnels.

serve for early suppression, while water spray and foam–water spray systems are used for hazardous material fires.

3.3 Fire Detection and Early Warning Technologies

Fire detection systems can be classified into six categories based on their detection principles: heat, smoke, flame, gas, image-based, and multi-sensor detectors. Heat detectors identify fires by sensing temperature changes using thermal elements or fiber optic sensors. Smoke detectors detect airborne particles through ionization or photoelectric methods. Flame detectors monitor ultraviolet or infrared radiation emitted by fire. Integrating multiple types of detectors improves accuracy, reduces false alarms, and enhances early fire detection in complex tunnel environments.

3.4 Emergency Evacuation Strategies

Environmental conditions significantly affect evacuation effectiveness during a tunnel fire. Voice alarm systems are crucial for alerting and directing individuals, especially at night when auditory signals may be the only means of warning. The effectiveness of voice-based alarms depends on factors such as age, background noise, sleep depth, gender, speech clarity, and medication use.

Emergency lighting ensures visibility when power is lost during a fire, enabling occupants to find exits quickly. The visibility of exit and evacuation signage—especially in low-light or smoke-filled environments—directly influences decision-making and evacuation time. In evacuation scenarios, individuals are advised to move against the smoke flow, cover their noses and mouths with a damp cloth to filter toxic gases, and follow tunnel walls to locate fireproof doors or exits. Vehicle management should follow the principle of "drive forward if ahead of the fire, abandon the vehicle if behind." When abandoning a vehicle, occupants should leave the door unlocked and the key inside to facilitate rescue operations.

4. Tunnel Structural Fire Safety Design

The principles behind tunnel safety design help us understand how to interpret information during tunnel fires and how passengers behave during evacuations. They also assist tunnel operators in making informed decisions [33]. For example, people familiar with the tunnel layout can quickly find cross passages in the event of a fire, while those who are unfamiliar with the design might instinctively try to escape through the tunnel's entrance or exit. In this system, different fire scenarios are considered to calculate the temperature distribution and fire development over time, which then inform the design of the tunnel's firefighting procedures. This approach allows engineers to test and improve various aspects of the firefighting plan, such as the tunnel's structure, layout, ventilation, and fire suppression systems.

4.1 Fire Simulation

By utilizing VR technology, designers can simulate fires originating from various sources (Figure 4), such as different types of vehicle fires, hazardous materials, or even fires caused by ventilation system failures. This flexibility provides a solid foundation for a comprehensive assessment of how different fire types and sources impact the environment.

This system uses smoke visualization technology and lightweight modeling technology to accurately simulate fire scenarios in underground spaces. A Fire Dynamics Simulator (FDS) model is created for simulating fires in underground stations, with fire data calculated in parallel. Figure 3 shows the visualization rendering results of FDS smoke data. Initially, voxel rendering is used to create a visual representation of how smoke spreads over time and space in the fire scenario.

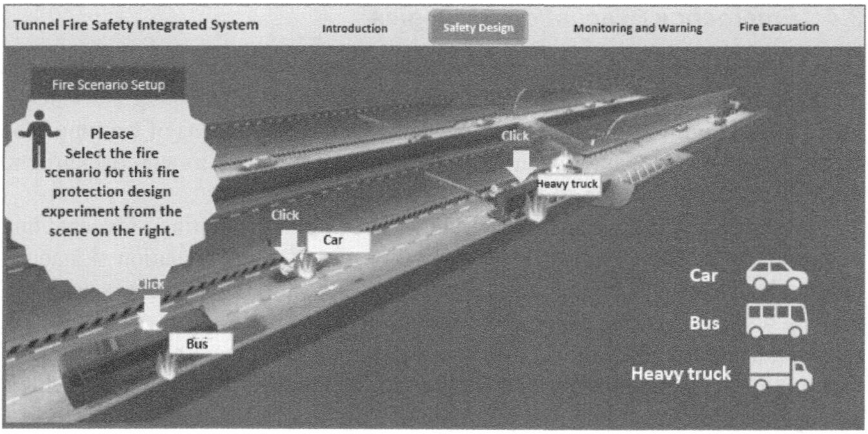

Fig. 4 Fire scenario setting within the VR system.

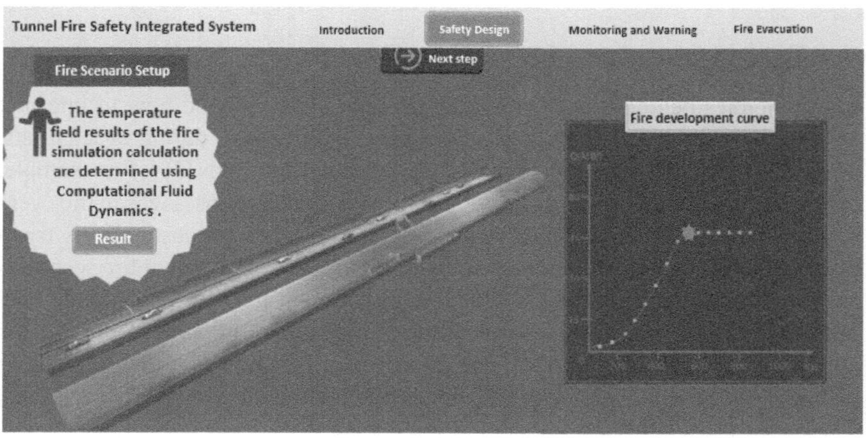

Fig. 5 FDS Smoke Data Visualization Rendering Technology.

Changes in smoke density are used as key information to represent the smoke in the voxel format. Extract smoke location data and temporal variation patterns from FDS calculations, and load them into a network-level model. Voxel rendering helps visualize smoke in space and time. The voxel rendering process requires the FDS model to correspond with the virtual reality environment, where each voxel represents part of the fire scenario. Through simulations and real tunnel fire tests, we study the fire development patterns, the rate of smoke spread along the tunnel, the thickness of the smoke layers, and how the temperature changes within these layers. Furthermore, by applying conservation principles for mass, momentum, and energy in tunnel fire scenarios, we develop models for visualizing smoke, including changes in temperature, density, and spread speed. We also use specific ignition source characteristics to quickly and efficiently model static flames and smoke.

4.2 Tunnel Structural Fire Safety Design

Maintaining the integrity of a tunnel's structure during a fire is essential. Concrete tunnel linings, for example, may crack under extreme heat. With VR, engineers can simulate different fire intensities and durations, allowing them to assess the fire resistance of tunnel structures and reduce the likelihood of concrete cracking. Using thicker linings or fiber-reinforced concrete can significantly improve the structure's resistance to heat-induced damage. Concrete segments reinforced with steel or polypropylene fibers offer excellent crack resistance. Additionally, insulating layers can reduce temperatures within the tunnel during a fire. When the insulation thickness is between 4 and 6 cm, it can provide optimal fire protection. New fire-resistant coatings can also enhance the structural integrity of tunnels, with different coatings requiring specific thicknesses for maximum effectiveness.

After determining the required fire resistance level, engineers select the appropriate tunnel lining type—whether standard or fire-resistant (Figure 4). Based on this choice, the lining thickness is adjusted to balance safety with cost-effectiveness. By using VR to simulate the entire lining installation process, engineers can finalize the fire protection design. They can also adjust the design based on how the structure might respond to fire, such as altering the lining type or thickness if needed.

4.3 Tunnel Building Fire Safety Design

Evacuation and rescue during a fire are among the most challenging aspects of tunnel operation. A well-designed tunnel layout can significantly improve the success of evacuation and rescue efforts. During a fire, there are three main types of evacuation and rescue methods used in road tunnels: transverse, longitudinal, and combined transverse-longitudinal modes.

Transverse Evacuation Mode involves using transverse passages between two parallel tunnels, allowing people to evacuate or be rescued from the affected tunnel to the adjacent one. Longitudinal Evacuation Mode uses escape passages located beneath the tunnel for evacuation and rescue. During a fire, people can

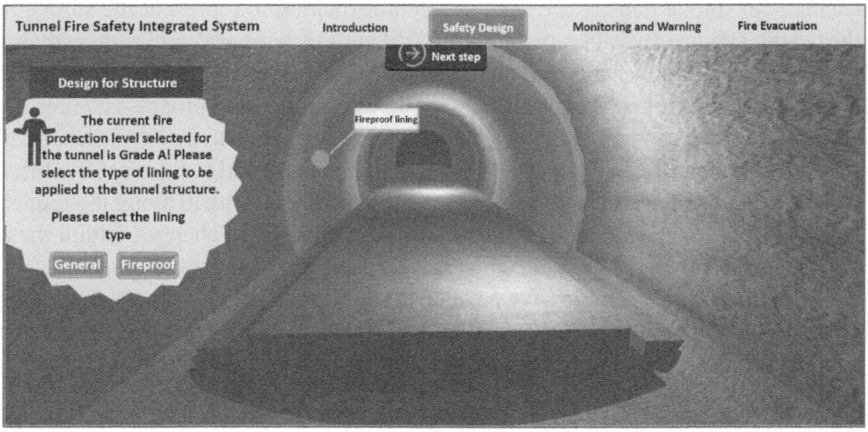

Fig. 6 Structural fire protection design within the VR system.

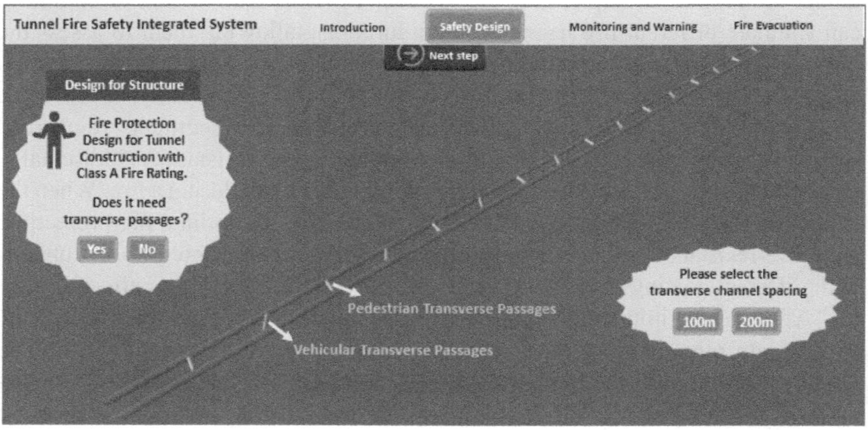

Fig. 7 Fire protection design of buildings within the VR system.

escape via escape slides or vertical ladders that lead to longitudinal passageways beneath the roadway slab. The combined transverse-longitudinal mode combines both transverse and longitudinal methods. In this case, wide transverse passages are installed between the two tunnels, and escape stairs or slides are set up within the tunnel. The longitudinal passage under the roadway slab serves as the evacuation and rescue route during a fire.

When VR is combined with Building Information Modeling (BIM), engineers can use it to determine whether transverse passages are needed based on fire resistance requirements and help decide the optimal spacing for these passages (Figure 7). After finalizing the fire protection design, engineers can adjust the tunnel layout to ensure that both pedestrian and vehicle escape routes are positioned efficiently for emergency evacuation and rescue.

4.4 Tunnel Ventilation and Fire Safety Design

Ventilation is a critical element of tunnel fire safety, as it directly impacts the movement of smoke, heat, and toxic gases during a fire. Different ventilation modes are used in tunnels, including longitudinal, transverse, and semi-transverse ventilation. Longitudinal ventilation is the most common, while transverse ventilation is used less frequently. In the longitudinal smoke extraction mode, jet fans or other ventilation devices installed at the tunnel ceiling create airflow in the direction of traffic, preventing smoke from flowing back toward the fire source. This method is particularly effective in one-way traffic situations, where it is assumed that vehicles downstream of the fire have already exited the tunnel.

Additionally, a dedicated smoke extraction duct should be installed longitudinally, with several smoke vents strategically placed. During a fire, remote control systems open vents near the fire source to quickly remove smoke from the tunnel. Normally, these vents stay closed, but during a fire, specific vents and smoke extraction fans activate to direct smoke toward the extraction duct for efficient removal.

By using VR, engineers can simulate how smoke and heat spread throughout the tunnel, depending on the fire's location and the ventilation system (Figure 6). The more advanced the tunnel's fire resistance, the denser the ventilation system will need to be. VR simulations allow engineers to determine the optimal ventilation configuration, such as the number and spacing of jet fans, and test how well the system works for smoke evacuation and emergency exit routes.

4.5 Tunnel Firefighting System Design

Tunnel firefighting systems are essential for slowing the spread of fire and minimizing damage to the tunnel's infrastructure. The most common firefighting equipment used in tunnels includes hydrant systems, fire extinguishers, water mist systems, and foam-water mist combined suppression systems.

Hydrant systems and fire extinguishers are the most reliable and widely used equipment, especially in road tunnels longer than 500 meters, where they are mandatory. The self-rescue fire hoses that come with these systems are convenient and easy for the public to use. Water mist systems include various types: standard water mist systems, foam-water mist combined suppression systems, and high-pressure fine water mist systems. The foam-water mist system first applies foam to extinguish the fire and then uses water mist for cooling and protection, effectively handling oil-based fires in tunnels in a short time.

VR technology can simulate how these firefighting systems would perform in real-life tunnel fire scenarios. Engineers can select the firefighting equipment they plan to use, such as hydrant systems or water mist systems, and design the layout of the equipment, including spacing (Figure 8). VR simulations allow them to test the effectiveness of these systems under different fire conditions, such as varying intensities or types of fuel involved. This enables engineers to optimize firefighting systems before they are installed in actual tunnels.

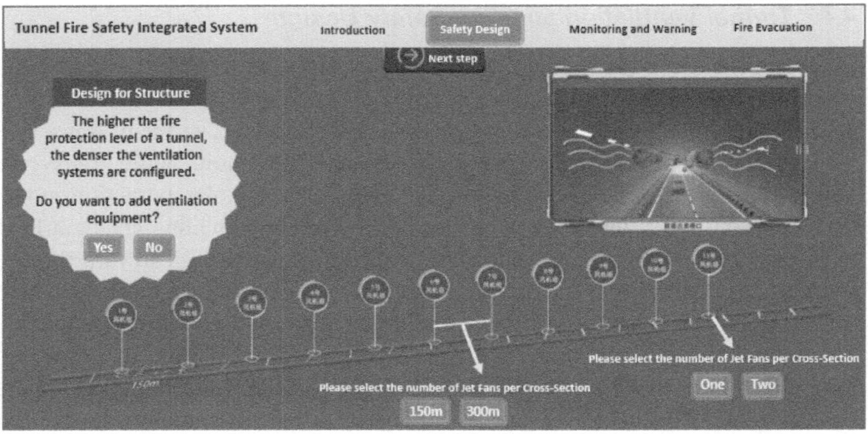

Fig. 8 Ventilation and fire protection design within the VR system.

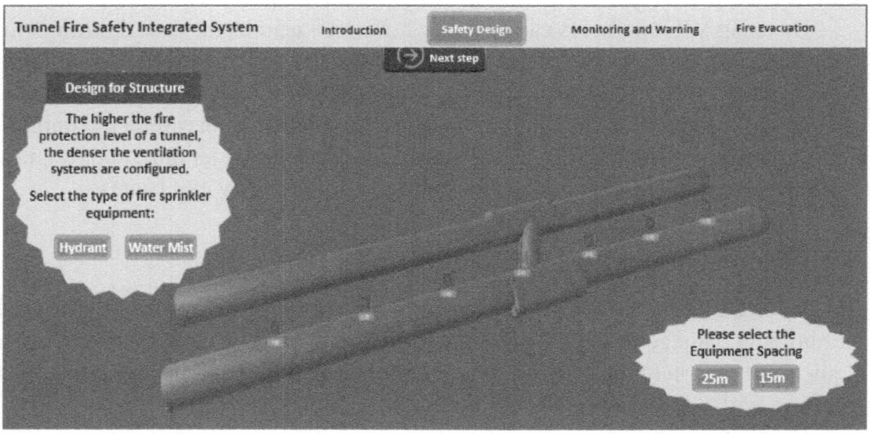

Fig. 9 Fire protection and fire prevention design within the VR system.

The integration of VR technology into tunnel fire prevention design marks a significant advancement in ensuring tunnel safety. Through customizable fire scenario simulations, VR provides a dynamic, detailed, and iterative platform for refining fire protection strategies across structural, ventilation, monitoring, and firefighting systems. By replicating the behavior of fire, smoke, and human response within a virtual environment, designers can identify weaknesses in their systems, optimize tunnel layouts, and improve fire safety protocols before actual implementation. This comprehensive approach to fire prevention design ensures that tunnels are equipped with the most effective safety measures to safeguard both occupants and rescue personnel during fire emergencies, ultimately leading to safer, more resilient tunnel infrastructures.

5. Monitoring and Warning

The monitoring system plays a crucial role in detecting fires early and initiating appropriate emergency responses. Although traditional fire monitoring systems are somewhat effective during a fire, they typically suffer from delays in response time, incomplete information, and challenges in simulation and prediction. Fire detection sensors can be categorized into six main types based on their detection principles: temperature, smoke, light (flame), gas, image, and composite detectors. Each type of detector can be further divided based on its detection target and principle. Common detectors include temperature, smoke, and light-based types. Temperature-based detectors can be further classified into thermal-sensitive, distributed fiber optic, and fiber grating types, depending on their temperature-sensitive components. Smoke detectors include ionization smoke detectors and photoelectric smoke detectors. Ionization smoke detectors detect smoke by measuring changes in the conductivity of air between electrodes in an ionization chamber, while photoelectric smoke detectors detect smoke based on how smoke particles absorb and scatter light. Flame detectors, a type of light-based detector, can be divided into ultraviolet (UV) flame detectors (which are sensitive to high-intensity UV light from flames), infrared (IR) flame detectors (which are sensitive to high-intensity IR light from flames), and composite UV-IR flame detectors.

A VR-based fire monitoring and early warning system primarily works by combining the above sensors to continuously monitor changes in the fire environment. The system's position labeling must be mapped to the 3D coordinate system in the VR environment. To achieve precise mapping of sensor data to the virtual environment, synchronous positioning technology is used to build a spatial digital twin model.

Engineers can analyze real-time data within the VR system to assess the potential risk of a fire. Cameras are installed inside tunnels to monitor the tunnel's real-time status (Figure 8). Engineers then develop a monitoring plan and review the coverage area of each camera to ensure the monitoring is both comprehensive and effective. When a fire occurs, the system provides key information, such as temperature, smoke, wind speed, temperature changes, and the distribution of CO concentration along the tunnel's length (Figure 11). Engineers can actively monitor the fire scene and, using the real-time data from the VR system, determine whether to trigger alarms, activate fire sprinklers, or initiate ventilation. If necessary, evacuation instructions can also be issued. Additionally, testing personnel must promptly identify the fire's source to ensure that the fire is detected as early as possible. Using VR technology, fire monitoring systems can create more accurate and comprehensive real-time simulations, making the data and dynamic changes during a fire more intuitive and easier to understand. This, in turn, improves the efficiency and accuracy of emergency responses.

6. Fire Evacuation

Tunnel fire protection design and monitoring systems create the basic environment needed for safe evacuation and smoke dissipation. To make fire evacuation

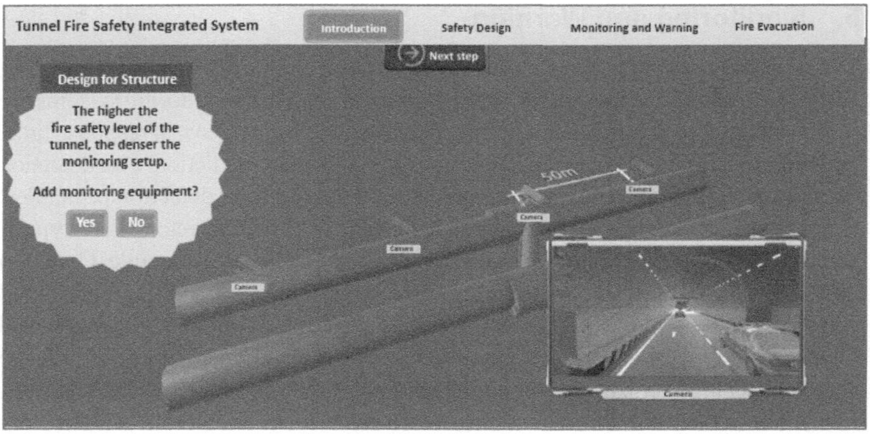

Fig. 10 Fire protection design for monitoring and control within the VR system.

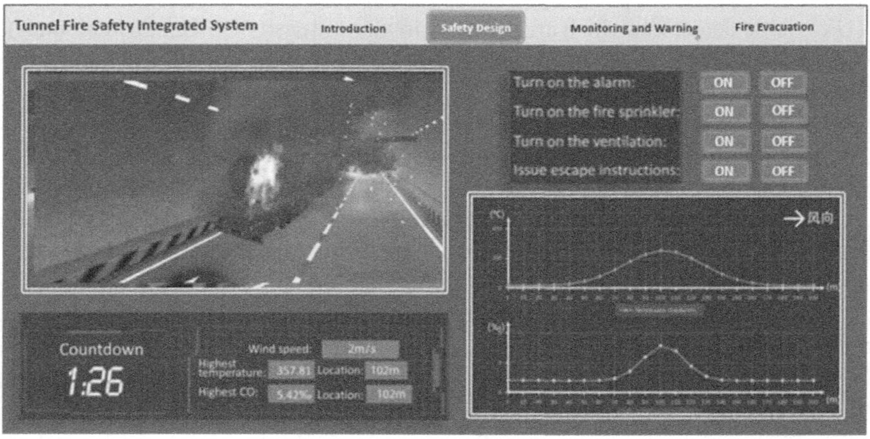

Fig. 11 Fire warning design within the VR system.

simulations feel more real, hardware systems are often used to enhance participants' sense of immersion. As shown in Figure 12, the systems include experimental cabins, thermal simulation devices, VR headsets, motion capture equipment, and heart rate monitors.

The experimental cabin is a semi-enclosed space made of transparent materials, designed to simulate heat and odors experienced during fire evacuations while keeping participants safe. A skylight at the top allows air exchange with the outside. During experiments, the thermal simulation device is activated. Participants stand on a motion capture platform and wear a heart rate sensor and an olfactory simulation device. The thermal simulation device uses a combination of heaters and fans to blow air at controlled speeds, creating the sensation of heat. Heating

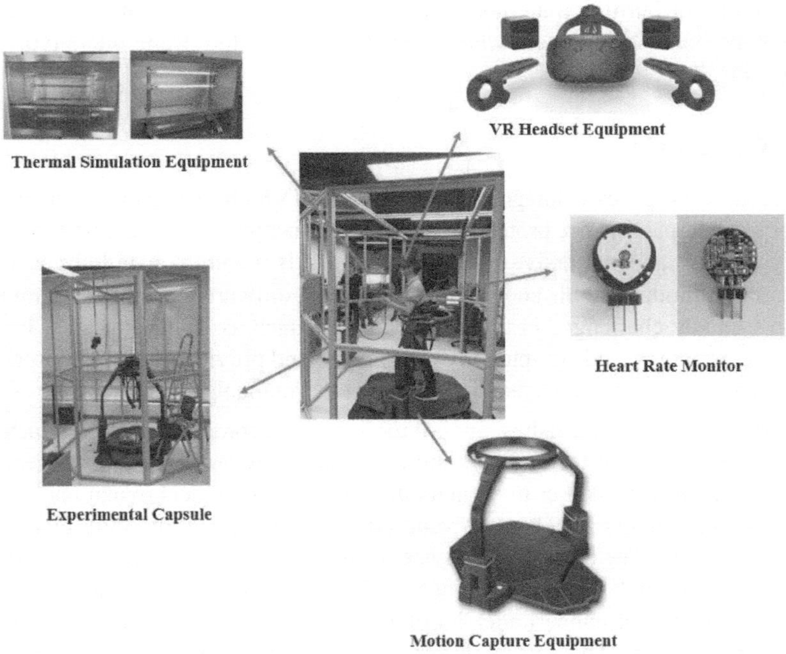

Fig. 12 Hardware composition of the integrated system.

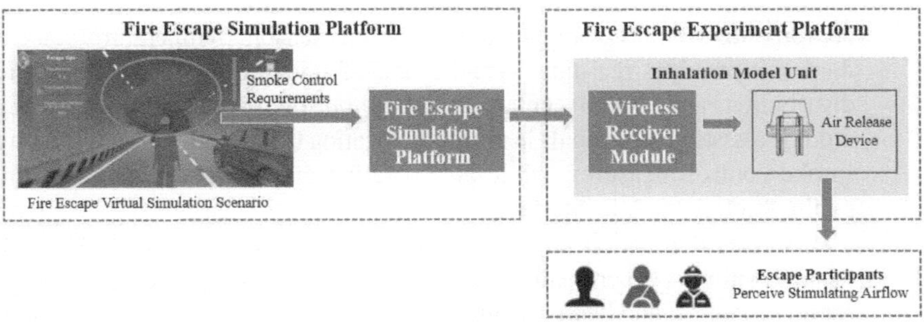

Fig. 13 Flowchart of olfactory unit realization.

levels are adjusted in real time based on data from temperature sensors to ensure realistic heat effects. The motion capture system includes a VR headset, handheld controllers, and a tracking headband, and serves three main purposes: creating a realistic visual and auditory fire scene, tracking and responding to participants' movements, and capturing participants' hand gestures. The Pulse Sensor is an optical device that measures heart rate, pulse waveforms, and heart rate variability. The olfactory simulation device is a small wearable unit that simulates smells like smoke or burning materials. It hangs just below the participant's nose and releases scents via Bluetooth when needed, recreating the strong odors typical of a fire (Fig. 13).

7. Conclusion

This chapter proposes an integrated Virtual Reality (VR)-based system for tunnel fire safety, encompassing fire protection design optimization, dynamic monitoring and early warning, and immersive evacuation training. By combining multidimensional interactive modules with coordinated hardware–software systems, this solution addresses key challenges in tunnel fire safety—such as enclosed space, limited evacuation routes, and complex fire dynamics—and provides a novel approach to enhancing fire safety management in tunnel environments.

1. The system establishes an interactive fire protection design platform. Leveraging VR technology, engineers can simulate various fire scenarios (e.g., car fires, tanker fires) and adjust tunnel parameters dynamically, such as lining material selection, ventilation layout, and firefighting equipment placement. For instance, the system supports design optimization of lining thickness and fireproof coatings, while also enabling validation of smoke control performance under longitudinal ventilation strategies with jet fan arrangements. This virtual design environment facilitates intuitive, iterative testing and ensures the rationality of both equipment distribution and evacuation route planning.
2. The system enables dynamic monitoring and multi-role early warning coordination. By integrating data from various sensors (temperature, smoke, flame), it provides real-time visualization of smoke diffusion, temperature distribution, and CO concentration throughout the tunnel. The monitoring module assists professionals in promptly locating the fire source and initiating smoke control measures, while also offering emergency guidance to the public through the VR interface.
3. The system offers a high-immersion evacuation training framework. Through the integration of experimental chambers, thermal simulation devices, motion capture systems, and heart rate monitors, it accurately replicates the sensory stimuli experienced in tunnel fire environments. Participants wearing VR headsets and using hand-held controllers can make real-time decisions within simulated tunnels—such as choosing between lateral passages or longitudinal escape routes. The use of responsive hardware enhances realism and improves behavioral training outcomes, enabling users to internalize emergency procedures.

Future research may further extend the system's capabilities by increasing simulation accuracy for complex scenarios such as multi-vehicle chain collisions, improving cross-platform compatibility of sensor data, or incorporating haptic feedback into rescue equipment training. Overall, this work provides a valuable technological tool for tunnel fire safety management and contributes to the advancement of intelligent transportation infrastructure and emergency preparedness.

References

[1] Wu, Y., Ahmed H. and Hafez, Y.N. CFD Simulations to study the Effect of Ambient Temperature on the Ventilation in a Metro Tunnel under Different Heat Release Rate[J]. *Journal of Electrical Systems*. 2024. 20-4s, 1998–2009.

[2] Manish Kumar, PMP., Manish Kumar and Jay Bardhan. Revolutionary Design and Optimization of Railway Electrification Overhead Equipment in High-Altitude Tunnels: A Case Study of the Magnificent T-80 Tunnel[J]. *International Journal for Multidisciplinary Research*, 2024. 6(3): May 7.

[3] Jianchun Sun, Heng Zhang and Fen Xiang. Challenges of New Energy Vehicles in Tunnel Fire and Discussion on Emergency Rescue Technology[C]. In *Proceedings of the 7th International Conference on Informatics, Environment, Energy and Applications (IEEA '18)*. 2018:88–92.

[4] Genovese, G.A. Numerical Evaluation for Estimating the Consequences on Users and Rescue Teams Due to the Fire of an Electric Bus in a Road Tunnel[J]. *Applied Sciences*, 2024, 14(20), 9191.

[5] McDaniel, A. New Methodology to Evaluate Critical Fire Life Safety and Emergency Preparedness in Vehicular Road Tunnels[D]. 2017, Colorado School of Mines.

[6] Fernandez Martin S., Ignacio, D.R.L. and Alberto, F.D.L. Tunnel Ventilation Analysis Using a Probabilistic Approach: Case Study, Fire in Road Tunnels with Longitudinal Ventilation[J]. Fire Technology, 2021, 57(3): 1115–1134. DOI:10.1007/s10694-020-01029-2.

[7] May, A., Mitchell, V. and Piper, J. A user-centered design evaluation of the potential benefits of advanced wireless sensor networks for fire-in-tunnel emergency response[J]. Fire Safety Journal, 2014, 63:79–88. DOI: 10.1016/j.firesaf.2013.11.007.

[8] QIU, Hanzhao and FANG Weining. Effect of high-speed train interior space on passenger evacuation using simulation methods[J]. *Physica A: Statistical Mechanics and Its Applications*, 2019, 528: 121322.]

[9] NAJMANOVÁ H., KUKLÍK L., PEŠKOVÁ V. et al. Evacuation trials from a double-deck electric train unit: experimental data and sensitivity analysis[J]. *Safety Science*, 2022, 146: 105523.

[10] Wu, Y. The Challenges of Effective Risk Assessment and Safety Management of Fire and Explosion Hazards[C]//International Seminar on Fire and Explosion Hazards.2017.

[11] Lemkowitz, S.M., Pasman, H.J. *Assessment and Control of Fire and Explosion Hazards and Risks of Particulates[J]*. Springer International Publishing, 2014. 19:97–151. DOI:10.1007/978-3-319-00714-4_4.

[12] Wang, Z., He, R., Rebelo, F., et al. A human-centered design approach: design a new evacuation alarm system for building fire emergency considering the influence of pre-emergency activity[J]. Theoretical Issues in Ergonomics Science, 2023. 25(2), 225–245. DOI:10.1080/1463922x.2023.2166144.

[13] WANG, Z.J., MAO, Z.X., LI, Y.X., et al. VR-based fire evacuation in underground rail station considering staff's behaviors: model, system development and experiment[J]. Virtual Reality, 2022, 27 (2):1145–1155.

[14] Xu, Z., Lu, X.Z., Guan, H., Chen, C. and Ren, A.Z. (2014). A virtual reality-based firetraining simulator with smoke hazard assessment capacity. *Advances in Engineering Software*, 68, 1–8.

[15] Skjermo, J., Moscoso, C., Nilsson, D., Frantzich, H., Hoem, A.S., Arnesen, P. and Jenssen, G. D. (2024). Analysis of visual and acoustic measures for self-evacuationsin road tunnels using virtual reality. *Fire Safety Journal*, 148, Article 104224.

[16] Zhang, X.C., Chen, L.J., Jiang, J.H., Han, S.Y., Zhu, T., Xu, W.B. and Tang, F. (2023). Risk analysis of people evacuation and its path optimization during tunnel fires using virtual reality experiments. *Tunnelling and Underground Space Technology*, 137, Article 105133.

[17] Ronchi, E., Kinateder, M., Müller, M., Jost, M., Nehfischer, M., Pauli, P. and Mühlberger, A. (2015). Evacuation travel paths in virtual reality experiments for tunnel safety analysis. *Fire Safety Journal*, 71, 257–267.

[18] Davide, C., Federico, D.L., Alberto, C., et al. Immersive virtual reality and passive haptic interfaces to improve procedural learning in a formal training course for first responders[J]. *Virtual Reality*, 2022, 27(2): 985–1012.

[19] Lovreglio, R, Duan, X.Y., Rahouti, A., et al.Comparing the effectiveness of fire extinguisher virtual reality and video train ing[J]. *Virtual Reality*, 2020, 25: 133–145.

[20] Artem, D.O., Mikhail, N.K., Denis, L.D., et al. The study of virtual reality influence on the process of professional training of miners. [J]. *Virtual Reality*, 2022, 27(2): 735–759.

[21] Lin, J., Cao, L. and Li, N. How the completeness of spatial knowledge infuences the evacuation behavior of passengers in metro stations:a VR-based experimental study!]. Automation in Construction, 2020, 113: 103136.

[22] Zhou, R., Cul, Y., Wang, Y., et al. A modified social force model with diferent categories of pedestrians for subway station evacuation!]. Tunnelling and Underground Space Technology, 2021, 110: 103837.

[23] Zhang, M., Xu, R., Slu, M.F., et al. Human decision change in crowd evacuation:a virtual reality-based study!. *Journal of Building Engineering*, 2023, 68: 106041.

[24] Koutitas, G., Smith, S. and Lawrence, G. Performance evaluation of AR/VR training technologies for EMS first responders [J]. *Virtual Reality*, 2020, 25(1): 83–94.

[25] Dang, P., Zhu, J., Cao, Y.G., et al. A method for multi-person mobile virtual reality fire evacuation drills based on pose estimation: Consistency of vision and perception[J]. *Safety Science*, 2024, 170: 106334.

[26] Tang, G., Shang, C., Qin, Y. and Lai, J. Current Advances in Flame-Retardant Performance of Tunnel Intumescent Fireproof Coatings: A Review[J]. *Coatings*. 2025; 15(1): 99.

[27] Maevski, Igor, Klein, et al. Impact of Tunnel Ventilation on Tunnel Fixed Fire Suppression System[J]. *ASHRAE Transactions*, 2011. 117(1): 629.

[28] SUN, Q. and TURKAN, Y.A. BIM-based simulation framework for fire safety management and investigation of the critical factors affecting human evacuation performance[J]. *Advanced Engineering Informatics*, 2020, 44: 101093.

[29] FU Meiqing, Liu Rui, Ragan, E. An immersive virtual reality experimental study of occupants' behavioral compliance during indoor evacuations[J]. *International Journal of Disaster Risk Reduction*, 2024, 107: 104420.

[30] Lin, J., Zhu, R., Li, N., et al. Do people follow the crowd in building emergency evacuation? A cross-cultural immersive virtual reality-based study[J]. *Advanced Engineering Informatics,* 2020, 43: 101040.

[31] Zhu, R., Lin, J. and Becerik-Gerber, B., et al. Human-building-emergency interactions and their impact on emergency response performance: a review of the state of the art[J]. *Safety Science,* 2020, 127(11): 104691.

[32] Li, Y. Z. and Ingason, H. (2018). Overview of research on fire safety in underground road and railway tunnels. *Tunnelling and Underground Space Technology*, 81, 568–589.

[33] Ling, J., Li, X., Shen, Y., et al. Human centric VR system development supporting fire emergency evacuation: A novel knowledge-data dual driven approach, *Expert Systems with Applications,* 2025, 273: 126895.

6 Research on the Theory and Technology of Equivalent Lighting in Tunnel Light Environments

He Shiyong[1], Niu Jia An[2*], Liang Bo[3] and Liu Hao[4]

China's tunnel construction, which is characterized by diverse types and large spans, has achieved remarkable accomplishments that have garnered global attention. However, ensuring tunnel lighting safety and low-carbon practices remains challenging in the post-"large-scale construction era." Traditional lighting technologies are inadequate for optimizing the tunnel tunnel light environment, necessitating advancements in lighting theory and technology under the coupled effects of human–vehicle–tunnel tunnel light environment interactions. To address the scientific challenge of non-equivalence between tunnel lighting loads and biological perception effects during tunnel operation, this study first identifies three key issues: difficulty in evaluating driver information perception states under tunnel lighting conditions, insufficient adaptability of existing tunnel light environment optimization theories and technologies, and lack of targeted strategies for tunnel light environment maintenance and management. Second, based on the mechanism of driver information perception, this paper proposes three levels of "equivalence" for tunnel light environments: physical quantity equivalence, biological quantity equivalence, and synergistic equivalence of physical and biological quantities, thereby establishing and validating the equivalent lighting theory. Subsequently, the paper introduces equivalent lighting technologies for tunnel light environments, advocating a paradigm shift from "meeting physical quantity standards" to "human-factor adaptation" through physical and biological equivalence principles. Finally, considering the evolution of driving demand patterns and evaluation methods in the context of intelligent transportation, future research directions are outlined, including the construction of large-scale traffic data models via multi-source data fusion and the development of digitalized, safe, and low-carbon tunnel lighting equipment.

[1] State Key Laboratory of Mountainous Bridge and Tunnel Engineering, Chongqing Jiaotong University.
[2-4] School of Civil Engineering, Chongqing Jiaotong University, Chongqing 400074.
* Corresponding author: ja_niu@163.com

DOI: 10.1201/9781003514480-6

1. Introduction

With the rapid development of China's transportation infrastructure, the total number of highway tunnels nationwide had reached 27,297 by the end of 2023, with a combined length exceeding 30.23 million linear meters, making them a critical component in ensuring the efficient operation of the transportation network[1]. However, as enclosed traffic spaces, tunnels exhibit photometric environmental characteristics that differ significantly from those of open road sections, leading to heightened risks to driving safety and energy consumption issues. According to statistics, in 2022, the accident and mortality rates per kilometer in highway tunnels were 1.62 times and 1.39 times higher than those on open road sections, respectively. Additionally, the annual carbon emissions during tunnel operation reached 8.04 million tons, with electricity costs exceeding 10 billion RMB[2-4]. Driven by the "Dual Carbon" goals, the construction of safe, energy-efficient, and comfortable tunnel lighting environments has become a pressing challenge in the transportation sector. Although current research has made progress in optimizing, evaluating, and maintaining tunnel lighting environments, three critical issues remain unresolved: difficulty in assessing driver information perception states, inadequacy of adaptive lighting optimization theories, and lack of targeted maintenance strategies, all of which require systematic breakthroughs.

Issue 1: Difficulty in Evaluating Driver Information Perception States in Tunnel Lighting Environments

Driver information perception is central to ensuring tunnel safety. Studies have shown that 80% of human-acquired information is visual, yet dynamic lighting variations in tunnels (e.g., light–dark adaptation at portals and differences in sidewall reflectivity) significantly affect the visual performance and physiological load of drivers[5,6]. Existing evaluation systems for tunnel lighting overemphasize physical metrics (e.g., pavement luminance and uniformity) while neglecting biological perception indicators (e.g., ECG, EEG, pupil area, and reaction time) [7]. For instance, standardized pavement luminance calculations exhibit errors of 14.9%–23.4%. While bidirectional reflectance distribution function (BRDF)-based driver luminance perception models reduce errors to <6%, they fail to integrate physiological and psychological parameters[8-11]. Furthermore, tunnel pollutants (e.g., exhaust fumes and smoke) reduce visibility, and the nonlinear relationship between target–background contrast and luminance exacerbates perception complexity[12,13]. Large-sample accident analyses revealed that 92.6% of tunnel accidents stem from human error, with visual distortion and psychological stress during light–dark adaptation posing significant risks. Current evaluation methods cannot quantify such dynamic biological metrics. Therefore, a multidimensional perception evaluation model that integrates physical and biological quantities must be urgently developed to accurately reflect the true perception states of drivers in complex lighting environments.

Issue 2: Insufficient Adaptability of Tunnel Light Environment Optimization Theories and the Need for Integrated Lighting Performance Enhancement Technologies

Current research on tunnel lighting optimization focuses on isolated improvements to single parameters. Examples include adopting high-efficiency LED luminaires (with luminous efficacy reaching 200 lm/W), optimizing lamp spacing and angles (spacing: 3–7 m, angles: 15–25°), and enhancing sidewall reflectivity (e.g., a 21% increase in road surface illuminance when using reflective materials with a reflectance coefficient of 0.65). Although these technologies individually achieve energy-saving goals (e.g., up to a 33% reduction in luminaire quantity) or safety improvements (e.g., a 26.2% increase in small-target visibility), their synergistic effects and adaptability remain systematically unverified. For instance, the combination of LED light sources and high-reflectivity sidewall materials significantly improves lighting uniformity; however, its applicability across varying tunnel cross-sections (e.g., dual vs. triple lanes), traffic volumes (e.g., peak vs. off-peak hours), and environmental conditions (e.g., day–night differences) lacks quantitative analysis. Moreover, existing optimization theories are primarily based on static models (e.g., discretized iterative lighting algorithms) and fail to adequately account for dynamic factors, such as traffic flow speed and weather variations. Studies have revealed significant differences in drivers' reaction times, heart rates, and pupil areas between daytime and nighttime. However, current technologies lack targeted adjustments to spectral composition and correlated color temperature (CCT) to mitigate day–night adaptation conflicts. Notably, the current digitization level of tunnel lighting systems remains low, and real-time data acquisition and analysis capabilities are lacking. This hinders precise adjustments based on real-time traffic data (e.g., traffic volume and vehicle speed) and driver biological parameters (e.g., eye movement metrics and ECG and EEG signals). Therefore, a dynamic, multidimensional theoretical framework for light environment optimization that enables efficient alignment between technical parameters and scenario-specific demands must be established.

Issue 3: Lack of Targeted Maintenance and Management Strategies for Tunnel Light Environments

The long-term maintenance and real-time control of tunnel light environments are critical for ensuring their sustainability. However, current maintenance strategies rely predominantly on generic standards (e.g., Guidelines for Highway Tunnel Lighting Design)[14-17] and lack differentiated solutions tailored to specific tunnel characteristics. For example, aging sidewall materials, luminous decay of luminaires, and pollution accumulation significantly degrade reflectance efficiency and lighting quality. However, existing maintenance cycles and cleaning protocols are not dynamically adjusted based on material properties, such as the durability differences between reflective coatings and cement mortar. In terms of control, while intelligent lighting systems can achieve on-demand lighting through dynamic dimming[18-20], their control logic is often based on fixed thresholds and fails to integrate real-time traffic data (e.g., traffic volume and vehicle speed) or driver physiological feedback (e.g., eye-tracking data and EEG signals). The introduction of natural light (e.g., solar light-guiding technologies) and visual guidance systems (e.g., reflective rings and colored pavement) can reduce the lighting load and enhance spatial perception. However, deploying these technologies requires customized designs that account for tunnel geometric parameters (e.g., length and slope) and regional daylight conditions. Therefore, a data-driven maintenance and management framework must be established to transition from "experience-based decision-making" to "precision regulation," ensuring adaptive and sustainable tunnel light environment operations. To address the aforementioned challenges, this study centers on the "equivalent lighting theory for tunnel light environments," and a research framework for the collaborative optimization of physical and biological quantities is proposed. By integrating luminaire parameters, sidewall materials, and driver perception data, this study established dynamic evaluation models, adaptive optimization technologies, and targeted maintenance and management strategies. The research outcomes aim to provide theoretical support for reducing tunnel accident rates and achieving energy-saving and emission-reduction goals, thereby advancing tunnel light environments from "single-indicator compliance" to "human–vehicle–environment synergistic enhancement" through leapfrog development.

2. Previous Research: Equivalent Lighting Theory for Tunnel Luminous Environments

To address the core scientific challenge of non-equivalent characteristics between tunnel lighting loads and drivers' biological perception effects and to create a safe and energy-efficient equivalent lighting environment for highway tunnels, our research team proposed the Equivalent Lighting Theory for Tunnel Luminous Environments. Under the premise of ensuring safe and comfortable lighting operation conditions for tunnel traffic, this theory achieves original lighting design standards or maintains necessary lighting quantities for traffic safety through technical interventions while reducing the load on lighting systems. This theory was developed from reflections on the limitations of traditional lighting evaluation systems, with its research evolution divided into two phases: initial exploration of dark adaptation and static perception and the proposal of dynamic perception and comprehensive equivalent theory.

To validate this theory, our team conducted matching studies on lighting fixture parameters (e.g., light source type, installation spacing, and mounting angle) and sidewall material parameters (e.g., material type, color, and reflectance coefficient) using lighting models, simulation experiments, and real-world vehicle tests. The feasibility of the theory was evaluated using both physical parameters (road surface illuminance, illuminance uniformity, and lighting efficiency enhancement) and biological parameters (visual performance, physiological indicators, and visual characteristics), progressing from static to dynamic perception analyses[21,22].

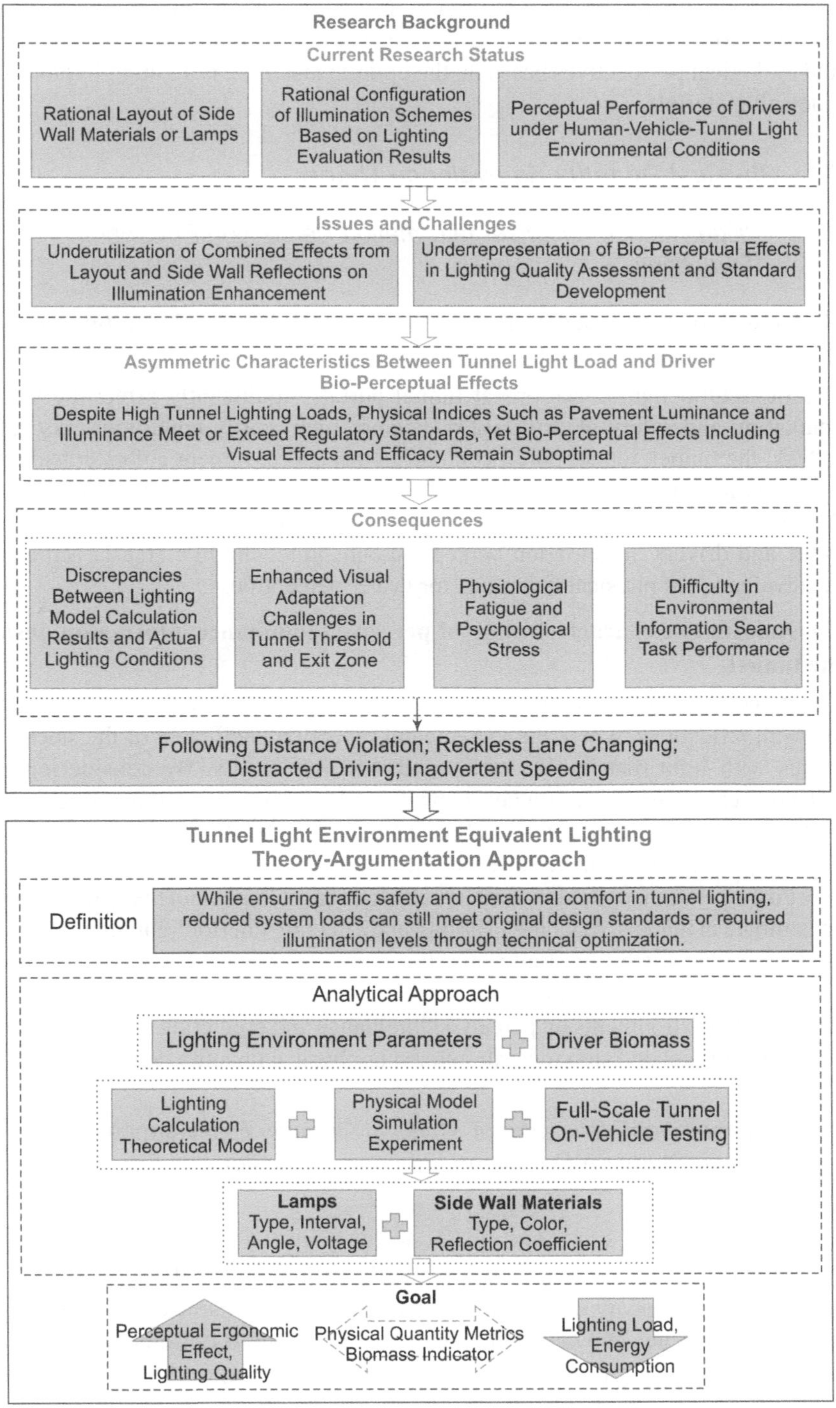

Fig. 1 Construction of equivalent lighting theory in a tunnel light environment.

The theoretical framework is illustrated in Figure 1. To further demonstrate the feasibility of the theory, our team conducted validation studies from both physical and biological perspectives based on the optimal matching of lighting fixtures and sidewall materials.

2.1 Physical Quantity Equivalence Theory

2.1.1 "Equivalence" of Driver Information Perception Physical Quantities

(1) Quantitative characterization of light environment information outside the tunnel portal.

Drastic natural light variations at tunnel portals significantly affect the visual adaptation and safety of drivers[23]. Factors such as sky conditions, sky ratio outside the tunnel, tunnel portal configurations, and pavement colors outside the tunnel influence the $L_{20}(S)$ value (external luminance at the tunnel entrance). Accurately predicting and characterizing the impact of these factors on the $L_{20}(S)$ value and drivers' information perception constitutes the first step in achieving "equivalence" of physical quantities for driver perception.

(2) Quantitative characterization of perceived luminance information inside the tunnel.

The perceived luminance of drivers is primarily provided by direct lighting and sidewall reflections, where the incremental reflection coefficient of the sidewalls varies with light distribution ratios and wall reflectivity. By considering the reflective characteristics of tunnel sidewalls and road surfaces, a theoretical model of driver visual luminance was established using the BRDF, which can control the driver's perceived luminance error to within 6%.

Post-operation, pollutants such as vehicle exhaust fumes and tire wear particles accumulate in tunnels, reducing visibility and affecting interior lighting luminance. Under varying visibility conditions, the target–background contrast increases nonlinearly with lighting luminance[24-28]. Accurately modeling the impact of tunnel light environments on perceived luminance and predicting pollution effects represents the second step towards achieving physical quantity "equivalence."

2.1.2 Theoretical Validation of Physical Quantity Equivalence—Case Study on Combined Luminaire and Sidewall Equivalent Lighting

A combined lighting and sidewall system was analyzed to theoretically validate physical quantity equivalence. A computational lighting model was developed based on light intensity distribution, sidewall reflection characteristics, and light reflection principles to evaluate the tunnel lighting load. By discretizing sidewall units, a discrete iterative illumination algorithm incorporating bidirectional reflectance increments was developed. This model accounts for sidewall material reflectivity and its enhancement of lighting quality, yielding more accurate illuminance values than standard specifications.

Next, simulation tests were performed to evaluate the lighting performance under various parameters: sidewall parameters (material type and reflectance coefficient), luminaire parameters (light source type, spacing, and installation angle), and tunnel structural parameters (tunnel segments and cross-sectional forms). The key metrics included illuminance enhancement, uniformity, and illuminance. The results demonstrated that optimal matching of the luminaire and sidewall parameters achieved remarkable outcomes. For example, the entrance and middle tunnel segments experienced 16% and 21% increases in pavement illuminance, respectively, and the total uniformities improved by 8% and 15%, respectively. Luminaire coverage expanded, reducing the required number of luminaires by up to 33% while maintaining lighting quality. This confirms the feasibility of the physical quantity equivalence theory. By rationally selecting and coordinating the luminaire and sidewall parameters, both lighting quality improvement and lighting load reduction can be achieved without compromising illumination requirements.

2.2 Biological Quantity Equivalence Theory

2.2.1 Equivalent Approach to Driver Information Perception Biomass

During tunnel driving, physiological parameters such as a driver's visual recognition ability, visual performance, visual load, and visual illusions most directly reflect their driving state. Furthermore, the interaction mechanisms of visual perception capability, attention resource allocation, cognitive activation level, and maximum utility value are significantly influenced by the tunnel lighting environment. By integrating drivers' eye movement patterns and cardiovascular–physiological indicators under tunnel lighting conditions, "equivalence" in driver information perception biomass can be achieved by constructing a quantitative evaluation model for drivers' visual perception capability and dynamic perception mechanisms.

2.2.2 Validation of Biomass Equivalence Theory—A Case Study on Driver Visual Performance

Static laboratory experiments and dynamic field trials were conducted to validate the biomass equivalence theory, using driver visual performance as an example. The key experimental variables included different tunnels (the Nanshan, Zhenwushan, Cimushan, and Yeyatang tunnels), time periods (daytime and nighttime), sidewall materials (cement mortar, reflective paint, and reflective materials), and lighting layouts (central and bilateral lighting). The variation patterns of the biomass indicators related to driver visual performance under these conditions were analyzed as critical evaluation criteria.

Reaction time analysis results indicated that the reaction times of drivers generally exhibited a "decrease followed by increase" trend, with the minimum reaction time observed in the tunnel middle section. For instance, the reaction times at 16:00 were lower and less variable than those at 12:00. In tunnels with reflective sidewall materials, the average reaction times at 12:00 and 16:00 were 0.95 and 0.93 times those in tunnels with fireproof materials and 1.01 and 0.99 times those in tunnels with cement mortar sidewalls, respectively. These findings highlight how higher sidewall reflectivity modulates driver reaction time variations across tunnel zones, indirectly demonstrating the effect of lighting environments on visual performance. Regarding small-target visibility, this study revealed that for every 0.1 increase in sidewall material reflectivity, the visibility differences for 1 cm, 2 cm, and 3 cm small targets increased by 26.2%, 9.9%, and 21.7%, respectively. These results confirm that optimized lighting and sidewall parameters enhance visual performance (shorter reaction times and higher target visibility), thus validating the biological quantity equivalence theory. By balancing safety and energy efficiency, this approach reduces the lighting load and energy consumption while meeting illumination standards. By combining physical and biological equivalence theories, the coordinated optimization of luminaire parameters (spectrum, color temperature, and light distribution), sidewall materials, and tunnel geometry can resolve the imbalance between lighting load and driver perception. This "one-up, one-down" strategy (improved light quality and perception with reduced energy demand) establishes a safe, energy-efficient, and perceptually optimized lighting environment for highway tunnels.

3. Important Findings: Tunnel Light Environment Equivalent Lighting Technology

Through nearly two decades of research, our team has summarized the achievements in tunnel light environment equivalent lighting technology into two components: physical quantity information equivalence technology and biological perception equivalence technology. The research process has evolved from physical quantity studies to biological perception studies, reflecting a paradigm shift in tunnel illumination from "meeting physical quantity standards" to "human factor adaptation."

3.1 Physical Quantity Information Equivalence Technology

This technology can be divided into tunnel sidewall coating material, equivalent illumination technology, and luminaire–sidewall combination equivalent illumination technology.

3.1.1 Tunnel Sidewall Coating Material Equivalence Illumination Technology

To investigate the optimization and energy-saving effects of sidewall coating materials, our team validated the feasibility of tunnel light environment equivalence illumination through theoretical modeling. Using a discrete element approach, the tunnel sidewall was divided into multiple reflective units (Fig. 2). By incorporating the BRDF of the sidewall and considering the luminaire's light intensity distribution, sidewall reflection characteristics, and light reflection principles (Fig. 3), we developed a discrete iterative illuminance calculation model (the model construction and equations are detailed in [29]). This model was used to simulate tunnel illumination under sidewalls coated with reflective materials versus cement mortar. As shown in Fig. 4 [30], the model results aligned closely with experimental measurements, with a relative error of only 0.63%, meeting the precision requirements for tunnel illumination calculations.

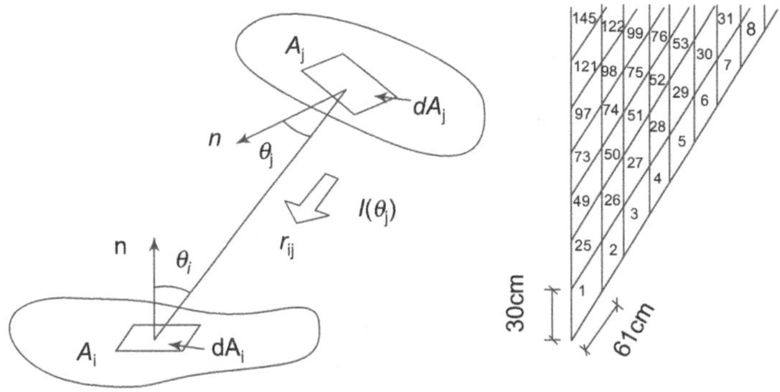

(a) Unit Discretization (b) Unit Distribution

Fig. 2 Discretization of tunnel sidewall elements.

Based on this model, we calculated the horizontal illuminance at road surface points while considering the reflective enhancement mechanism of sidewall materials. Comparisons with existing tunnel lighting standards and on-site measurements (using the four-corner grid method according to GB/T 5700-2008 [31]) are shown in Fig. 5. For a unidirectional two-lane tunnel, 28 measurement points (7 transverse × 4 longitudinal) were established.

The results indicate that conventional lighting standards deviate from measured data by 26.3–31.5 lx (14.9–23.4% error), while the model's calculations deviate by only 2.9–4.8 lx (2.6–4.7% error). This confirms that the model, which accounts for sidewall reflectance, significantly improves the illumination quality and accuracy.

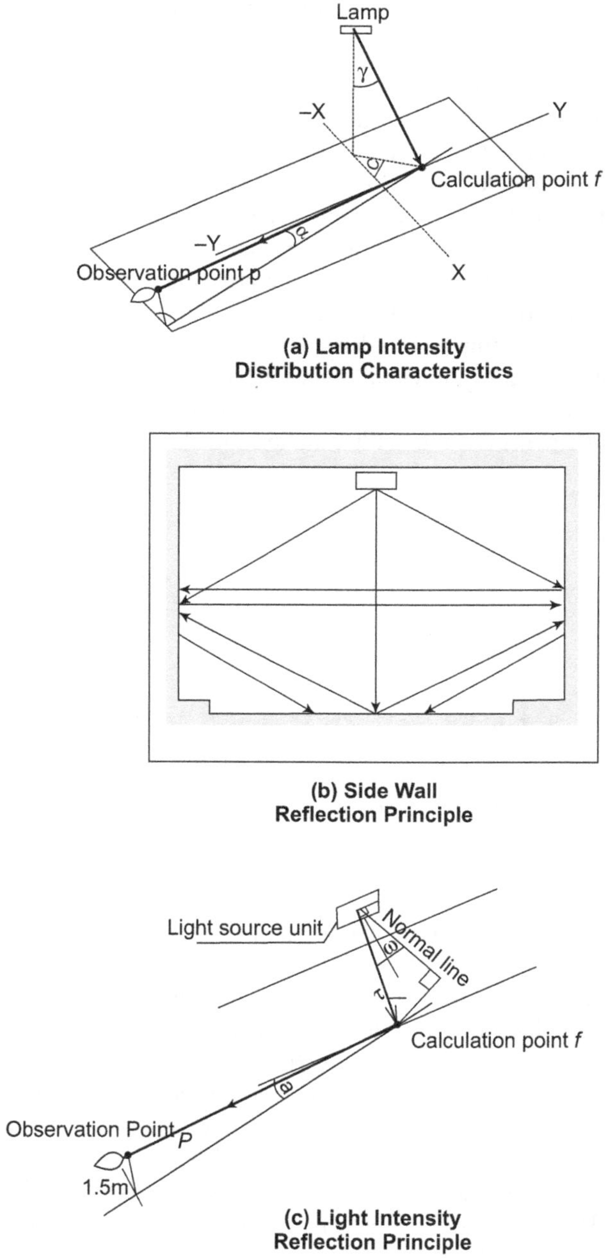

Fig. 3 Luminaire light intensity distribution characteristics and tunnel sidewall light reflection principles.

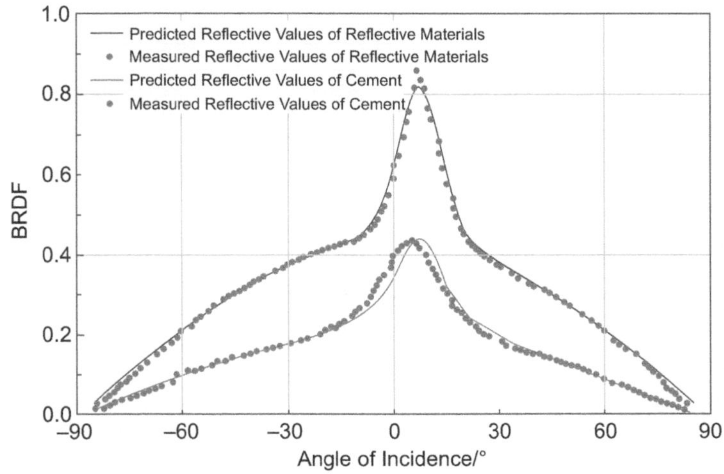

Fig. 4 Model fitting results vs. experimental measurements.

3.1.2 Luminaire–Sidewall Combination Equivalence Illumination Technology

Through simulated and real-world vehicle tests, our team evaluated the feasibility of combining luminaires with reflective sidewalls to optimize lighting load. The key findings were as follows:

- Sidewall reflectance exhibits a linear relationship with road surface illuminance and uniformity.
- Reducing the luminaire spacing from 70 to 50 cm increases illuminance by 41.4% and uniformity by 15.6%.
- Adjusting luminaire angles from -9° to 18° enhances illuminance by 10.5–11.9%.
- A 0.1 increase in sidewall reflectance increases illuminance by 1.93 lx and uniformity by 0.0025 (Fig. 6).

Owing to constraints in field experimental conditions, the installation height and angle of the luminaires in the physical tunnel were fixed. Our team conducted crossover experiments by controlling different lighting zones and sidewall materials in the physical tunnel. The test results are shown in Figures 7 and 8.

As shown in Figure 7, in the tunnel entrance zone, the illuminance of reflective paint and retroreflective materials was 1.09–1.14 and 1.16–1.25 times that of cement mortar, respectively. In the tunnel middle zone, the illuminance values of reflective paint and retroreflective materials reached 1.18–1.14 and 1.08–1.21 times that of cement mortar, respectively. Figure 8 demonstrates that in the entrance zone, the overall uniformity of the reflective paint and retroreflective materials was approximately 1.07 and 1.09 times that of cement mortar, while in the middle zone, these values increased to 1.09 and 1.15 times that of cement mortar, respectively.

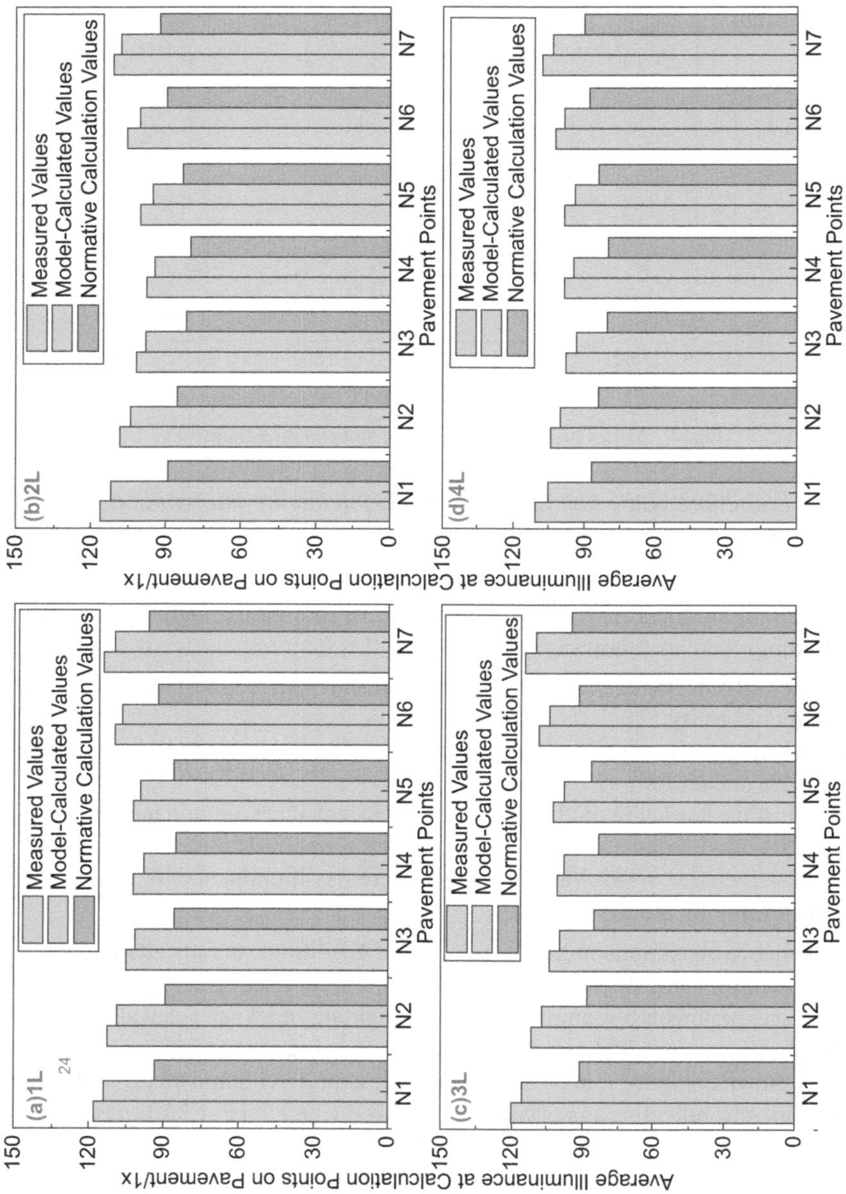

Fig. 5 Horizontal illuminance comparison at road surface points.

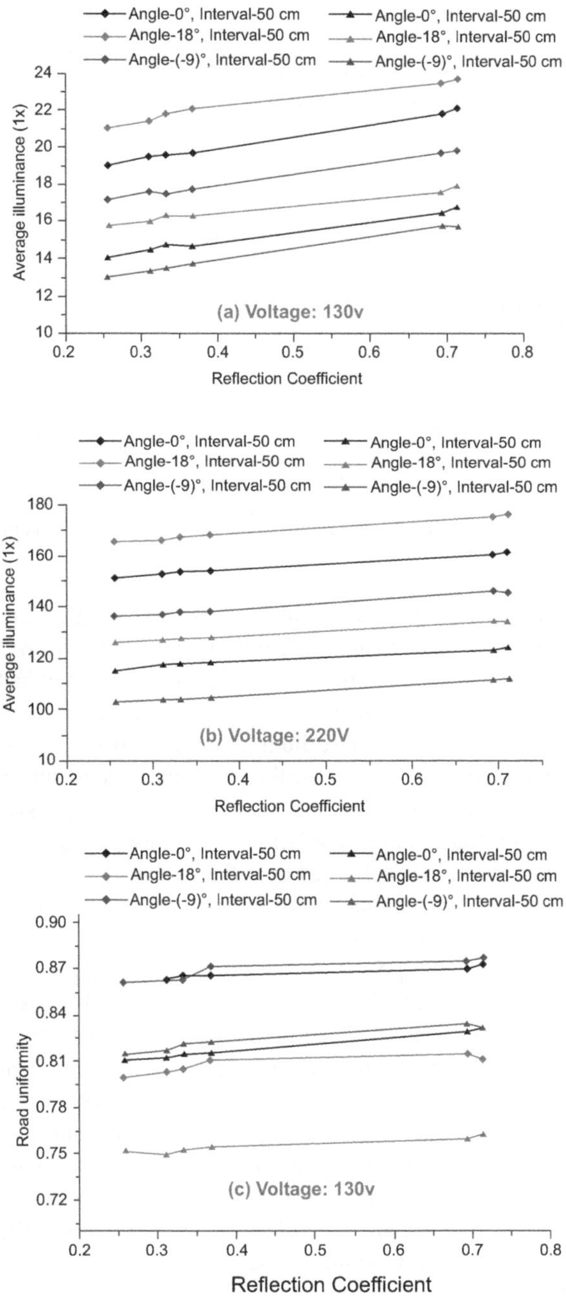
Angle-0°, Interval-50 cm
Angle-18°, Interval-50 cm
Angle-(-9)°, Interval-50 cm
Angle-0°, Interval-50 cm
Angle-18°, Interval-50 cm
Angle-(-9)°, Interval-50 cm
Average illuminance (1x)
(a) Voltage: 130v
Reflection Coefficient
Angle-0°, Interval-50 cm
Angle-18°, Interval-50 cm
Angle-(-9)°, Interval-50 cm
Angle-0°, Interval-50 cm
Angle-18°, Interval-50 cm
Angle-(-9)°, Interval-50 cm
Average illuminance (1x)
(b) Voltage: 220V
Reflection Coefficient
Angle-0°, Interval-50 cm
Angle-18°, Interval-50 cm
Angle-(-9)°, Interval-50 cm
Angle-0°, Interval-50 cm
Angle-18°, Interval-50 cm
Angle-(-9)°, Interval-50 cm
Road uniformity
(c) Voltage: 130v
Reflection Coefficient

Fig. Contd.

Fig. Contd.

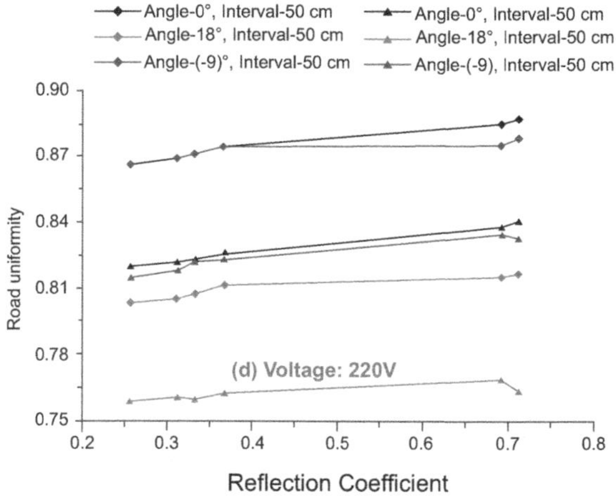

Fig. 6 Impact of reflectance on illuminance and uniformity under varying conditions.

As shown in Figure 9, under all luminaire spacing conditions, the road surface illuminance for different sidewall materials follows the descending order of reflective material > reflective paint > cement mortar. The illumination enhancement effect of various luminaires on road surfaces also decreased sequentially as LED lamps > metal halide lamps > high-pressure sodium lamps. Additionally, the figure reveals that as the luminaire spacing decreases, the disparity in road surface illuminance between different sidewall materials and luminaire types progressively intensifies, reaching maximum differentiation at a 3-m spacing configuration. Specifically, the maximum illuminance values achieved by reflective paint and reflective material measured 1.21 and 1.47 times that of cement mortar, respectively. Similarly, the illumination enhancement ratios of metal halide lamps and LED lamps registered 1.18 and 1.55 times that of high-pressure sodium lamps, respectively.

Based on the above analysis, high-reflectivity sidewall materials have been demonstrated to enhance road surface illuminance in two critical areas: the lateral zones of the tunnel mid-section and the central zone of the tunnel entrance section. Meanwhile, field test results from physical tunnels show general consistency with laboratory experimental data from simulated tunnel environments, collectively validating the illuminance enhancement and uniformity improvement achieved through high-reflectivity sidewall materials. Specifically, under identical lighting conditions, both the illuminance levels and uniformity coefficients exhibit proportional increases corresponding to the elevation of the sidewall materials' reflective coefficient.

The experimental results indicate that tunnel lighting quality is influenced by factors such as sidewall reflectance, luminaire spacing, tilt angle, and luminaire

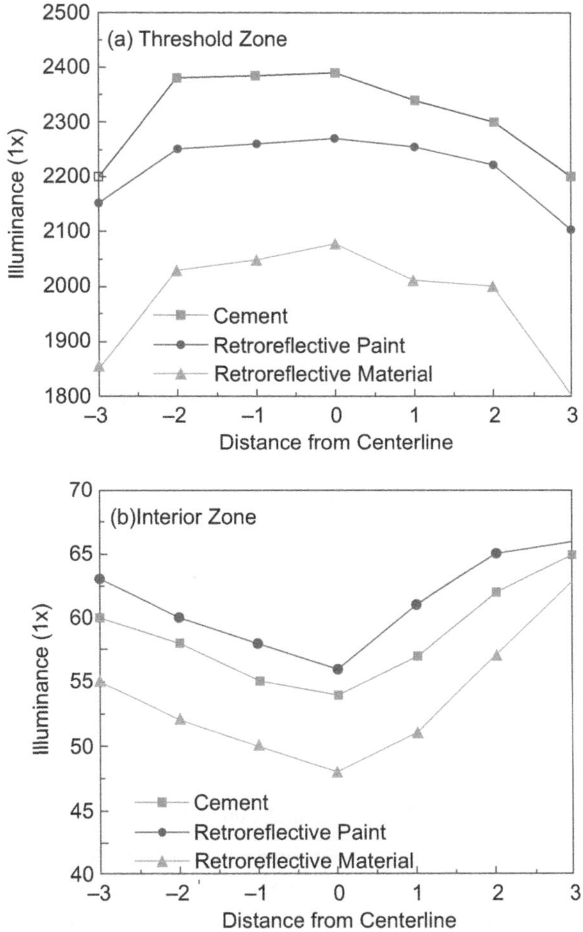

Fig. 7 Road surface illuminance in entrance/interior zones.

type. By adopting high-diffuse-reflectance materials for sidewalls, selecting LED luminaires, and configuring appropriate lighting parameters (3 m installation spacing, 0 m offset from tunnel centerline, and 18° tilt angle), the road surface illuminance in tunnel entrance zones can be enhanced by at least 16%, with a 9% improvement in total uniformity, while the intermediate zones demonstrate a 25% increase in illuminance and 15% enhancement in uniformity. Consequently, through rational coordination between sidewall materials and luminaire layout design, it becomes feasible to expand lighting coverage while maintaining the original illumination requirements. This integrated approach effectively reduces tunnel lighting energy consumption by reducing lighting load demands.

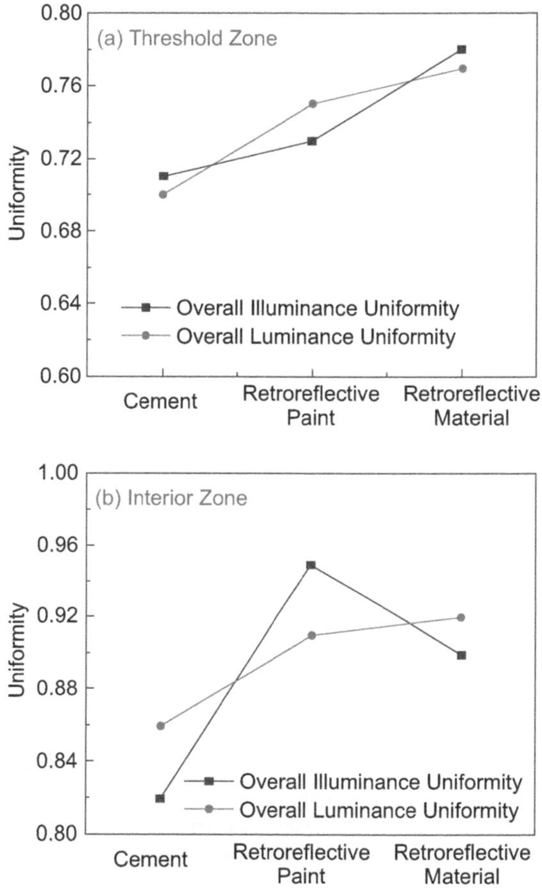

Fig. 8 Total uniformity in entrance/interior zones.

3.2 Biomass Perception Equivalence Technology

We summarize biomass perception equivalence technology in two dimensions: visual performance-based biomass perception equivalence technology and physiological parameter-based biomass perception equivalence technology.

3.2.1 Biomass Perception Equivalence Technology Based on Visual Efficacy

(1) Reaction Time

Our research team's study indicates that driver reaction times generally exhibit a trend of first decreasing and then increasing, with the minimum reaction time occurring in the middle section of tunnels. As shown in Figure 10, compared to 12:00, the reaction time at 16:00 is lower and exhibits smaller fluctuations. Notably,

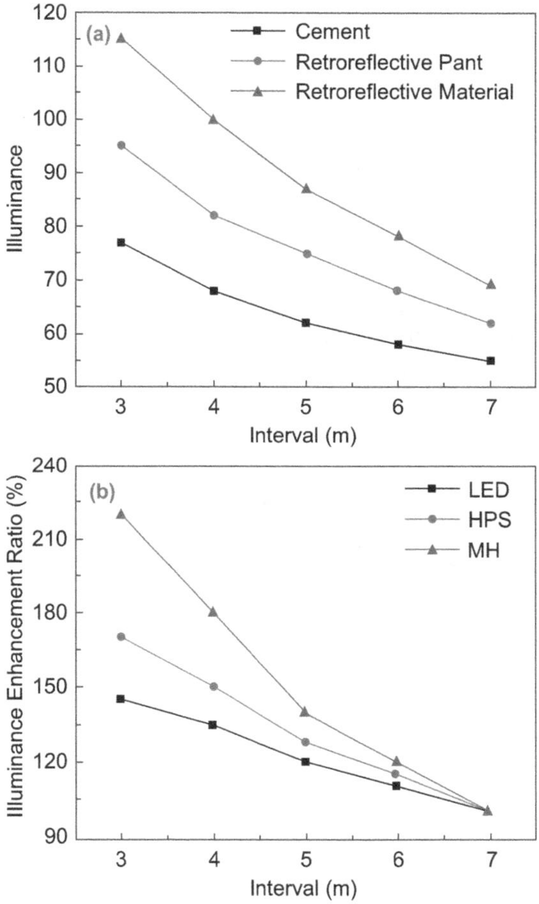

Fig. 9 Illuminance under different sidewall materials and luminaires.

however, driver reaction times at 16:00 show an upward trend. This is attributed to the low reflectivity of the sidewall materials in the Zhenwu Mountain and Cimu Mountain Tunnels, which creates a significant luminance contrast between the tunnel interior and exterior, thereby increasing the reaction times of drivers.

Figure 10 reveals that in tunnels with reflective sidewall materials, drivers' average reaction times at 12:00 and 16:00 are 0.95 and 0.93 times and 1.01 and 0.99 times those in tunnels with fireproof materials and cement mortar sidewalls, respectively. This indicates that, in tunnels with higher sidewall reflectivity, drivers exhibit longer reaction times near tunnel portals and shorter reaction times in the tunnel interior. When approaching the tunnel entrance, drivers are affected by the "black hole effect," which requires time for visual adaptation to luminance changes, resulting in prolonged reaction times compared to other sections. As drivers move further into the tunnel, they gradually adapt to the tunnel environment, leading to

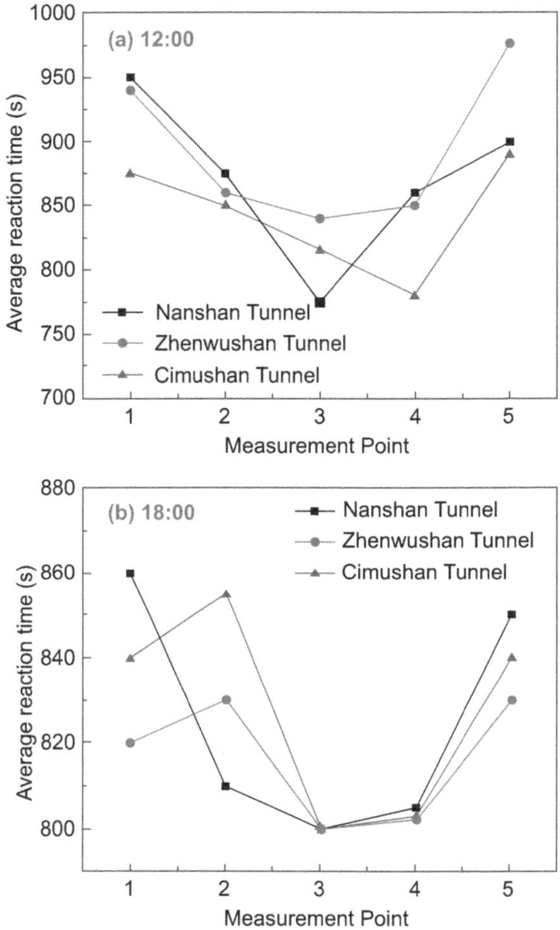

Fig. 10 Driver reaction time under varying conditions.

reduced reaction times, particularly in the middle section (As shown in Figure 11 (a)). At the tunnel exit, the "white hole effect" causes temporary visual distortion and blurring owing to abrupt luminance changes (As shown in Figure 11 (b)). The lower reaction time at 16:00 suggests that reduced luminance contrast at the portal sections enhances driver adaptability to light variations, resulting in relatively shorter reaction times. These findings further confirm that external luminance levels influence obstacle detection time of drivers, providing theoretical support for the development of equivalent lighting technologies in luminous tunnel environments[32–37].

(2) Small Target Visibility

The research findings regarding small target visibility are shown in Figure 12. For every 0.1 increase in the reflectance of sidewall materials, the visibility contrast

(a) Black hole effect (b) White hole effect

Fig. 11 Black hole and white hole effects at tunnel entrances and exits.

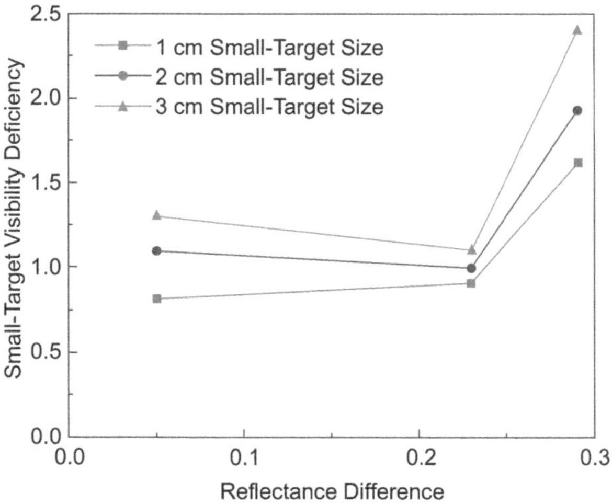

Fig. 12 Reflectance and small-target visibility.

differences for small targets measuring 1, 2, and 3 cm increase by 26.2%, 9.9%, and 21.7%, respectively. Furthermore, the results in Figure 12 demonstrate that as the size of small targets increases, the corresponding enhancement in visibility contrast becomes more pronounced.

3.2.2 Physiological Parameter-Based Biomass Sensing Equivalence Technology

(1) Respiratory Rate, Heart Rate, and Electroencephalogram (EEG)

As shown in Figure 13, the physiological parameters of drivers, including heart rate, respiratory rate, and EEG signal, exhibit significant variations under different temporal and spatial conditions. Near the tunnel entrance, the heart and respiratory rates of drivers increased by 1.31% and 0.86%, respectively, whereas EEG power

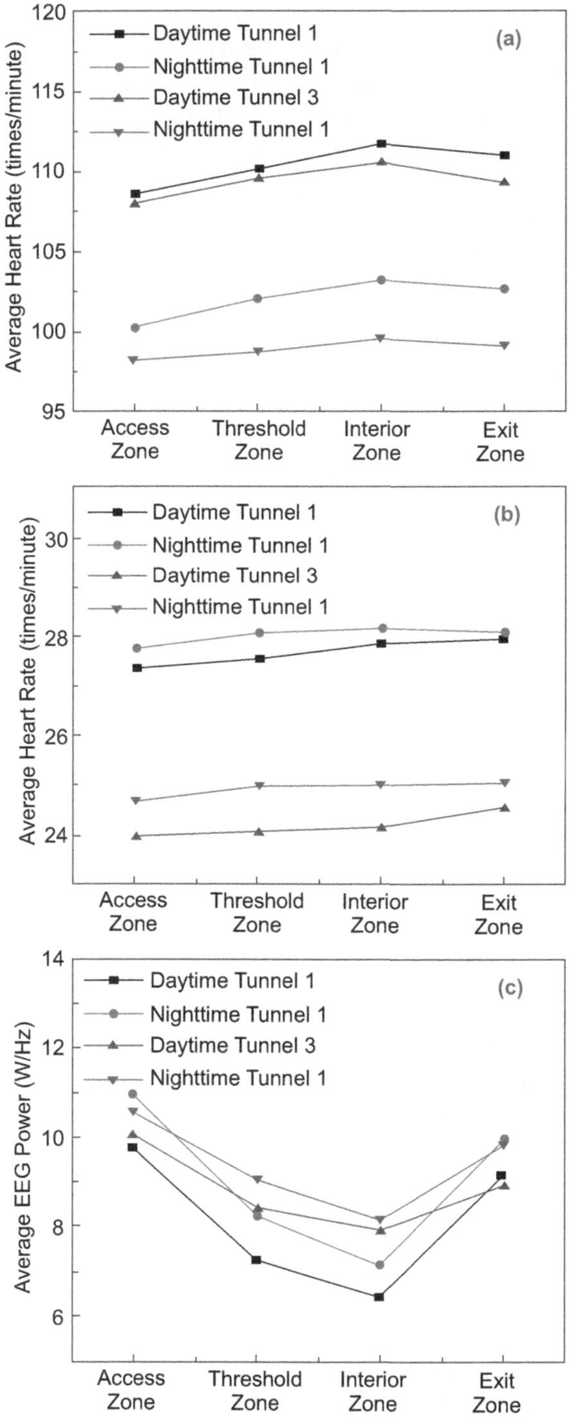

Fig. 13 Physiological parameters (day and night).

decreased by 20.01%. In the middle section of the tunnel, the heart rate reached its maximum value (106.3 beats per minute, bpm), EEG power dropped to its minimum (7.5 W/Hz), and respiratory rate continued to rise. When approaching the tunnel exit, heart rate declined, EEG power increased, and respiratory rate peaked at 26.3 breaths per minute.

Figure 13 also indicates that, compared to daytime conditions, drivers at night exhibited lower respiratory rates and EEG power (reduced by 2.1% and 7.9%, respectively) but higher heart rates (increased by 9.3%). These physiological trends align with the findings of Feng et al. [38–40]. Additionally, Qiao et al. observed that, near tunnel entrances, the illuminance variation rate surged to 104%, correlating with a peak heart rate growth rate of 31.95% in drivers. Upon entering the tunnel, as the illuminance variation rate gradually decreased, the heart rate growth rate also decreased [41–42].

(2) Fixation Time Percentage, Saccade Amplitude, and Pupil Area

As shown in Figure 14, when driving into tunnels during daytime, the fixation time percentage and pupil area increase sharply (26.85% and 52.11%, respectively), while the saccade amplitude decreases slowly (3.06%). In the middle section, Tunnel 1 shows a gradual increase in fixation time percentage (3.89%), whereas Tunnel 3 exhibits a slow decline in fixation time percentage and saccade amplitude (8.33% and 7.88%, respectively), with a sharp rise in pupil area (20.83%). At the exit section, the fixation time percentage and pupil area drop sharply (10.38% and 32.66%, respectively), while saccade amplitude remains largely unchanged. During nighttime tunnel entry, the fixation time percentage increases slowly (4.16%), saccade amplitude decreases gradually (4.35%), and pupil area declines sharply (40.17%). In the middle section, fixation time percentage and saccade amplitude rise slowly (4.01%), while saccade amplitude and pupil area increase rapidly (27.27% and 26.02%, respectively). At the exit section, fixation time percentage, saccade amplitude, and pupil area all decrease gradually (2.57%, 7.15%, and 5.34%, respectively).

Figure 14 also reveals that, compared to daytime drivers, nighttime drivers exhibit lower values and smoother variations in fixation time percentage, saccade amplitude, and pupil area. These visual characteristic trends align with findings from studies by He [9], Feng [43], and Liang et al. [44]. Shi further observed high cumulative fixation time percentages in extra-long tunnel entrance, driving, and exit sections: 81.9%, 86.2%, and 75.9%, respectively. Driver pupil areas in the approach, entrance, middle, and exit sections were measured as 1224, 1581, 2518, and 1520 px, respectively [45]. Zhang also found that lower tunnel lighting quality correlates with higher proportions of small-amplitude saccades among drivers [46].

Combining Figures 13 and 14, compared to Tunnel 1 (coated with fireproof material), drivers in Tunnel 3 (coated with reflective material) exhibited reductions in daytime and nighttime fixation time percentage (37.5% and 2.39%, respectively), saccade amplitude (27.2% and 2.3%, respectively), pupil area (25.2% and 15.9%, respectively), heart rate (0.9% and 3.1%, respectively), and respiratory rate

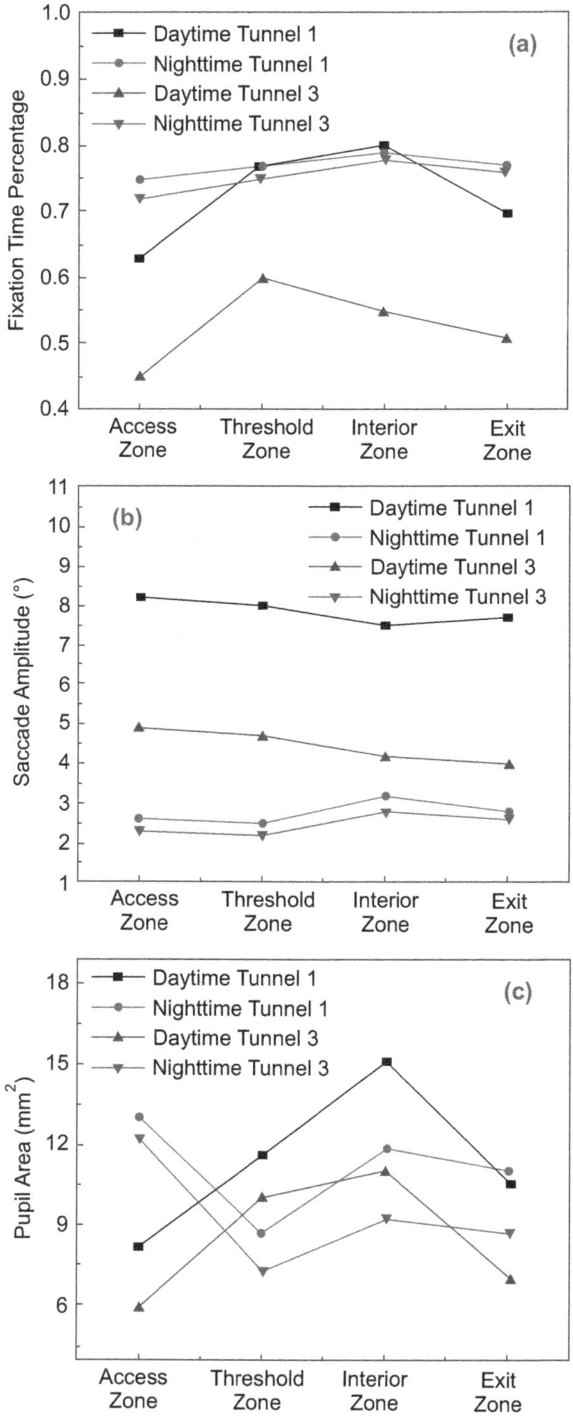

Fig. 14 Visual behavior metrics (day and night).

(12.6% and 11.0%, respectively), alongside increased EEG power (8.6% and 3.6%, respectively). These results are consistent with those reported by Qin [9] and He et al. [47]. Research indicates that reflective materials, a type of diffuse reflective sidewall material rich in yellow-green light, significantly enhance tunnel lighting quality and improve ability of drivers to distinguish sidewalls from road surfaces[48–52]. Consequently, drivers in tunnels with reflective sidewalls experience reduced anxiety, visual search difficulty, risky driving behavior frequency, and light–dark adaptation challenges, enabling calmer psychological states and more stable pupil adjustments for environmental observation[53–59].

The findings demonstrate that reflective sidewall materials under various luminaire configurations effectively enhance driver bio-perception, manifested through shorter reaction times, higher small-target visibility, and lower, more stable heart rate, respiratory rate, fixation time, saccade amplitude, and pupil area. This indicates that optimized coordination between lighting fixtures and sidewall materials improves driver obstacle recognition and environmental perception, alleviates psychological stress and physiological fatigue during driving, and significantly enhances driving safety and comfort.

Based on comprehensive research findings, it can be concluded that the rational matching of lighting fixtures and sidewall materials can effectively reduce tunnel lighting load while enhancing drivers' biological visual perception, further demonstrating the feasibility and reliability of equivalent lighting technology for tunnel luminous environments. The research outcomes also reveal that this equivalent lighting technology successfully addresses two critical issues: the rational matching of composite luminous environments formed by multiple light sources or multidimensional combinations, and the existing lighting quality evaluation standards that neglect biological perception effects. Consequently, it resolves the inherent non-equivalence between tunnel lighting load and driver biological perception characteristics. This technological advancement achieves the dual objectives of energy conservation and traffic safety enhancement in tunnel lighting systems, marking significant progress in both sustainable engineering practices and transportation safety management.

Future research

1. In the context of intelligent transportation, the traditional "human–vehicle–tunnel" model is gradually transitioning to emerging transportation modes and scenarios, such as "new energy vehicles," "intelligent driving," and "extended tunnel clusters," resulting in significant changes in operational environmental factors and operating conditions. Therefore, there is an urgent need to research driving safety demand patterns for these emerging transportation modes and scenarios in intelligent transportation contexts. For instance, studies should focus on aspects such as the information perception capabilities of drivers in human–machine co-driving vehicles within intelligent transportation systems. In this regard, VR technology can be utilized to create immersive virtual tunnel environments that simulate

various lighting conditions and scenarios. This allows researchers to study how different lighting environments affect drivers' information perception and reaction times in a safe and controlled setting.

2. Current tunnel operation environments generate substantial data volumes, yet face challenges in terms of organic integration and low digitalization levels. Consequently, it is imperative to establish a large-scale tunnel traffic operation data model through multi-source data fusion (including traffic flow, driver factors, and tunnel lighting environment data). This initiative aims to clarify the driving safety requirements under spatiotemporal environmental characteristics of tunnel locations, integrate these requirements with tunnel lighting energy-saving technologies, and ultimately enhance the digitalization and intelligent regulation capabilities of tunnel operations. In the process of establishing this data model, VR can be employed to visualize the complex interactions between tunnel lighting and other operational factors. By creating virtual models of tunnels with different lighting configurations, researchers can better understand the impact of lighting on driving safety and energy efficiency, and optimize the data model accordingly[13].
3. Given the current lack of targeted maintenance and control strategies for tunnel lighting environments, it is essential to utilize the tunnel traffic operation data model to analyze evaluation methods for the impact of traffic pollutants (resulting from increased "new energy vehicles") on tunnel driving safety. This process should incorporate safety and carbon reduction indicators while advancing the development of digital lighting equipment tailored to extra-long tunnels. Meanwhile, VR can be used to simulate the effects of traffic pollutants on tunnel lighting and visibility. Researchers can create virtual scenarios with varying levels of pollutants and observe how they affect the performance of different lighting systems. This can help in developing more effective maintenance strategies and designing digital lighting equipment that can better cope with the challenges posed by traffic pollutants.

References

[1] Ministry of Transport of the People's Republic of China 2024. Statistical Bulletin on the Development of the Transport Industry in 2023.

[2] National Bureau of Statistics, 2023. China Statistical Yearbook. Beijing: China Statistics Press.

[3] Ministry of Public Security of the People's Republic of China, 2017. Annual Statistical Report of Road Traffic Accidents of the People's Republic of China. Beijing: China Communications Press.

[4] Peña-García, A. (2018). The impact of lighting on drivers well-being and safety in very long underground roads: New challenges for new infrastructures. Tunn. Undergr. Space Technol. (80) 38–43.

[5] Peña-García, A. (2017). The SLT equation: A tool to predict and evaluate energy savings in road tunnels with sunlight systems. Tunn. Undergr. Space Technol. (64)43–50.

[6] Peña-García, A. (2019). Optical coupling of grouped tunnels to decrease the energy and materials consumption of their lighting installations. Tunn. Undergr. Space Technol. (91)103007.

[7] Amundsen, F., Ranes, G., 2000. Studies on traffic accidents in Norwegian road tunnels. *Tunn. Undergr. Space Technol.* 3–11.

[8] Zhu, H., Y. Deng, Y. Shen, et al. (2022). Application analysis of life-cycle green index of highway tunnel light environment. *China J. Highw. Transp.* 35(1):13.

[9] He, S., Liang, B., Pan, G., et al. (2017). Influence of dynamic highway tunnel lighting environment on driving safety based on eye movement parameters of the driver. *Tunn. Undergr. Space Technol.* 57:52.

[10] He, S., Ren, H. and Liu, et al. (2022). A novel tunnel lighting method aided by highly diffuse reflective materials on the sidewall: Theory and practice. *Tunn. Undergr. Space Technol.* 122:1.

[11] He, S., Liu, H., Niu, J. et al. (2023). Driver's visual luminance model considering the reflection characteristics of a tunnel's inner wall: Theory and practice I. *Tunn. Undergr. Space Technol.* 131:1.

[12] Niu, J., Liang, B., He, S, et al. (2022). Long tunnel lighting environment improvement method based on multiple-parameter intelligent control: Considering dynamic changes in luminance difference. Tunn. Undergr. Space Technol. 128, 104637.

[13] Niu, J., Liang, B. and He, S, et al. (2024). Assessment and optimization of tunnel lighting quality based on drivers' visual comfort: From methodology to application. Tunn. Undergr. Space Technol. 143, 105487.

[14] Guide for the lighting of road tunnels and underpasses: CIE88: 2004. Vienna: CIE Central Bureau, 2004.

[15] Calculation of Tunnel Lighting Quality Criteria: CIE189: 2010. Vienna: CIE Central Bureau, 2010.

[16] Guidelines for design of lighting of highway tunnel: JTG/T D70/2- 01-2014. Beijing: China Communications Press, 2014.

[17] Urban road lighting design standard: CJJ 45-2015 . Beijing: China Architecture Publishing and Construction Press, 2015.

[18] Li, N., Wang, X., Wu, J. et al. (2013). Application of genetic algorithm in intelligent optimization of highway tunnel lighting. *J. Cent. South Univ. (Sci. Technol.)* 44(S1):342.

[19] Qin, L., Shi., X., Leon, A. et al. (2020). Dynamic lighting control in road tunnels: A framework for energy efficiency and safety. Building and Environment, 106844.

[20] Liang, B., Dong, Y., Yan, Z. et al. (2021). A sensitivity analysis of lighting condition parameters in the Middle section of tunnel based on biomass index. *J. Transp. Inf. Saf.* 39(6):91.

[21] Liang, B., Liang, J., He, S. et al. (2020). Study on light environment information perception and experiment method of highway tunnel. *Tunn. Constr.* 40(9):1251.

[22] He S., Zhou, Y., Liang, B. et al., (2022). Research status and progress of drivers information perception representation methods in long tunnel light environment . *Tunn. Constr.* 42(02):176.

[23] Mayeur, A., Brémond, R. and Bastien, C. (2010). The effect of the driving environment on drivers' visual behavior. *Transportation Research Part F: Traffic Psychology and Behaviour*, 13(6): 376–386.

[24] Zhang, Y., He, S., Sun, Y., et al. (2023). Study on Variation of Driver's Field of View Contrast in Tunnel Light Environment and Its Influence on Driving Safety . *Mod. Tunnell. Technol.* 60(5):30.

[25] He, S., Li, X., Liu, Q., et al. (2023). Research on the Influence of Longitudinal Spacing within Tunnels Group on Drivers' Physiological Load . *Highway* 68(11):192.

[26] Feng, S., Mei, J., Mao, W. et al. (2021). Study on light environment in tunnels decorating with multifunctional energy-storage and luminescent material . *Chin. J. Undergr. Space Eng.* 17(6):2030.

[27] Liang, B., Niu, J., Li, S. et al. (2023). Dynamic optimization and intelligent control of highway tunnel lighting considering visibility effects. *Control Theory Appl.* 40(10):1783.

[28] Peña-García, A. (2022). Sustainable tunnel lighting: One decade of proposals, advances and open points. Tunn. Undergr. Space Technol. 104227.

[29] Ren, Y., He, S., Pan, S. et al. (2020). Calculation method of highway tunnel lighting based on bidirectional reflectance distribution function and energy saving measures. Tunn. Constr. 40(12):1748.

[30] Pan, G. (2014). Systematic experiment and research on sidewall coating material assisting tunnels energy-saving lighting. Chongqing: Chongqing Jiaotong Univ.

[31] Lighting measurement method: GB/T 5700-2008. Beijing: Standards Press of China, 2008.

[32] Niu, J., Liang, B., Wong, Y.D, et al. (2024). Dynamic traffic safety risk assessment in road tunnel entrance zone based on drivers' psychophysiological perception states: Methodology and case-study insights. Tunn. Undergr. Space Technol. 147, 105677.

[33] Liu, N., Liang, B., Xiao, Y. et al. (2020). Driving safety of extra-long tunnel based on reaction time characteristics of drivers. *Sci. Technol. Eng.* 20(26):10927.

[34] Bassan, S. (2016). Sight distance and visibility analysis in road tunnels. Transportation Research Part C: Emerging Technologies, 71: 76–90.

[35] He, S., Liang, B. and Luo, H. (2017). On a drivers visual reaction to different coatings on tunnel sidewalls. *Mod. Tunn. Technol.* 54(1):48.

[36] Liang, B., Li, X., Wei, Q.H., et al. (2018). Research on the influence of diffuse reflectance of light-storing reflective material on the quality of tunnel lighting. *Chin. J. Undergr. Space Eng.* 14(2):565.

[37] Liang, B., Niu, J., He, S., et al. (2023). Tunnel lighting calculation model based on bidirectional reflectance distribution function: considering the dynamic changes in light transmittance in road tunnels. Tunn. Undergr. Space Technol. 140, 105313.

[38] Feng, Z., Yang, M., Ma, C., et al. (2018). Effect of slope and velocity on drivers heart rate growth in urban underpass tunnel longitudinal section. *China J. Highw. Transp.* 31(4):66.

[39] Ding, G. (2011). Study on the impact of freeway tunnel environment on drivers' physiology and psychology. Chang'an Univ.

[40] Wu, L. (2018). Study on driver risk characteristics in extra-long highway tunnels based on physiological and behavioral indicators. Chang'an Univ.

[41] Torre, F., Meocci. M., Domenichini, L. et al. (2019). Development of an accident prediction model for Italian freeways. Accident Analysis & Prevention, 1–11.

[42] Luo, J., Lu, B., Xu, J. et al. (2019). Effect of light and velocity on driver workload in mountain city tunnel. *Journal of Chongqing University of Technol. (Nat. Sci.)* 33(6):13.

[43] Yang, F., (2014). Study on effects of non-sufficient lighting environment on driver's physiology in the extra-long tunnel. Inner Mongolia: Inner Mong. Agric. Univ.

[44] Liang, B., Qin, C., Niu, J. et al. (2024). Visual load in threshold zone of highway tunnel based on Topsis improved factor analysis method. *China Saf. Sci. J.* 34(1):77.

[45] Shi, L. (2011). Study on human eye motion and fixation characteristics of drivers in extra-long highway tunnels. Chang'an Univ.

[46] Zhang, Y., Wei, Q., Liang, B. et al. (2018). Influence of short highway tunnel lighting environment on driver's physiology. *ZHAOMING GONGCHENG XUEBAO* 29(2):80.

[47] Peña-García, A. (2018). The impact of lighting on drivers' well-being and safety in very long tunnels: New challenges for new infrastructures. *Tunn. Undergr. Space Technol.* 272: 38–44.

[48] Peña-García, A. and Cabeza-Lainez, J. (2023). Daylighting of road tunnels through external ground-based light-pipes and complex reflective geometry. Tunn. Undergr. Space Technol. 104788.

[49] Luo, H., Shi, L., Tu, Y., et al. (2016). Feasible analysis of the application of solar fiber lighting in tunnel. *ZHAOMING GONGCHENG XUEBAO* 27(4):55

[50] Wu, K., Du, Z., Han, L. et al. (2023). Influence of shading shed at connection section between two tunnels on visual load of drivers . *Tunn. Constr.* 43(10):1760.

[51] Yu, Li., Wang, S., Luo, X. et al. (2023). A study on geometric design parameters of grille sunshades in highway tunnels. *Mod. Tunn. Technol.* 60(1):34.
[52] Liang, B., Li, J. N., S. et al. (2023). Dynamic optimization and intelligent control of highway tunnel lighting considering visibility effects . *Control Theory Appl.* 40(10):1783.
[53] He, X., et al. (2022). A real-time tunnel lighting control method based on vehicle speed and traffic flow. IEEE Transactions on Intelligent Transportation Systems, 23(7): 6812–6822.
[54] Wang, C., et al. (2021). Visual guidance system for tunnel interiors using LED contour lights. Lighting Research & Technology, 53(2): 167–184.
[55] Liang, B., Qin, Y., Zhang, H. et al. (2024). Study on information perception of visual illusion speed control marking in approach section of expressway tunnel. *J. Highw. Transp. Res. Dev.* 41(4):146.
[56] Ye, F., Wen, X., Zhang, X. et al. (2023). Contribution of gradual change design in alleviating visual abrupt change in tunnel landscape . *China J. Highw. Transp.* 36(11):375.
[57] Liu, C., Zang, Y. and Sun, X. (2019). Analysis of the impact of urban tunnel midsection ceiling landscapes on driving safety based on visual characteristics . *Mod. Tunn. Technol.* 61(3):228.
[58] Gibbons, R.B. and Hankey, J.M. (2012). Effect of pavement marking retroreflectivity on traffic crash frequency. Transportation Research Record, 2272(1): 113–120.
[59] Xu, S., Yang, Y., Zhang, Y., et al. (2017). Research on the effectiveness of colorful pavement based on driver's visual behavior in long downhill. *Sci. Technol. Eng.* 17(33):349.

7 Immersive Perception of the Tunnel Environment in VR/AR Platform: from a multi-sensory perspective

Shaobin Zhou[1], Haihang Hu[2], Mun On Wong[3] and Tianhang Zhang[4*]

Tunnel environments present unique challenges for immersive simulations due to their complex sensory characteristics. This study proposes an Input-Feedback-Evaluation (IFE) framework to systematically structure human-XR interactions in tunnel VR/AR applications. Building on a comprehensive review, the study examines multi-modal input methods (e.g., controllers, gestures, motion tracking), multi-sensory feedback mechanisms (primarily visual and auditory), and multi-dimensional evaluation techniques (such as eye tracking and EEG signals). Results highlight that physical controllers and visual feedback currently dominate existing XR tunnel research, while emerging modalities like voice input, haptic feedback, and physiological evaluations offer significant potential for enhancing realism and user experience. By consolidating the state-of-the-art and identifying future research opportunities, this work provides a methodological foundation for advancing XR-based tunnel simulation systems to improve safety, training, and operational efficiency across the tunnel lifecycle.

1. Introduction

Virtual reality (VR) and augmented reality (AR) technology offer significant advantages throughout the full lifecycle of tunnel construction, operation, maintenance, and emergency management. At each stage, VR/AR provides distinct benefits by enabling immersive environments, realistic scenario simulations, and safe interactions, leading researchers to extensively integrate these technologies into tunnel-related applications.

During the construction phase, to address the risks associated with respirable coal and silica dust, researchers developed an immersive educational tool that visualizes computational fluid dynamics (CFD) modeling datasets of ventilation and dust flow characteristics within an underground VR environment[13]. In the operation phase, VR technology has been utilized to establish an intelligent decision-support system for highway tunnel environments. For example, driving

1,3 Department of Civil and Environmental Engineering, Faculty of Science and Technology, University of Macau, Macau SAR, China.
2,4 Research Center for Urban Fire Safety Engineering, Zhejiang University, Hangzhou, China.
* Corresponding author: thz@zju.edu.cn

DOI: 10.1201/9781003514480-7

simulation experiments with various main lighting color temperatures were conducted to develop a metric model linking attention growth rate to light source color temperature, along with response time distribution patterns. [24]). Another study proposed a novel framework based on digital twins to optimize tunnel lighting environments, integrating virtual world design (numerical simulation and VR experiments) with real-world validation (tunnel models and on-site experiments) [34]. In the field of emergency response, scholars have employed VR technology to study tunnel evacuation strategies. A VR-based experiment involving 96 participants provided recommendations for emergency exit flashing light designs in highway tunnel evacuations [32]. Additionally, researchers examined the effects of VR agents, emergency evacuation signage, and fire extinguishers on evacuation behavior during tunnel fires. Using VR simulations, they analyzed evacuation risks and optimized escape routes in tunnel fire scenarios [44]. Beyond the extensive research on visual perception in tunnel VR/AR applications, other perceptions like auditory, haptic, and temperature are equally crucial, as shown in Fig. 1. Studies have shown that in low-visibility tunnel conditions, the role of auditory cues becomes increasingly significant [1,8,48,46,47]. Additionally, haptic feedback technology has also been explored to enhance the immersive simulations of underground utility tunnel environments [20]. Besides, physically thermal temperature feedback has also been replicated using heating wires and fans to mock up realistic fire evacuation experiences in virtual tunnel simulations [25]. While haptic and thermal cues also play important roles in multisensory perception, this study concentrates on visual and auditory guidance, which are most critical for drivers and evacuees responding rapidly in tunnel environments [19,40].

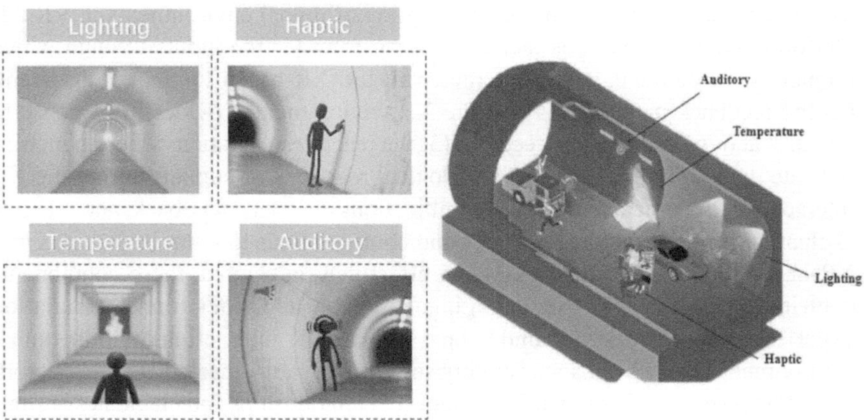

Fig. 1 Multi-form AR/VR platforms for enhancing human perceptions in tunnel environment simulations.

To sum up, effective tunnel simulations require multi-modal input methods to enable intuitive human interactions within virtual environments, robust sensory feedback for immersive and realistic experiences, and comprehensive evaluation mechanisms to accurately measure psychological and physiological responses. However, most existing studies are dominated by conventional controller inputs and visual feedback, accompanied by relatively limited evaluation methods. Meanwhile, emerging XR interaction modalities such as gesture control, voice commands, eye tracking, haptic and thermal feedback, and advanced evaluation techniques, including EEG, physiological, and emotional metrics, represent substantial untapped potential. Exploring these advanced VR/AR forms can significantly enhance the realism, effectiveness, and overall utility of XR-based tunnel environment simulations.

To address these opportunities and existing research gaps, this chapter proposes a novel input-feedback-evaluation (IFE) framework, aiming to establish a structured approach to integrating multi-modal inputs, diverse sensory feedback, and multi-dimensional evaluation systematically. The primary motivation behind proposing the IFE framework is to better understand the different forms of XR and how they could be effectively utilized for various tunnel environment-related research. The chapter structure begins with a detailed explanation of the IFE framework, followed by a systematic literature review that elaborates on state-of-the-art developments. Critical findings from this review are summarized and discussed concerning inputs, feedback, and evaluations. Finally, future research directions are suggested to advance the practical application and effectiveness of XR technologies in tunnel lifecycle management.

2. Framework

The natural and immersive human perception of tunnel environments in VR/AR platforms hinges on various factors, ranging from (1) the inputs through which humans send commands and interact with the XR-based tunnel environment, (2) the feedback mechanisms that the XR environment provides in response to humans' actions and sensory needs, to (3) the reactive evaluations wherein metrics indicate the psychological and physiological reactions of humans during the XR interactions. Specifically, different combinations of inputs, feedback, and reactive evaluation bring distinct experiences and configurations to immerse humans into XR-based tunnel environments. When effectively integrated, these components contribute to more realistic, engaging, efficient, and safer tunnel simulation experiences [30]. To better understand human-XR interactions within tunnel environment research, this study proposes a novel multi-faceted IFE framework (Fig. 2) that structuralizes their interconnected relationships and elucidates various forms of inputs, feedback, and reactive evaluations within the interaction process.

Multi-modal inputs enable humans to intuitively explore and interact with the XR platform as if they were in real-world tunnel environments. The most typical input forms are physical controllers, including keyboards, steering wheels, braking pedals, and joysticks (Brown et al., 2010), which allow users to touch and manipulate physical objects to gain realistic experiences. Graphical interfaces are also widely

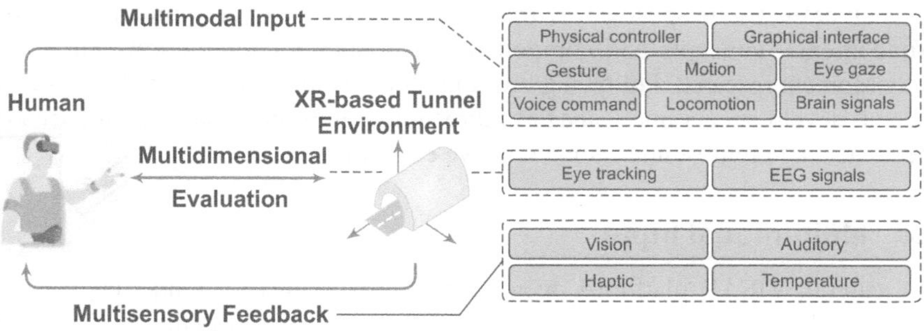

Fig. 2 The input-feedback-evaluation framework for human-XR interactions in tunnel environment simulation.

used to facilitate human-XR interactions, particularly for triggering functions such as menu selections, tool activation, or information retrieval. In addition, human-centric inputs have gained prominence in recent years, including eye gaze, voice commands, gesture recognition, body motion tracking, and locomotion [29]. These human-centric inputs reflect instinctive human behaviors, such as communicating with other evacuees during tunnel emergencies and navigating through the tunnel by physically walking [44]. The careful selection of appropriate input forms can significantly enhance user presence and immersion, thereby empowering realistic simulations of various tunnel scenarios, such as driving or escaping.

Another crucial component constituting the cycle of human-XR interactions for tunnel environment research is multi-sensory feedback, which refers to the responsive configurations designed to enhance human perception by emulating real-world conditions. It includes visual, auditory, haptic, and temperature feedback. Visual feedback responds to environmental elements such as lighting or landscape layout [24][50], while auditory feedback enhances spatial awareness and immersion [6]. Haptic feedback simulates touch sensations, allowing users to feel the resistance of virtual objects [51]. Temperature feedback replicates real environmental conditions, such as the heat experienced during a tunnel fire, to further increase immersion of users [25]. Together, these feedback mechanisms stimulate multiple senses, making the virtual tunnel environment more convincing and engaging for users.

In addition to the input-feedback cycle of human-XR interactions for tunnel environment research, reactive evaluation serves as an independent component that measures human perception levels within XR-based tunnel environments. Unlike the inputs and feedback, which are directly involved in enhancing human perceptions, reactive evaluation is an indirect method that quantifies psychological and physiological behaviors of humans, such as through electroencephalography (EEG) and eye trackers. These methods enable the monitoring of reaction time, cognitive loads, attention, and other metrics to provide insights into how users perceive and react to various situations within the virtual tunnel environments [9].

Overall, the proposed IFE framework comprehensively incorporates the fundamental concepts and components in human-XR interactions for tunnel environment research. Following this, this study would further elucidate state-of-the-art developments among the three directions: multi-modal input, multi-sensory feedback, and multi-dimensional reactive evaluation.

3. Multi-modal Input

In general, VR/AR platforms open up new possibilities for tunnel engineering research by offering a risk-free, cost-effective, and high-fidelity environment. These platforms enable a wide range of scenarios, including tunnel environmental design, operational inspection, driving simulation, emergency training, and human behavior analysis. While promising, ensuring that the XR environment accurately replicates real-world interactions between humans and tunnel environments remains one of the most challenging yet essential issues that deserves systematic investigation. Within this context, multi-modal inputs are critical for facilitating communication and control from humans to the XR environment. These multi-modal inputs, ranging from physical controllers, gesture detection, motion tracking, voice commands, to location detection, serve to create a more immersive, realistic, and human-in-the-loop tunnel simulation. Therefore, this chapter endeavors to provide an in-depth review of the application of multi-modal inputs in XR-based tunnel environments, highlighting both current achievements and future research directions.

Among the various interaction modalities, physical controllers have emerged as the most widely applied solution. According to Chinese scholars [10], when users are driving in a simulated virtual tunnel environment, the DSC-BV-TI system provides an HTC Vive handle and steering wheel kit for enhanced immersive experiences, as shown in Fig. 3(a). Users manipulate vehicles in the virtual tunnel environment using not only a keyboard but also a steering wheel and brake pedal, allowing them to experience and discuss tunnel designs under various weather or operational circumstances. In addition to developing a driving system, [11] He et al. (2020) also utilize a more immersive on-board diagnostic (OBD) virtual cockpit for driving. Drivers respond to various tunnel facilities and landscapes, adjusting the speed of the car accordingly. Another application is about risk analysis of evacuation. Users control a joystick to move or escape during a tunnel fire, responding to emergencies and deciding their escape routes based on the scene they observe in VR glasses [44]. It is important to note that the VR agent's shouting and emergency signals influence the user's judgment in choosing the correct escape path [52], demonstrating the interaction effect between users' input and sensory feedback. The effectiveness of emergency response training can be increased by integrating inputs with XR tunnel simulation systems, as this allows trainees to gain safety knowledge through immersive experiences [39].

Gesture detection is also vital in tunnel emergency training and on-site inspection. Recent studies [39] simulated emergency scenarios, such as fires and

(a) Physical controllers (b) Gesture detection (c) Motion detection

Fig. 3 The application of multi-modal inputs within XR-based tunnel environments.

vehicle accidents, guiding users to take on different roles through instructions presented in VR glasses. For those participants trained as central operators or coach passengers, the VR system enhances response efficiency and reduces reaction time by providing virtual training environments with different gestures like picking up, holding, pointing, and releasing. These interactions add tangible elements to the experience, further boosting training effectiveness. Besides, gesture detection is applied in AR technologies. Melnyk et al. [28] introduce air tapping gestures as a pivotal function that enables inspectors to select buttons and set virtual markers within real tunnel environments when using AR glasses, facilitating prompt and unobstructed anchoring to overlay digital contents onto the real tunnel environments precisely (Fig. 3(b)).

Furthermore, motion detection is widely applied in different areas. For fire emergency evacuation, motion tracker equipment is used to follow subjects' movements, such as escaping from a fire, covering their mouth, or bending their body [25]. This is particularly valuable in tunnel environments where hostile conditions, such as low light, smoke, or dust, can compromise single-sensor detection capabilities [37]. When it comes to the hazard identification training, Isleyen et al. [16] propose a method to track users' head and hand motion using infrared cameras and attached sensors for achieving high precision motion tracking. In this environment, motion tracking is performed at a real-scene scale, which not only heightens the participants' sense of immersion but also mitigates motion sickness, a common issue in VR experiences (Fig. 3(c)). Furthermore, a prior study indicates that AR-based inspection processes begin with inputting precise localization and virtual markers to ensure accurate anchor positioning [28]. This contrasts with conventional methods that depend heavily on manual judgment, leading to more precise and reliable results.

To summarize, this chapter critically examines the application of multi-modal inputs within XR-based tunnel environments, with a focus on physical controllers, gesture detection, and motion detection. The use of these diverse input approaches across different phases of tunnel environments, such as design, inspection, and training, offers immersive and efficient experiences while contributing to decision-making and advanced preparedness. However, current applications primarily concentrate on mechanical (physical controllers and motion) or optical (gestures) input modalities, with limited exploration of acoustic and biophysical options [27]. In the robot control field, Szczurek et al. [35] have developed a multimodal input system that incorporates voice commands, enabling remote control through voice recognition in hazardous environments. In the medical field, Hussain et al. [14] have mentioned a brain-computer interface (BCI) system to translate the intentions of disabled individuals from brain signals to control electric-powered wheelchairs. These examples illustrate the potential of voice command and brain signal research in expanding the diversity of input modalities, providing valuable insights for advancing tunnel environment simulation research.

4. Multi-sensory Feedback

Tunnels, as an essential component of modern transportation infrastructure, provide a unique perceptual environment where lighting and auditory conditions significantly impact human experience. Unlike open roads, tunnels feature a semi-enclosed structure that alters the way individuals perceive and adapt to their surroundings. Key perceptual changes include transitioning from a bright external environment into the relatively dim interior of a tunnel, triggering the visual adaptation process known as "dark adaptation", as well as the "light adaptation" process, where the driver's visual system must adjust to the rapid increase in lighting when transitioning from the dim tunnel environment to the bright external light [12]. Improper tunnel lighting design has been shown to prolong visual reaction times, increase psychological stress, and impact overall perceptual comfort [2]. However, when considering human perception in virtual reality (VR) and augmented reality (AR) simulations, the challenge extends beyond safety—ensuring an immersive and authentic multi-sensory experience is key to making simulation results generalizable and applicable to real-world environments.

The realism of the lighting environment plays a crucial role in immersive perception. A VR/AR simulation must accurately replicate lighting effects, including luminance levels, uniformity, glare, and temporal changes in brightness, to ensure that the simulated experience aligns with real-world conditions. The effectiveness of tunnel lighting cannot be assessed solely through safety metrics but must also account for human visual adaptation and comfort under different lighting parameters. Therefore, VR/AR-based studies should focus on how individuals respond to a range of lighting conditions, assessing the equivalence of visual effects rather than just driver performance alone.

Furthermore, multi-sensory feedback is a crucial aspect of human perception in tunnel environments. While visual stimuli dominate tunnel perception, auditory cues also contribute significantly to situational awareness and spatial orientation. The interaction between vision and hearing plays a fundamental role in how individuals process their surroundings, especially in environments where visual perception is limited. The ability to integrate both visual and auditory inputs in VR/AR simulations is vital for creating a realistic and immersive experience.

This chapter provides a review of previous research on human eye responses to different lighting environments, including the effects of luminance adaptation, color temperature, and luminance uniformity on visual perception. Additionally, it investigates the role of auditory feedback in tunnel environments, examining how individuals process and respond to sound cues when visual information is constrained. By combining insights from both lighting and auditory research, this study aims to provide a comprehensive review of multi-sensory perception in tunnels and its implications for VR/AR-based environmental simulations.

First, luminance level is one of the core parameters in tunnel lighting design. According to Pieron's law, luminance is inversely related to visual reaction time [5]. Research has shown that as the average luminance (L_{av}) inside the tunnel increases, drivers' reaction speed improves, and overall visual performance is enhanced [43]. However, the visual environment inside a tunnel is monotonous, and during driving, continuous stimuli from light-dark fluctuations can create psychological pressure on the driver. If lighting uniformity is poor, the "zebra effect" on the tunnel road surface exacerbates this psychological burden [15], potentially impacting tunnel driving safety. Therefore, increasing luminance also leads to greater light-dark contrast stimulation, especially when the luminance longitudinal uniformity (U_1) is low. In such cases, an increase in luminance may result in higher visual load and psychological stress, which, in turn, affects the driver's overall visual performance [19]. Thus, luminance longitudinal uniformity (U_1) has a significant impact on drivers' visual performance. Japanese scholars [21]) found through subjective assessments that, compared to a tunnel road surface with a luminance of 3.5 cd/m^2 and low uniformity, drivers in a tunnel lighting environment with a luminance of 2.3 cd/m^2 and uniformity above 0.9 had the same or better visual impression and comfort. This suggests that for drivers' subjective perception, there is a complementary relationship between luminance uniformity and luminance level. Jia [17] conducted subjective evaluation experiments based on small target recognition and found that, at the same luminance uniformity, when the luminance level reached 1.5 cd/m^2 or higher, participants found it easier to detect target objects.

As shown in Fig. 4(a), higher U_1 can effectively reduce drivers' psychological stress and visual fatigue. For example, when U_1 increased from 0.5 to 0.8, drivers' reaction time decreased from 410 ms to 330 ms, and the pupil area variation rate (U) decreased from 18% to 8% [19]. Moreover, low uniformity can cause the "flicker effect", increasing driver discomfort and potentially leading to visual oscillation, which can affect driving stability [15]. Further research has found that the effect of luminance uniformity on driver visual performance varies at different average luminance (L_{av}) levels. When U_1 is low, increasing L_{av} tends to result in longer reaction times. However, when U_1 is higher, increasing L_{av} it can actually improve visual performance. For example, when U_1 is 0.5, increasing L_{av} from 1 cd/m^2 to 6 cd/m^2 causes drivers' reaction times to increase from 330 ms to 410 ms, and the pupil area variation rate (U) increases from 11.75% to 18.82% [19], as is shown in Fig. 4(b).

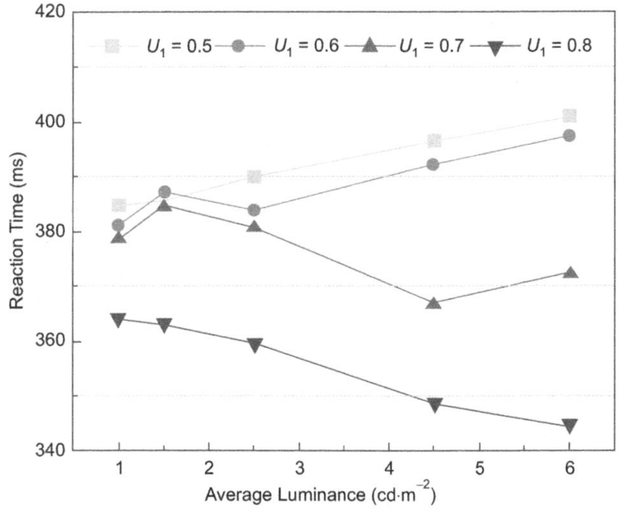

(a) Average luminance

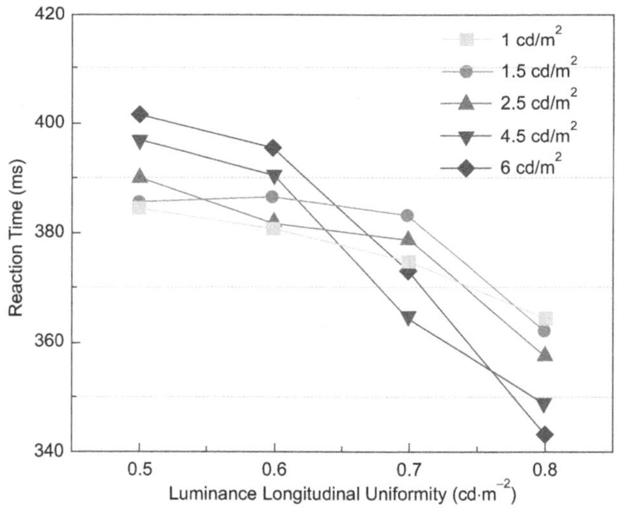

(b) Luminance longitudinal uniformity

Fig. 4 Influence of luminance characteristics on reaction time [19].

Luminance variation rate (V) is one of the key factors influencing drivers' visual performance, and the stepped change in luminance mainly occurs at the tunnel entrance. Due to the tunnel's unique tubular structure, a "black hole effect" is easily generated at the tunnel entrance, and drivers undergo a visual "dark adaptation" process, leading to a temporary visual blind spot. To ensure driving safety, a common approach in many countries and regions is to provide enhanced lighting at the tunnel entrance, dividing the lighting into multiple characteristic lighting segments, each with gradually decreasing luminance according to a certain ratio. This multi-step luminance distribution helps alleviate the negative effects of the dark adaptation process on the driver's visual performance.

In the 1960s, the International Commission on Illumination (CIE) proposed the tunnel lighting adaptation curve based on basic research results, such as the small-target visibility experiments conducted by European and Japanese scholars [33], and pointed out that this curve could be replaced by a stepped curve. He et al. [13] conducted indoor visual performance experiments based on the driving process in the enhanced lighting section at the tunnel entrance and found that due to the proportional decrease in luminance across each lighting segment, stepped luminance transitions occurred at the junctions between adjacent segments. This caused drivers to experience slower reactions when driving through these junctions. [49] defined the phenomenon of visual oscillation based on changes in the driver's pupil area and pointed out that during the driving process from the tunnel entrance to the transition section, although the frequency of visual oscillation decreased compared to the tunnel opening, certain safety risks still remained.

Studies show that an increase in V leads to longer reaction times, increased pupil area variation rates, and exacerbated psychological pressure and visual fatigue [41]. Specifically, when the luminance variation rate increases from 3 cd/($m^2 \cdot s$) to 7 cd/($m^2 \cdot s$), the driver's reaction time increases from 520 ms to 640 ms, and the pupil area variation rate increases from 7% to 15%. A theoretical model was established to correlate luminance variation rate with the driver's safety reaction threshold under different initial luminance conditions (20–80 cd/m^2). The study found that when the tunnel lighting luminance changes too quickly, it results in decreased driver perception, affecting safe driving. Furthermore, research shows that smoother luminance change strategies, such as continuous dimming, are more effective at reducing driver discomfort and visual load compared to abrupt dimming strategies, such as stepped dimming. Indoor dynamic experiments show that when the enhanced lighting at the tunnel entrance decreases in a stepped manner, participants experience higher visual workload and psychological tension, with average reaction times exceeding 300 ms and average pupil diameter change rates around 8.6%. Particularly during short periods of sudden luminance decrease, the instantaneous pupil diameter variation rate exceeds 20%. In contrast, under continuous smooth dimming, the luminance variation per unit time is limited, and the average pupil diameter change rate is only around 4.6%, allowing the human eye to better adapt to light environment stimuli and resulting in higher visual performance [18]).

Color temperature is also an important factor influencing tunnel lighting environments. Due to the photobiological effects, the color temperature of the light source directly impacts drivers' reaction times and pupil variations, making it a key indicator in evaluating tunnel lighting quality. Zhang et al. [42], using a custom reaction time testing device, compared the differences in drivers' visual performance under high and low color temperature CFL light sources. He et al. [13], focusing on four different color temperature LED light sources, analyzed the

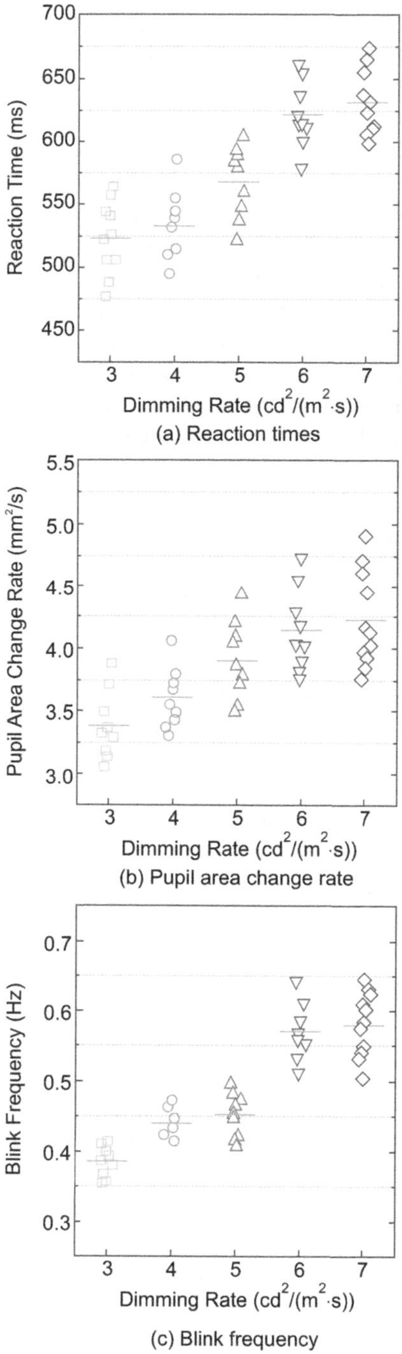

Fig. 5 The impact of dimming rate on visual performance when the luminance decreases through dimming [41]

changes in reaction times at the junctions of various tunnel lighting sections. Zhao, using small-target recognition experiments, studied the impact of background light source color temperature and target color differences on drivers' reaction times [45]. Studies have shown that low color temperature (around 2800 K) and high color rendering index (close to 100) LED light sources are more suitable for tunnel entrance lighting. Under lower color temperature conditions, drivers' visual reaction times are approximately 12% shorter compared to those under higher color temperature conditions. Furthermore, low color temperature designs shorten adaptation time in light-dark transition zones by about 15%, effectively reducing drivers' visual fatigue. Specifically, in a low color temperature environment, drivers' reaction time upon entering the tunnel decreases from an average of 450 ms to 400 ms, and the pupil variation rate drops from 15% to 10% [19]). Changes in color temperature affect drivers' visual adaptation ability, particularly in light-dark transition zones at tunnel entrances and exits. A well-designed color temperature can effectively reduce drivers' visual fatigue and improve overall visual comfort [26].

In situations where visual perception is limited, auditory signals provide more effective information [4] because sound propagation is not affected by smoke or visual obstructions. Based on existing indoor auditory guidance research, it can be inferred that gentle, soothing, and informative guiding sounds are more effective than noisy, irritating, and meaningless sounds. For example, a soft female voice is preferable to a male voice; speech signals are more effective than alarm sounds; and tonal whistles are better than beeping sounds [7,36]. These findings highlight the importance of sound characteristics in environments with limited visual cues, such as tunnels, where auditory signals can significantly enhance the overall sensory experience.

In such cases, the multisensory interaction between auditory and visual stimuli in virtual reality/augmented reality (VR/AR) platforms becomes crucial. When visual signals are limited, auditory signals can support or even replace visual cues, improving the safety and navigability of tunnel environments. Additionally, considering physiological differences between individuals, aging leads to a decrease in the human ear's ability to transmit sound signals, meaning that elderly individuals require higher sound pressure levels to perceive them effectively [38]. Therefore, when designing tunnel environments, it is important to account for age-related sensory differences, particularly in VR/AR platforms aimed at providing personalized and accessible experiences. Increasing the sound pressure level for older populations can bridge sensory gaps, ensuring that auditory guidance is effective for people of all ages.

In conclusion, the design of tunnel lighting environments needs to comprehensively consider factors such as brightness, luminance uniformity, color temperature, and brightness variation rate to ensure visual comfort and safety for drivers. The integration of multi-sensory feedback from both auditory and visual elements is critically important in extreme environments, and the interaction between different sensory modalities can significantly enhance user perception.

Future research can further explore the interactions between these factors and, by incorporating VR and AR technologies, simulate real-world tunnel environments to more comprehensively assess the impact of multi-modal information on user perception.

5. Multi-dimensional Evaluation

Apart from the input and feedback mechanisms, the multi-dimensional evaluation is critical in interpreting human responses and interactions within XR-based tunnel environments. Specifically, eye tracking and EEG signal analysis are widely used in existing studies. He, [11] employed SMI eye tracker to investigate users' visual focus and reactions during simulated emergency hazard events. Their study observed that various tunnel facilities and landscape elements could affect pupil dilation, which is closely associated with mental tension and psychological strain. Similarly, Li et al. [22] utilize VIVE Pro eye tracking technology to monitor users' pupil behaviors under various illumination conditions in virtual tunnels. Based on the eye closure patterns and eye movements, the eye states are divided into blinking, observing, and saccadic movements, which reflect participants' fatigue levels and concentration states, as shown in Fig. 6(a). It is noted that eye-tracking devices can be used to collect specific metrics such as saccade count, saccade duration, pupil diameter, and blinking frequency, thereby delivering comprehensive evaluations on tunnel construction workers, particularly when they are operating excavators for material handling.

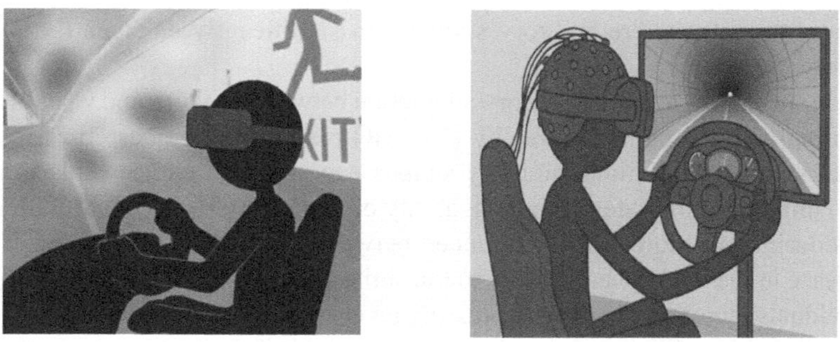

(a) Eye tracking (b) EEG signals

Fig. 6 The application of multi-dimensional evaluations within XR-based tunnel environments.

Additionally, EEG signals serve as valuable indicators of cognitive loads or emotional states among tunnel workers during safety training scenarios [9]. EEG systems record brainwave activities across multiple training scenarios, yielding insights into cognitive and emotional responses. This evaluation helps in understanding the level of immersion achieved and the usability of the XR simulation system, providing valuable insights for optimizing the user experiences.

Beyond the safety training contexts, prior studies also utilized EEG signals to evaluate driver fatigue in extra-long tunnels influenced by landscape layouts during the design phase [23], as shown in Fig. 6(b). The results highlighted that EEG signals not only serve as an effective indicator for fatigue monitoring but also help to reveal the influence of visual environments on drivers' attention, which provides a scientific basis for better tunnel design aimed at enhancing driving safety.

Overall, multi-dimensional evaluation methods offer essential insights for optimizing XR-based tunnel assessments, elevating training and design effectiveness, enhancing realism, and better preparing workers for real-world emergencies (Fig. 6). By assessing both attentional focus and cognitive responses, these evaluation methods are critical for improving the outcomes of emergency safety training and tunnel design programs. However, current research pays predominant attention to evaluating the signal information of users' cognitive responses, like brain activities and eye tracking signals, which only covers a limited range of human reactions in tunnel simulation situations. For example, emotional states such as nervousness, anxiety, and panic experienced during emergencies substantially impact workers' behaviors and performances. Therefore, future multi-dimensional evaluations should integrate emotional and physiological indicators, including heart rate variability and galvanic skin response, to achieve a more comprehensive and realistic assessment of human responses in tunnel emergency simulations.

6. Discussion

VR and AR technologies have been extensively applied to tunnel engineering research due to their immersive, safe, and interactive capabilities. Among the entire lifecycle of tunnel engineering, VR and AR are employed across phases differently depending on their inherent features, as shown in Fig. 7. VR technology is predominantly adopted in the design, training, and evaluation phases, highlighting its significant strength in creating immersive, controlled, and interactive scenario-based environments. Specifically, VR has been extensively applied during the design phase to visualize various tunnel configurations (e.g., lighting and ambient color patterns), environmental layouts, and operational conditions in a fully immersive virtual setting. This immersive visualization supports better-informed decision-making, iterative refinement of designs, and effective stakeholder engagement, leading to optimized tunnel construction and operational plans.

In the training phase, VR has been widely used to simulate diverse scenarios realistically, including emergency response drills, vehicle operation practices, and hazard identification sessions. By providing risk-free yet authentic environments, VR-based training significantly enhances workers' preparedness, situational awareness, and response efficiency during real emergencies. The adoption of VR in the evaluation phase further demonstrates its capability to systematically measure and analyze human cognitive responses, fatigue, attention levels, and behaviors in

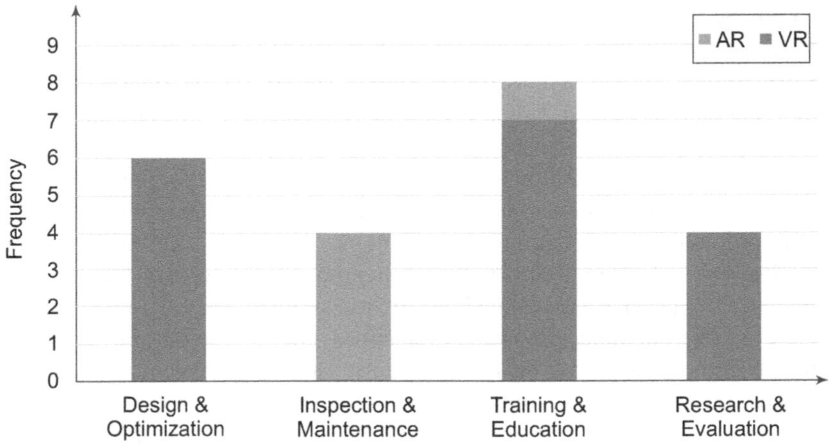

Fig. 7 The utilization of VR and AR platforms in various stages of the tunnel lifecycle.

immersive simulations, facilitating comprehensive assessments of design efficacy and training effectiveness.

AR, on the other hand, demonstrates substantial potential primarily within the inspection phase, where its capacity to overlay digital information onto the real-world environment proves particularly advantageous. AR enables inspectors to conduct accurate and efficient on-site verifications, defect identifications, and maintenance assessments by seamlessly integrating virtual annotations, instructions, or spatially-anchored markers into the physical context. Moreover, AR for on-site operational training for tunnel management and maintenance gradually received attention in recent years, which showcases its additional potential for enhancing real-time interactive guidance, procedural assistance, or supplemental information during hands-on tasks.

Both VR and AR platforms for tunnel engineering research involve multiple forms of inputs, feedback mechanisms, and evaluations to achieve better human-XR interactions. Figure 8 illustrates the current statuses and trends of existing XR-based tunnel environment simulations using various forms. Regarding the multi-modal inputs, physical controllers currently dominate in existing studies, given their intuitive usability and reliability for controlling virtual equipment, machinery operation, and navigation within complex tunnel environments. Other input methods also emerge, including gesture detection, motion detection, locomotion tracking, and eye-gaze interaction. These input methods are increasingly being explored, suggesting significant potential for more naturalistic, immersive, and hands-free user experiences, especially valuable for training, inspection, and emergency response scenarios.

For feedback mechanisms, visual feedback remains the most extensively applied form as it is crucial for realism, immersion, spatial orientation, and accurate representation of tunnel conditions and designs, such as lighting effects for tunnel

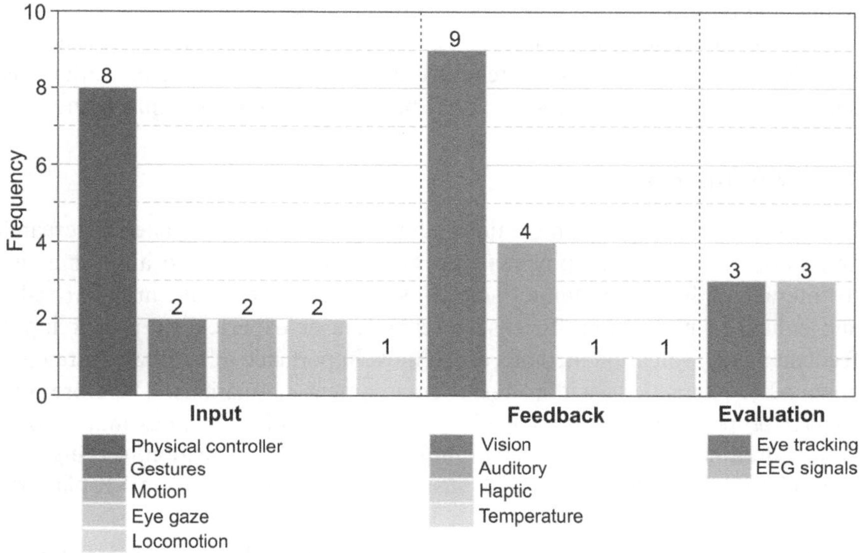

Fig. 8 The frequency analysis of multi-form XR in tunnel engineering research.

driving safety. Auditory feedback also plays an important complementary role in enhancing situational awareness, signaling hazards, and providing realistic environmental cues such as alarms or equipment sounds. Other sensory feedback methods, such as haptic feedback (simulating vibrations or resistance when interacting with virtual tools or equipment) and temperature feedback (replicating conditions like heat during tunnel fires), appear more recently and present exciting opportunities to further enhance sensory realism, thus significantly benefiting emergency response training, hazard perception, and overall simulation authenticity.

In the evaluation, EEG signals and eye-tracking methods have demonstrated effectiveness for assessing cognitive load, attentional responses, fatigue levels, and user engagement during simulated tunnel operations, thereby supporting improved training effectiveness, optimized tunnel design, and enhanced safety assessments. Moving forward, incorporating emerging physiological and emotional indicators, such as heart rate variability and galvanic skin response, could provide deeper insights into user experiences, stress responses, emotional states, and decision-making processes during critical tunnel emergency scenarios. Embracing these broader evaluation metrics will substantially enhance the comprehensiveness, effectiveness, and realism of XR-based tunnel simulations, ultimately advancing tunnel safety, design quality, and operational preparedness [53][54].

Overall, multiple forms of inputs, feedback, and evaluations are increasingly employed in XR-based tunnel environment simulations to enhance the intuitiveness, sensation, and immersion of human-XR interactions. Untapped opportunities exist for using combinations of various XR components within the IFE framework to further bridge the gap between reality and virtuality, ultimately enabling seamless replication of actual tunnel experiences across diverse applications. Additionally, systematic comparative experiments are essential to validate the effectiveness of these XR approaches and to comprehensively understand the strengths, limitations, and practical implications of various XR-based tunnel simulation platforms.

7. Conclusion

VR and AR technologies bring unique advantages to tunnel engineering research and simulation by generating realistic virtual environments to allow humans to interact with various tunnel scenarios with high freedom, minimal risks, and enhanced perception. To ensure realistic user experiences, proper input, feedback, and evaluation methods are of most importance to facilitate instinctive human-XR interactions within tunnel environment simulations. Therefore, this chapter proposes an IFE framework that systematically integrates multi-modal inputs, multi-sensory feedback, and multi-dimensional evaluation, serving as a methodological reference to provide insights for further multi-form XR platform developments in tunnel engineering.

Following the structure of the IFE framework, a systematic review is conducted to provide details and examples of state-of-the-art developments in the multi-form XR-based tunnel environment simulation. Several critical findings are observed and summarized as follows:

1. Physical controllers, gesture detection, and motion tracking dominate current multi-modal inputs, providing intuitive and reliable interaction mechanisms essential for tasks such as tunnel inspection, equipment operation, and emergency response training. Emerging input methods such as voice commands and eye-gaze tracking indicate potential pathways toward more natural and immersive user experiences, particularly useful in scenarios requiring hands-free operations such as working within confined tunnel spaces.
2. Multi-sensory feedback, notably visual and auditory, has been effectively utilized to improve spatial awareness and realism during tunnel navigation, maintenance, and emergency response exercises. Promising opportunities exist in emerging areas such as haptic and temperature feedback, particularly for tunnel emergency scenarios.
3. Multi-dimensional evaluations using EEG and eye-tracking methodologies effectively measure cognitive load and attention, supporting improved tunnel design and emergency training outcomes. Future developments should integrate broader physiological and emotional metrics, including heart rate variability and galvanic skin response, offering comprehensive insights into human responses.

Overall, the proposed IFE framework establishes a structured basis for advancing XR applications in tunnel engineering, driving forward research that prioritizes safety, user experience, and operational efficiency throughout all phases of tunnel lifecycle management.

Reference

[1] An, W., Hu, M., Gao, J., Xie, Y. and Yu, S. (2022). Experiment on Evacuation Behavior in Full-Scale Tunnel with Different Guiding Pattern. *International Journal of Intelligent Transportation Systems Research, 20*(3). https://doi.org/10.1007/s13177-022-00327-z.

[2] Bassani, M., Catani, L., Salussolia, A. and Yang, C.Y.D. (2019). A driving simulation study to examine the impact of available sight distance on driver behavior along rural highways. *Accident Analysis and Prevention, 131*. https://doi.org/10.1016/j.aap.2019.07.003.

[3] Brown, M., Kehoe, A., Kirakowski, J. and Pitt, I. (2010). *Beyond the Gamepad: HCI and Game Controller Design and Evaluation*. https://doi.org/10.1007/978-1-84882-963-3_12.

[4] Bryan, J.L. (1977). *Smoke as a Determinant of Human Behavior in Fire Situations:(Project People)*. US Department of Commerce, National Bureau of Standards.

[5] Burr, D.C. and Corsale, B. (2001). Dependency of reaction times to motion onset on luminance and chromatic contrast. *Vision Research, 41*(8). https://doi.org/10.1016/S0042-6989(01)00019-0.

[6] Chen, L., Zhao, H., Shi, C., Wu, Y., Yu, X., Ren, W., Zhang, Z. and Shi, X. (2024). Enhancing Multi-Modal Perception and Interaction: An Augmented Reality Visualization System for Complex Decision Making. *Systems, 12*(1). https://doi.org/10.3390/systems12010007.

[7] Dong, L., Zhao, E., Chen, Y., Qin, G. and Xu, W. (2020). Impact of LED Color Temperatures on Perception Luminance in the Interior Zone of a Tunnel considering Fog Transmittance. *Advances in Civil Engineering, 2020*. https://doi.org/10.1155/2020/3971256.

[8] Fridolf, K., Ronchi, E., Nilsson, D. and Frantzich, H. (2013). Movement speed and exit choice in smoke-filled rail tunnels. *Fire Safety Journal, 59*. https://doi.org/10.1016/j.firesaf.2013.03.007.

[9] Guo, Z., Tao, H., Yang, Y. and Lin, W. (2024). Evaluating the Effectiveness and Efficiency of a Training Program for Tunnel Construction Workers. *Buildings, 14*(12), 3733. https://doi.org/10.3390/buildings14123733.

[10] Hao, Z., Zhang, W. and Zhao, Y. (2021). Integrated BIM and VR to implement IPD mode in transportation infrastructure projects: System design and case application. *PLoS ONE, 16*(11 November). https://doi.org/10.1371/journal.pone.0259046.

[11] He, J., Zhang, C., Zhang, F. and Liu, L. (2020). Tunnel traffic facilities and landscape evaluation technology based on virtual reality. *Journal of Tongji University* (Natural Science), *48*(02), 215–222. https://doi.org/10.11908/j.issn.0253-374x.18338

[12] He, S., Liang, B., Tähkämö, L., Maksimainen, M. and Halonen, L. (2020). The influences of tunnel lighting environment on drivers' peripheral visual performance during transient adaptation. *Displays, 64*. https://doi.org/10.1016/j.displa.2020.101964.

[13] He, S.Y., Tähkämö, L., Maksimainen, M., Liang, B., Pan, G.B. and Halonen, L. (2017). Effects of transient adaptation on drivers' visual performance in road tunnel lighting. *Tunnelling and Underground Space Technology, 70*. https://doi.org/10.1016/j.tust.2017.07.008.

[14] Hussain, S.A.H., Raza, I., Hussain, S.A., Jamal, M.H., Gulrez, T. and Zia, A. (2025). A mental state aware brain computer interface for adaptive control of electric powered wheelchair. *Scientific Reports, 15*(1), 9880. https://doi.org/10.1038/s41598-024-82252-7

[15] International Commission on Illumination. (2004). CIE 88–2004 Guide for the Lighting of Road Tunnels and Underpasses. In *Commission internationale de l'éclairage*.

[16] Isleyen, E. and Duzgun, H.S. (2019). Use of virtual reality in underground roof fall hazard assessment and risk mitigation. *International Journal of Mining Science and Technology, 29*(4). https://doi.org/10.1016/j.ijmst.2019.06.003.

[17] Jia, H. (2017). *Study on average luminance and overall luminance uniformity index of urban overpass road lighting*. Tianjin University.

[18] Kang, C., Du, Z., Lin, G., Zhang, C., Huang, Z. and Wu, K. (2022). Luminance stepped decreasing ratio of the tunnel entrance enhanced lighting based on visual performance. *Applied Optics, 61*(22), 6553. https://doi.org/10.1364/ao.457524.

[19] Kang, C., Lin, G., Ye, L., Zhang, C., Huang, Z. and Wu, K. (2023). Influence of luminance and luminance longitudinal uniformity of tunnel lighting on drivers' comprehensive visual performance. *Tunnelling and Underground Space Technology, 140*. https://doi.org/10.1016/j.tust.2023.105282.

[20] Kang, N., Seo, S.-W., Jung, W.-S., Lee, J. and Lee, Y.-T. (2025). Effective Underground Facilities Management in Virtual Reality by Applying a Cybersickness Reduction Technique Based on a Four-Degree-of-Freedom Motion Simulator. *IEEE Access, 13*, 19598–19611. https://doi.org/10.1109/ACCESS.2025.3532597.

[21] Kimura, M., Hirakawa, S., Uchino, H., Motomura, H. and Jinno, M. (2014). Energy savings in tunnel lighting by improving the road surface luminance uniformity – A new approach to tunnel lighting. *Journal of Light and Visual Environment, 38*. https://doi.org/10.2150/jlve.IEIJ120000487.

[22] Li, J., Zhu, J. and Guan, C. (2024). Assessing illumination fatigue in tunnel workers through eye-tracking technology: A laboratory study. *Advanced Engineering Informatics, 59*. https://doi.org/10.1016/j.aei.2023.102335.

[23] Li, X. jun, Ling, J. xin and Shen, Y. (2021). Visual fatigue relief zone in an extra-long tunnel using virtual reality with wearable EEG-based devices. *Journal of Central South University, 28*(12). https://doi.org/10.1007/s11771-021-4882-8.

[24] Li, X., Ling, J., Shen, Y., Lu, T. and Zhu, H. (2021). Effect of Color Temperature of Light Source in Tunnel on Driving Safety Based on Virtual Reality Technology. *Tongji Daxue Xuebao/Journal of Tongji University, 49*(2). https://doi.org/10.11908/j.issn.0253-374x.20303.

[25] Ling, J., Li, X., Shen, Y., Chen, C., Yan, Z., Zhu, H. and Li, H. (2025). Human centric VR system development supporting fire emergency evacuation: A novel knowledge-data dual driven approach. *Expert Systems with Applications, 273*, 126895. https://doi.org/10.1016/J.ESWA.2025.126895.

[26] Liu, Y., Peng, L., Lin, L., Chen, Z., Weng, J. and Zhang, Q. (2021). The impact of LED spectrum and correlated color temperature on driving safety in long tunnel lighting. *Tunnelling and Underground Space Technology, 112*. https://doi.org/10.1016/j.tust.2021.103867.

[27] Ma, G., Xiao, M. and Chan, C.T. (2019). Topological phases in acoustic and mechanical systems. In *Nature Reviews Physics* (Vol. 1, Issue 4). https://doi.org/10.1038/s42254-019-0030-x.

[28] Melnyk, O., Huymajer, M., Fenzl, D., Huemer, C., Wenighofer, R. and Mazak-Huemer, A. (2024). Augmented reality for enhanced documentation and anchor inspection reporting in conventional tunnelling. *Tunnelling and Underground Space Technology, 153*, 106040. https://doi.org/10.1016/j.tust.2024.106040.

[29] Oviatt, S. and Cohen, P. (2000). Perceptual user interfaces: multimodal interfaces that process what comes naturally. *Communications of the ACM, 43*(3). https://doi.org/10.1145/330534.330538.

[30] Pan, M., Wong, M.O., Lam, C.C. and Pan, W. (2024). Integrating extended reality and robotics in construction: A critical review. *Advanced Engineering Informatics, 62*, 102795. https://doi.org/10.1016/J.AEI.2024.102795.

[31] Ren, T., Qiao, M., Roberts, J., Hines, J., Chow, Y.W., Zong, W., Sugden, A., Shepherd, M., Farrelly, M., Kennedy, G., Hai, F. and Susilo, W. (2024). Development and evaluation of an immersive VR-CFD-Based tool for dust exposure mitigation in underground tunnelling operations. *Tunnelling and Underground Space Technology, 143*. https://doi.org/10.1016/j.tust.2023.105496.

[32] Ronchi, E., Nilsson, D., Kojić, S., Eriksson, J., Lovreglio, R., Modig, H. and Walter, A.L. (2016). A Virtual Reality Experiment on Flashing Lights at Emergency Exit Portals for Road Tunnel Evacuation. *Fire Technology, 52*(3). https://doi.org/10.1007/s10694-015-0462-5.

[33] Schreuder, D.A. (1964). *The lighting of vehicular traffic tunnels* [Industrial Engineering and Innovation Sciences]. https://doi.org/10.6100/IR88368.

[34] Shen, Y., Ling, J., Li, X., Li, H., Feng, S. and Zhu, H. (2022). Holistic digital-twin-based framework to improve tunnel lighting environment: From methodology to application. *Building and Environment, 224*. https://doi.org/10.1016/j.buildenv.2022.109562.

[35] Szczurek, K.A., Prades, R.M., Matheson, E., Rodriguez-Nogueira, J. and Castro, M. Di. (2023). Multimodal Multi-User Mixed Reality Human-Robot Interface for Remote Operations in Hazardous Environments. *IEEE Access, 11*. https://doi.org/10.1109/ACCESS.2023.3245833.

[36] Tronstad, T.V., Jenssen, G.D., Moscoso, C., Södersten, C. and Zaikonnikov, E. (2021). Sound signals to improve evacuation in road tunnels. *Fire Safety Journal, 125*. https://doi.org/10.1016/j.firesaf.2021.103431.

[37] Wang, X., Sun, Z., Chehri, A., Jeon, G. and Song, Y. (2024). Deep learning and multi-modal fusion for real-time multi-object tracking: Algorithms, challenges, datasets, and comparative study. *Information Fusion, 105*. https://doi.org/10.1016/j.inffus.2024.102247.

[38] Wong, L.T. and Leung, L.K. (2005). Minimum fire alarm sound pressure level for elder care centres. *Building and Environment, 40*(1). https://doi.org/10.1016/j.buildenv.2004.05.004.

[39] Yu, X., Yu, P., Wang, C., Wang, D., Shi, W., Shou, W., Wang, J. and Wang, X. (2022). Integrating Virtual Reality and Building Information Modeling for Improving Highway Tunnel Emergency Response Training. *Buildings, 12*(10). https://doi.org/10.3390/buildings12101523.

[40] Zhang, C., Kang, C., Hu, H., Ye, L., Huang, Z. and Wu, K. (2025). Time-varying characteristics of the luminance of components near the tunnel opening and luminance L20 in the access zone. *Tunnelling and Underground Space Technology, 155*, 106214. https://doi.org/10.1016/J.TUST.2024.106214.

[41] Zhang, C., Kang, C., Ye, L., Weng, J., Huang, Z. and Wu, K. (2023). The influence of dimming-induced luminance change on driving safety in tunnels. *Heliyon, 9*(1). https://doi.org/10.1016/j.heliyon.2023.e12928.

[42] Zhang, Q., Chen, Z., Hu, Y. and Yang, C. (2008). Study on the effect of light source color temperature on visual performance in tunnel and road lighting. *China Illuminating Engineering Journal, 19*(2), 24–29.

[43] Zhang, Q., Yang, C., Chen, Z. and Hu, Y. (2009). A research of new light source for road lighting based on vision function method. *Tongji Daxue Xuebao/Journal of Tongji University, 37*(6).

[44] Zhang, X., Chen, L., Jiang, J., Ji, Y., Han, S., Zhu, T., Xu, W. and Tang, F. (2023). Risk analysis of people evacuation and its path optimization during tunnel fires using virtual reality experiments. *Tunnelling and Underground Space Technology, 137*. https://doi.org/10.1016/j.tust.2023.105133.

[45] Zhang, X., Hu, J., Wang, R., Gao, X. and He, L. (2017). The comprehensive efficiency analysis of tunnel lighting based on visual performance. *Advances in Mechanical Engineering, 9*(4). https://doi.org/10.1177/1687814017696449.

[46] Zhang, Y., Kinateder, M., Huang, X. and Warren, W.H. (2025). Modeling competing guidance on evacuation choices under time pressure using virtual reality and machine learning. *Expert Systems with Applications, 262*, 125582. https://doi.org/10.1016/J.ESWA.2024.125582.

[47] Zhang, Y., Kinateder, M., Templeton, A. and Warren, W.H. (2025). A Virtual Reality Experiment on Visual and Auditory Guidance for Egress in Road Tunnel Fires. *Fire Technology*. https://doi.org/10.1007/s10694-024-01668-9.

[48] Zhang, Y., Zhu, H., Guo, Q., Carvel, R. and Yan, Z. (2021). The effect of technical installations on evacuation performance in urban road tunnel fires. *Tunnelling and Underground Space Technology, 107*. https://doi.org/10.1016/j.tust.2020.103608.

[49] Du, Z., Zheng, Z., Zheng, M., Ran, B. and Zhao, X. (2014). Drivers' visual comfort at highway tunnel portals: A quantitative analysis based on visual oscillation. *Transportation Research Part D: Transport and Environment, 31*, 37–47.

[50] Wu, K., Hu, H., Sun, H., Zhu, K., Yu, X., Jin, Y. and Zhang, T. (2026). Experimental Study on Human Visual Response to Safety Signage under Emergency Lighting Conditions. Developments in the Built Environment, 100852.

[51] Liu, Y., Han, L., Lv, S., Jiang, T., Duan, M., Guo, H., ... and Qian, X. (2025). Emotional and directional enabled programmable flexible haptic interface for enhanced cognition in disabled community. Research, 8, 0714.

[52] Sun, H., Zhu, K., Zhang, W., Ke, Z., Hu, H., Wu, K. and Zhang, T. (2025). Emergency path planning based on improved ant colony algorithm. *Journal of Building Engineering*, 100, 111725.

[53] Wu, K., Sun, H., Zhu, Z., Hu, H., Xu, J., Zhu, K., ... and Zhang, T. (2025). Experimental study on the emergency evacuation behavior in building with bottleneck group. *Journal of Building Engineering*, 106, 112576.

[54] Sun, H., Zhu, K., Wang, G., Hu, H., Guo, P., Wu, K. and Zhang, T. (2025). Experimental study on pedestrian evacuation characteristics through building bottleneck group structure. Developments in the Built Environment, 100734.

8 Tunnel Lighting in Driving Performance and Energy Saving

Xin Huang[1], Yi Rui[2*] and Guanlin Huang[3]

This chapter investigates the impact of tunnel lighting on driving behavior and traffic carbon emissions using a virtual reality (VR)-based experimental platform. First, we review the current research on tunnel lighting and driving behavior. In the field of tunnel lighting, existing studies have demonstrated that luminance and correlated color temperature (CCT) significantly influence driver fatigue and comfort through experimental manipulations. In driving behavior research, microscopic traffic models are widely employed to analyze car-following behaviors within tunnels. Meanwhile, studies on VR-based tunnel environment design reveal that optimizing tunnel lighting can enhance driving safety and comfort while effectively reducing energy consumption and carbon emissions. Subsequently, we introduce a VR driving experiment platform developed using Unity3D to simulate tunnel lighting environments and driving behaviors. The experiment explores changes in driving behavior under different combinations of luminance and CCT in a car-following driving mode. Results indicate that during car-following driving, driver behavior is influenced not only by the leading vehicle but also by the lighting environment, particularly at tunnel entrances and interior zones. The study further analyzes the impact of tunnel lighting on traffic carbon emissions by calibrating the Intelligent Driver Model (IDM) and integrating it with the MOVES model. Findings show that lower luminance and CCT (e.g., 1 cd/m^2, 5000 K) can reduce carbon emissions and improve traffic flow stability. In conclusion, this chapter reveals the comprehensive effects of tunnel lighting on driving behavior and traffic carbon emissions through VR experiments and model analysis, providing scientific evidence for optimizing tunnel lighting design and contributing to the development of low-carbon transportation.

1. Research Status on Tunnel Lighting and Driving Behavior

1.1 Research Status of Tunnel Lighting

In the domain of tunnel lighting, research about driving behavior predominantly relies on experiments that manipulate the luminance and correlated color temperature (CCT) of tunnel lighting [1–4]. By modifying factors such as luminance levels,

[1-3] Department of Geotechnical Engineering, Tongji University, Shanghai, 200092, China.
* Corresponding author: ruiyi@tongji.edu.cn

DOI: 10.1201/9781003514480-8

uniformity, spatial distribution, and CCT, researchers collect psychophysiological data from drivers, including EEG readings, heart rate measurements, and eye movement patterns, to assess driver fatigue and comfort, thereby identifying optimal tunnel lighting conditions [5–7]. For example, Liang et al. investigated driver response times under varying luminance and CCT settings and identified an optimal combination of 10 cd/m^2 luminance and a CCT of 5000 K[8]. Niu et al. collected eye movement data from drivers across diverse lighting conditions, including variations in luminance, luminance uniformity, and flicker frequency, to develop metrics for quantitatively assessing the impact of tunnel lighting on visual comfort [9].

The interior zone of a tunnel typically exhibits the lowest luminance levels and imposes greater visual demands on drivers compared to other tunnel sections [6, 10]. This is particularly evident in long tunnels where the duration spent navigating the interior zone far exceeds that of other zones. Therefore, analyzing how lighting settings in the interior zone affect driving behavior is crucial. Researchers generally categorize luminance and CCT into multiple levels and compare driving behavior characteristics under different parameter combinations to determine the most effective configuration [3, 8, 11]. In accordance with the Chinese Guidelines (JTG/T D70/2-01-2014), 1 cd/m^2, 3 cd/m^2, and 10 cd/m^2 correspond respectively to normal scenarios of low speed with low traffic flow, low speed with high traffic flow, or high speed with low traffic flow, and high speed with high traffic flow. Furthermore, based on findings by Li et al., significant distinctions in driver attention levels were observed at CCTs of 2000 K, 5000 K, and 8000 K[12].

1.2 Research Status of Driving Behavior

Vehicle behavior in tunnels necessitates taking into account how other vehicles interact with one another and how drivers are affected by their surroundings. Given that macroscopic traffic models fail to represent individual vehicles, and mesoscopic models overly simplify vehicle interactions (such as in queueing models, which only consider parameters like free-flow speed, road length, road capacity, and lane count without accounting for vehicle heterogeneity or interaction specifics), only microscopic simulation models are capable of providing a detailed representation of driving behavior patterns. These models accurately reflect the actual geometric conditions of traffic networks, the influence of the surrounding environment on drivers, and vehicle-to-vehicle interactions. Therefore, it is necessary to employ microscopic traffic models to analyze driver behavior in tunnels.

Typically, a complete driving behavior can be divided into two components: longitudinal and lateral behavior. The longitudinal behavior is further subdivided into free driving (when the vehicle is at a considerable distance from the preceding vehicle and not influenced by it) and car-following (when the vehicle is close to the preceding vehicle and its behavior is affected by it). Considering that lane changes are generally not allowed within tunnel highways, emphasis should be placed on analyzing the longitudinal behavior of traffic flow.

In microscopic car-following models, a fundamental assumption is that drivers adjust their behavior based on the actions of the leading vehicle[13]. Each vehicle responds according to its position, instantaneous speed, acceleration, or a combination of these factors. Existing microscopic car-following models can primarily be categorized into four types: safety distance models, stimulus-response models, cellular automata models, and psycho-physical models.

(1) The core of safety distance models is based on car-following rules that maintain a safe following distance. The earliest safety-distance model was proposed by Pipes[14] in 1953. He suggested that vehicle movement should be governed by an idealized "Law of Separation," where each vehicle should maintain a certain "Following Distance" from the vehicle ahead. The central formula, shown as Equation 1.1, expresses the position of the leading vehicle as a function of the following vehicle's characteristics.

$$x_{n-1} = x_n + d + Tv_n + l_{n-1}^{veh} \qquad ...(1.1)$$

where, d represents the statutory minimum safety distance between stationary vehicles, l_{n-1}^{veh} denotes the length of the preceding vehicle, and Tv_n indicates the legal safety gap based on the speed difference between vehicle n and vehicle $n-1$.

In 1981, Gipps enhanced the safety distance model by assuming that each driver sets limits on their desired deceleration and acceleration rates, thus developing a new following vehicle response model[15]. This model includes two primary constraints: ①The vehicle's speed does not exceed the driver's desired speed. It's free acceleration initially increases with engine torque, but gradually decreases to zero as the vehicle approaches the desired speed. ②During emergency braking, the vehicle will avoid colliding with the preceding vehicle, accounting for an additional reaction delay of the following vehicle's driver in responding to the actions of the preceding vehicle.

Furthermore, Newell introduced his own safety distance models in two separate publications: the first in 1961 [16] and the second in 2002[17].

(2) The stimulus-response model is an analytical framework that assumes each driver reacts to a given stimulus according to the relationship described in Equation 1.2 [18]:

$$Reaction = Stimulus \times Sensitivity \qquad ...(1.2)$$

Among the most classic models is the GHR model, proposed by Gazis [19] in 1961. Its core formula is shown in Equation 1.3:

$$a_n(t) = \alpha V_n(t)^{\beta} \frac{\Delta V_n(t-\tau_n)}{\Delta X_n(t-\tau_n)^{\gamma}} \qquad ...(1.3)$$

where, $V_n(t)^{\beta}$ serves as the stimulus term, and $\frac{\Delta V_n(t-\tau_n)}{\Delta X_n(t-\tau_n)^{\gamma}}$ serves as the sensitivity parameter.

Bando [20] proposed the Optimal Velocity model in 1995. As a model approximating a constant traffic flow state, the core idea is that vehicles will approach a specific optimal speed. This is specifically manifested by the vehicle's acceleration being determined by a sensitivity parameter and the difference between the current speed and the optimal speed. The core formulas are shown in Equations 1.4 and 1.5:

$$a_n(t) = \alpha[V_n^*(\varDelta X_n(t)) - V_n(t)] \qquad ...(1.4)$$

$$V_n^*(\varDelta X_n(t)) = v_{\max}(\tanh(s - c_1) + c_2) \qquad ...(1.5)$$

where, α is a constant sensitivity parameter, and $V_n^*(\varDelta X_n(t))$ is the optimal speed.

The IDM (Intelligent Driver Model) was proposed by Treiber[21] in 2000. This model posits that vehicles are influenced by both driving force and resistance, which ultimately determine the vehicle's state. The core formulas are shown in Equations 1.6 and 1.7 [22]:

$$a_n(t) = a_{\max}^{(n)}\left[-\left(\frac{V_n(t)}{V\ (t)}\right) -\left(\frac{S_n(t)}{S\ (t)}\right) \right] \qquad ...(1.6)$$

$$\tilde{S}_n(t) = S_{jam}^{(n)} + \max\left(0, V_n(t)\tilde{T}_n(t) - \frac{V_n(t)\Delta V_n(t)}{\sqrt[2]{a_{\max}^{(n)}\ a_{comf}^{(n)}}}\right) \qquad ...(1.7)$$

where, $a_{\max}^{(n)}$ is the maximum acceleration/deceleration of the following vehicle n, $\tilde{V}_n$ is the desired speed, V_n is the speed of the following vehicle, β is a parameter, S_n is the bumper-to-bumper inter-vehicle clearance, and $\tilde{S}_n$ is the desired following distance. $S_{jam}^{(n)}$ is the vehicle-vehicle clearance in stand-still situations, $\tilde{T}_n$ is the desired time headway, $\varDelta V_n$ is the speed difference, $a_{comf}^{(n)}$ is the comfortable deceleration.

(3) The Cellular Automaton model differs significantly from other car-following models. While it also models vehicle movement, its main distinction from the previous three types of models is that its modeling space is discrete [13]. Roads are typically divided into cells that are 7.5 meters long. Vehicles can either be present or absent in a cell. The operation of vehicles is driven by a series of rules that determine when a vehicle moves to the next cell.

The *NaSch* model is the simplest cellular automaton model [23], reproducing traffic flow with just four rules. Removing any of these rules would distort the simulation results, and more detailed vehicle behavior requires additional rules. During the transition from time t to $t + 1$, the four rules of the *NaSch* model are as follows [23]:

Rule 1: Acceleration Rule – $v_j(t) \rightarrow v_j(t + 1/3) = \min\{v_j(t) + 1, v_{\max}\}$

Rule 2: Deceleration Rule – $v_j(t + 2/3) = \min\{d_j(t), v_j(t + 1/3)\}$

Rule 3: Random Deceleration Rule (probability p) – $v_i(t + 2/3) \rightarrow v_i(t + 1) = \max\{0, v_i(t + 2/3) - 1\}$

Rule 4: Position Update Rule –Move forward by $v_j(t + 1)$ cells

To further refine the granularity of traffic flow modeling, several modified models have been derived from the *NaSch* model, primarily focusing on modifications to Rule 1: The T^2 model [24] posits that vehicles with exactly one empty cell ahead have a certain probability of accelerating, while other vehicles will definitely accelerate. The *BJH* model [25] suggests that the following vehicles have a certain probability of moving forward after the lead vehicle brakes. The *VDR* model [23]distinguishes acceleration probabilities for vehicles in different speed states. These three models differentiate the acceleration conditions for vehicles based on different spatial positions, times, and speeds.

Additionally, to improve simulation accuracy, Yeldan [26] introduced fuzzy rules into the cellular automaton model and discretized vehicles into cells, using floating-point numbers in car-following rules to control vehicle behavior, thereby proposing the Continuous Cellular Automaton model.

(4) In 2014, Saifuzzaman classified car-following models into engineering models and psychological-physical models, a categorization that has gradually gained widespread acceptance. Among psychological-physical models, the Action Point Model is the most successful [27].

The Action Point Model posits that drivers' vehicle operations are not continuous actions (a point validated by Wagner's 2011 study [28]). During acceleration (when the driver presses the accelerator or brake pedal), the process typically lasts between 0.3 and 2 seconds, while changes in acceleration often occur over very short periods (usually less than 0.3 seconds). This brief period of acceleration change is termed the Action Point [28]. The Action Point Model describes the stimuli perceived by drivers from the preceding vehicle as several threshold curves, thereby determining the driving behavior process [29]. The Action Point Model has also undergone a development process.

Early versions of the Action Point Model originated from Barbosa's 1961 study. Using driving simulators for extensive experiments, observations were made in a two-dimensional Cartesian coordinate system (with the x-axis representing the relative speed between the following and leading vehicles, and the y-axis representing the gap between the rear of the leading vehicle and the front of the following vehicle). It was found that certain parts of the trajectories roughly followed a parabolic shape [30]. The parabola indicates that the second derivative of the gap with respect to relative speed is piecewise constant. From this observation, Barbosa proposed the Decision Point Model, where under car-following conditions, drivers make decisions to accelerate or decelerate at a constant rate, resulting in trajectories that oscillate around an approximate equilibrium point [30].

In 1963, Michaelis [31] advanced this model theory by introducing the concept that drivers perceive speed differences through the rate of change of the visual angle of the preceding vehicle's apparent size within their field of view. Drivers have a certain threshold for perceiving the rate of change of the visual angle; actions are only taken when the rate exceeds a specific value, which is approximately $d\theta/dt \approx \Delta v/D_v^2 \approx 6 \times 10^{-4}$. Once this threshold is exceeded, drivers perceive the deceleration or acceleration of the preceding vehicle and take actions to adjust their speed accordingly [27]. When approaching a slower-moving vehicle ahead, drivers recognize some action points, while other action points are completed unconsciously.

Based on previous research, Wiedemann proposed the most classic Action Point Model in 1974 [30]. In a two-dimensional $\Delta x - \Delta v$ coordinate system, Wiedemann's Action Point Model identifies five threshold curves, each with specific meanings and mathematical formulas:

AX: The desired distance between adjacent vehicles in stopped traffic, composed of the length of the leading vehicle plus the desired following distance of the trailing vehicle.

ABX: The minimum distance between adjacent vehicles at low speed differences, which consists of AX plus a speed-dependent additional term: $ABX(v) = AX + BX\sqrt{v}$.

SDX: The threshold for maximum following distance. When this threshold is reached, the driver realizes they are moving away from the lead vehicle and consciously accelerates: $SDX(v) = AX + BX\sqrt{v} \cdot EX$.

CLDV: During a decreasing gap, when this threshold is reached, the driver perceives that they are too close to the lead vehicle and decelerates: $CLDV(Dx) = -D^2x/CX^2$.

OPDV: During an increasing gap, when this threshold is reached, the driver perceives that they are too far from the lead vehicle and accelerates: $OPDV(Dx) = D^2x/OP^2$.

Since driving behavior is a complex process influenced by and interacting with various factors such as environment, physiology, and psychology, the Action Point Model, as a physiological-psychological model, is more realistic in its modeling approach and best describes the car-following processes of most common driving behaviors [32].

2. Research Status on VR-based Tunnel Environment Design

2.1 The Impact of Tunnel Lighting on Driving Performance

Virtual Reality (VR) is a technology that generates a three-dimensional virtual environment through computer systems, combining multi-sensory interaction methods such as vision, hearing, and touch to immerse users in a virtual world where they can interact with their surroundings in real time[33].VR driving simulators are systems developed based on virtual reality technology. They utilize tools such as VR headsets, motion capture devices, and simulation hardware to provide users with a highly realistic driving experience. These systems are widely used for driver training, traffic safety research, and the analysis of driving behaviors in specific scenarios[34]. VR driving simulators offer several significant advantages:

- By using VR headsets or similar devices, drivers can experience a more realistic driving environment, as if they were in actual road conditions. This level of immersion allows for better simulation of complex traffic situations and emergency scenarios, thereby improving the authenticity and effectiveness of training.
- VR technology is not constrained by physical space, enabling training simulations to be conducted anywhere. Whether at a driving school, home, or other locations, learning or training can be carried out anytime and anywhere as long as VR equipment is available.

Thanks to these advantages, VR-based scene design has been widely applied to research on tunnel visual environments, including studies on sidewall design styles [35], pavement markings [36], sidewall texture characteristics [37], color usage [38], and lighting intensity levels[39].

Currently, related research focuses on two main directions: one is evaluating driver safety within tunnels based on different tunnel lighting environment designs, including aspects such as reaction time and fatigue levels[11, 12, 40]. The other involves optimizing and improving tunnel lighting environments based on the results of these safety or comfort assessments[9, 41, 42]. These studies not only promote the scientific development of tunnel design but also provide important references for enhancing road traffic safety.

In the context of evaluating driving safety for tunnel lighting design, Kircher and Ahlstrom[11] constructed a $2 \times 3 \times 2$ experimental framework encompassing wall color depth, lighting intensity, and task load. The experimental conditions included walls with either dark or light colors, three levels of lighting intensity (1.1 cd/m^2, 1.7 cd/m^2, and 2.5 cd/m^2), and two driving tasks (overtaking and car-following). Driver distraction was analyzed by assessing attention levels. The findings demonstrated that light-colored tunnel walls were more effective than high-intensity lighting in directing drivers' visual focus toward the road ahead. Furthermore, Domenichini et al.[43] developed two tunnel scenarios employing LED and high-pressure sodium vapor (HPSV) lighting, simulating potential driving failure situations. Driving safety was evaluated using metrics such as vehicle speed, lane position, and the standard deviation of lateral position. The results indicated that under LED lighting, drivers were able to detect obstacles earlier, enabling more controlled and relaxed avoidance maneuvers. Additionally, drivers reported higher levels of perceived driving comfort in tunnels illuminated by LED lighting. Li et al.[12] utilized Unity to render a tunnel model under seven distinct correlated color temperatures (CCT) ranging from 2000 K to 8000 K. A rear-end collision scenario was simulated in the middle of the tunnel to investigate the influence of CCT on driver behavior. The results revealed that the CCT of tunnel lighting significantly impacted driver fatigue during driving tasks. Moreover, when drivers encountered rear-end collision scenarios, CCT also exhibited a significant effect on their visual performance. Shao et al.[4] examined the coupling effects of slope gradients and lighting conditions in underwater tunnels through virtual driving experiments. Three lighting conditions (bright, red-tinted, and dim) and four longitudinal slope gradients (2.5%, 3.0%, 3.5%, and 4.0%) were tested. The findings indicated that slope gradients had a pronounced effect on driving performance, while lighting conditions primarily influenced eye movements and heart activity. Notably, the risk of rear-end collisions increased significantly under conditions of dim lighting combined with steep slopes. Lim et al.[40] designed five tunnel scenarios based on two internal brightness levels (9 cd/m^2 and 3.6 cd/m^2) and two information delivery systems: Variable Message Signs (VMS) and Head-Up Displays (HUD). A control scenario featuring a bright tunnel without any information system was also included. Driving behavior was assessed using metrics such as average speed, maximum deceleration, time to collision, average inter-vehicle distance, braking reaction time, and speed deviation. The results demonstrated that information delivery systems significantly enhanced driving safety. Specifically, drivers who received advance tunnel information via these systems reduced their reaction time to the vehicle ahead by 51.29%.

In studies aimed at optimizing tunnel lighting environments through driving safety and comfort, Shen et al. [41] designed three configurations for the placement of top and side luminaires to optimize the internal lighting layout of tunnels. Through VR-based driving experiments, they found that bilateral lighting arrangements significantly enhanced driving safety. The study indicated that the power of lower-side luminaires and main luminaires should be at least 8 W and 12 W, respectively. Additionally, the use of anti-collision lower-side luminaires effectively improved the brightness levels of both the road surface and sidewalls. Subsequently, the researchers applied the optimized lighting scheme to real-world tunnel scenarios, resulting in a significant reduction in tunnel accident rates. Yan et al. [42] explored the impact of self-explanatory environmental design on safety from the perspective of drivers' situational awareness. They designed eight experimental tunnel scenarios by combining two types of slope patterns, two types of colored pavements, and two types of sidewall patterns. Driving safety was analyzed using longitudinal and lateral speed metrics as well as eye-tracking parameters. The results demonstrated that visual illusion markings on colored pavements and light-colored decorative sidewalls provided superior self-explanatory performance, significantly enhancing drivers' cognitive understanding of the driving environment. This highlights the critical importance of self-explanatory tunnel environment design in improving driving conditions and safety. Furthermore, Niu et al.[9] conducted driving simulation experiments based on virtual reality (VR) technology to evaluate changes in visual comfort before and after optimizing tunnel lighting quality. The findings revealed that drivers' visual comfort in the internal zone of the tunnel increased significantly with higher brightness reduction coefficients and lower vehicle speeds. Additionally, drivers' visual comfort in the internal tunnel zone improved as pavement illuminance and visibility increased.

In summary, designing reasonable comparative driving experiments based on tunnel environmental factors is a fundamental step in assessing and optimizing tunnel lighting environments. These studies underscore the significant value of virtual reality (VR) technologies in simulating tunnel lighting environments for research purposes.

2.2 The impact of tunnel lighting on energy saving

One of the main sources of energy use and carbon emissions in tunnels is tunnel lighting. Numerous studies have focused on optimizing lighting system designs to reduce carbon footprints, such as the adoption of green luminaires, intelligent controls, and auxiliary lighting[44–46]. Additionally, researchers have proposed green evaluation metrics for lighting environments from a lifecycle perspective, encompassing the construction, operation, and maintenance phases of tunnels[47, 48]. By comprehensively considering lighting performance and carbon emission levels across these phases, these studies have achieved significant energy-saving and emission-reduction benefits[49]. However, there is limited literature exploring the impact of lighting energy consumption using VR-based equipment. Only Niu et al.established a visual comfort index and demonstrated through VR experiments that an optimized tunnel lighting environment could save 12.03% of the energy consumption of the lighting system.

Given the influence of tunnel lighting on driving behavior[2, 50], the carbon emissions generated by vehicles within tunnels should not be overlooked. Previous studies have shown that reducing the maximum speed limit in highway tunnels can simultaneously decrease carbon emissions from both lighting systems and passing vehicles[51]. Additionally, prior research has demonstrated that driving style significantly affects vehicle carbon emissions, with normal driving producing far less CO_2 than aggressive driving[52]. Therefore, future research could focus more on the potential impact of driving styles on traffic-related carbon emissions under varying tunnel lighting parameters.

3. A Case Study Based on VR Driving Experiments

3.1 VR Tunnel Scene Design

This section uses Unity3D as the development platform to implement a virtual driving experiment. The experimental platform is shown in Figure 1 below.

According to the Guidelines for Highway Tunnel Lighting Design (JTG_ TD70_2-01-2014), the tunnel should be divided into four sections (entrance section, transition section, interior zone, and exit section) in the experimental design, each with its own light environment parameters. This section focuses on

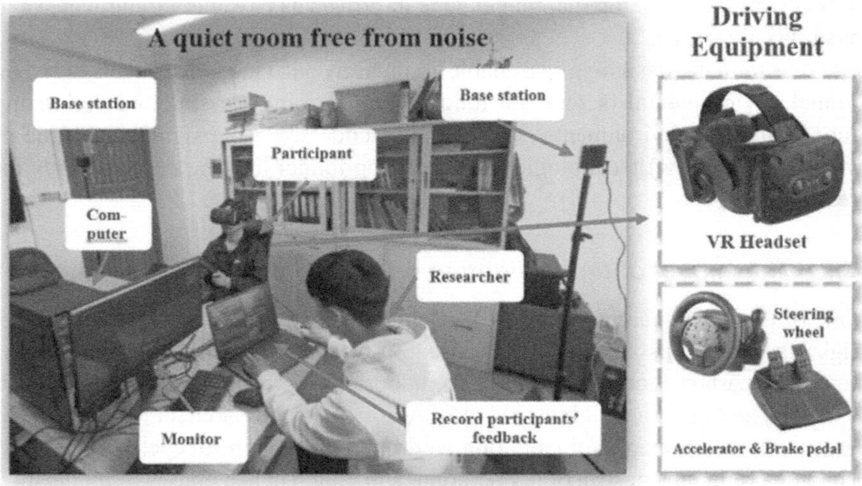

Fig. 1 The experiment platform.

the following behavior in the interior zone of the tunnel (a stable light environment area).

According to the Guidelines for Highway Tunnel Lighting Design (JTG_TD70_2-01-2014), for simplification , the optical parameters (luminance and CCT) along the longitudinal axis of the tunnel are set as piecewise constant functions. The specific parameter design is shown in Figure 2. Based on research in dynamic vision, three typical values for luminance and CCT are selected, resulting in a total of nine sets of experimental lighting environment parameters. The specific scene parameters are illustrated in Figure 3.

The procedure for the VR driving experiment is outlined below:

1. Before Testing: Participants received a brief introduction and detailed instructions on the driving simulator. They then put on the necessary equipment, which was adjusted with the help of researchers to ensure a comfortable and immersive driving experience. Next, each participant completed a practice trial to familiarize themselves with the simulator.
2. During Testing: Participants sequentially completed three sets of driving under a constant speed limit of 60-100 km/h. Overtaking, collisions, or lane changes were not allowed. After each test, participants had a 5-minute break during which the equipment was readjusted. Researchers monitored and recorded the participants' performance throughout the experiment.
3. After Testing: After completing all tests, each participant filled out a questionnaire providing their basic information and feedback on the experiment, including their driving experience and whether they felt fatigued. The duration of the experiment for each participant ranged from 50 to 80 minutes. None of the participants reported feeling dizzy or uncomfortable due to visual lag or other reasons.

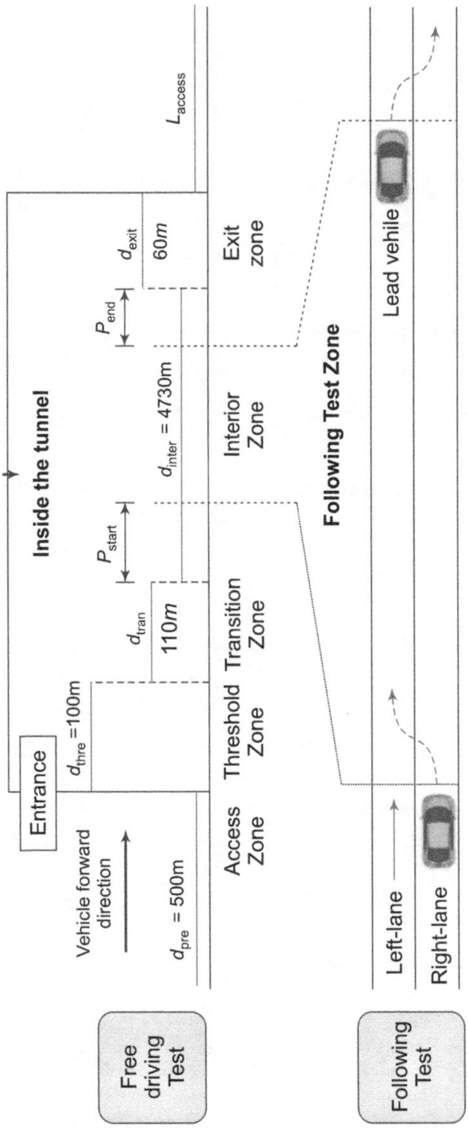

Fig. 2 Schematic diagram of experimental setup.

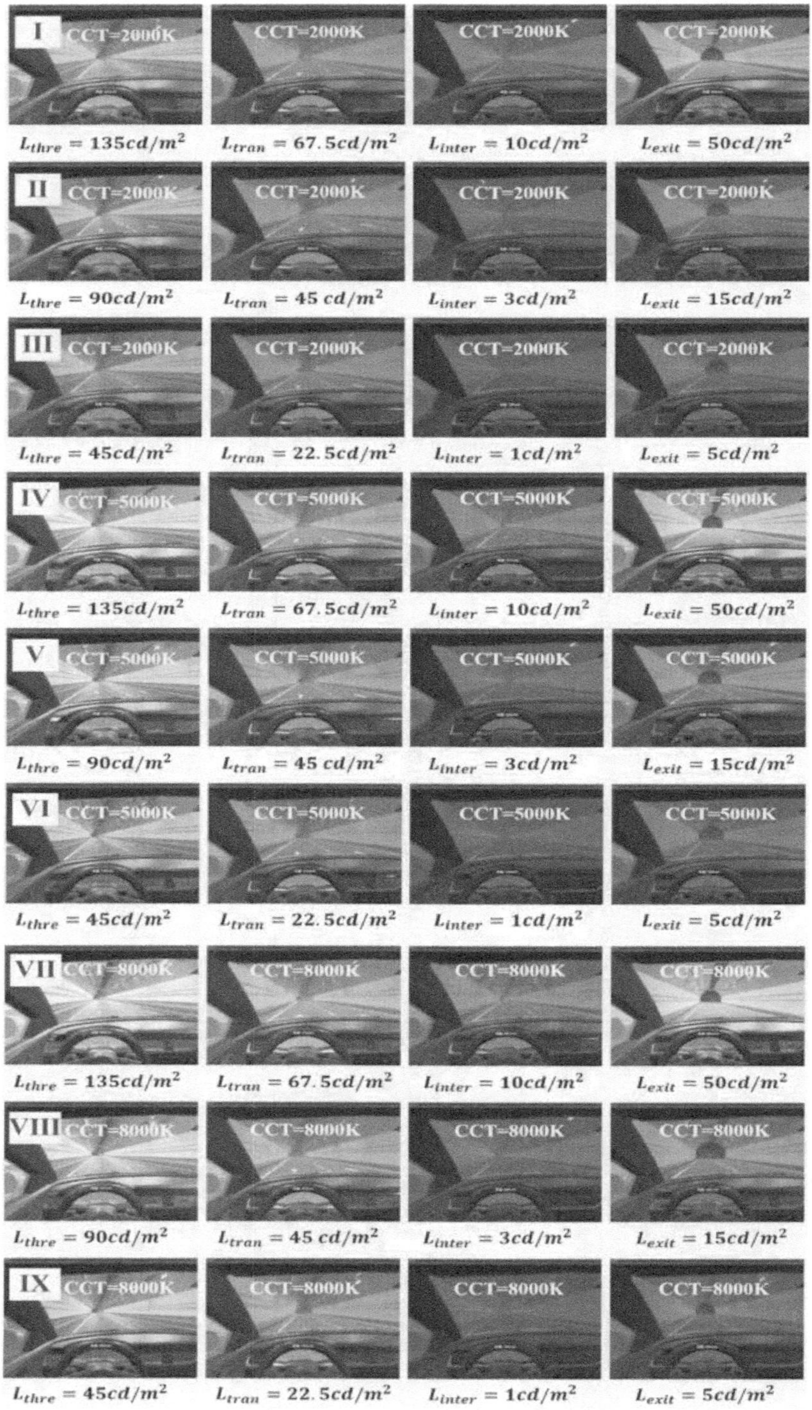

Fig. 3 Nine lighting environments in the virtual reality tunnels.

3.2 Analysis of Car-following Behavior Patterns

3.2.1 Tunnel Entrance

This section analyzes the data from nine tunnels. Specifically, the focus is on the 300-meter range before and after the tunnel entrances. The action points of drivers within each tunnel are binned at 100-meter intervals for analysis, with the results shown in Figure 4. Figure 4 displays the average magnitude of these action points under different lighting conditions.

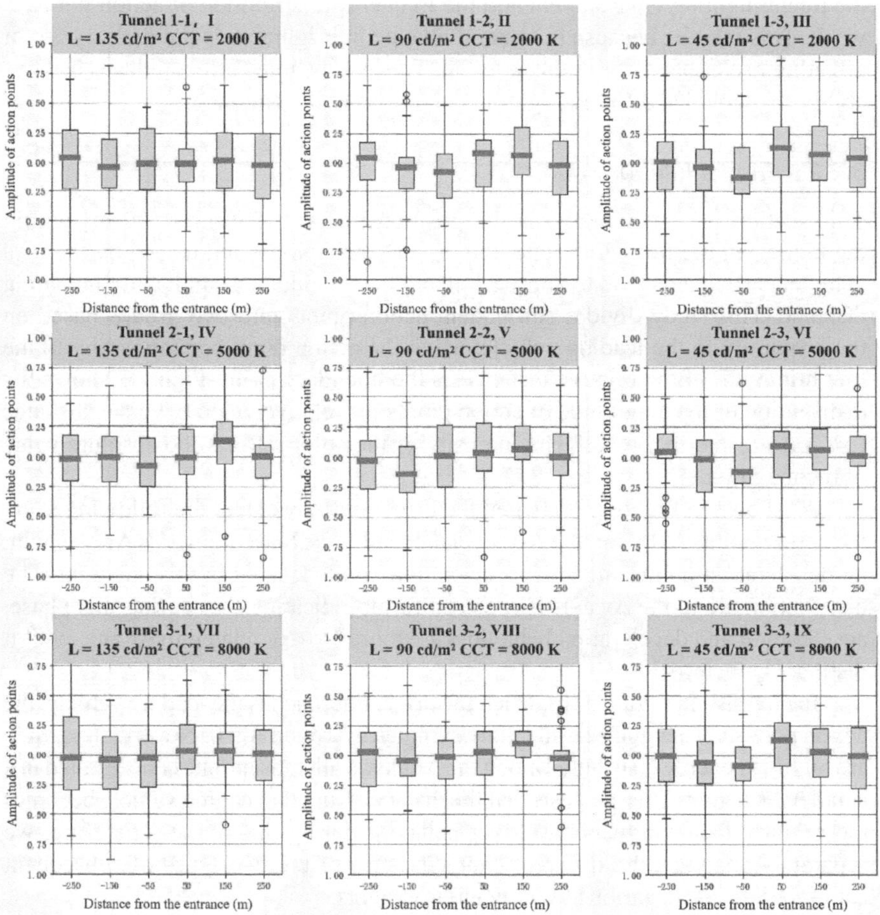

Fig. 4 Position distribution and the average value of the magnitude of drivers' APs around the tunnel entrance.

Under different tunnel lighting environments, a consistent behavior pattern is observed at the tunnel entrance: drivers tend to decelerate within the 200 meters before the entrance and then accelerate over the subsequent 200 meters. Vehicle behavior stabilizes during this transition zone, indicating that drivers have begun to adapt to the changes in lighting conditions.

Differences in tunnel lighting environments result in variations in the magnitude and position of driver action points around the tunnel entrance. As brightness decreases, the difference in brightness between inside and outside the tunnel diminishes, leading drivers to decelerate more as they approach the entrance. The greatest deceleration is caused by a CCT of 8000 K, with the impact of CCT on driving behavior decreasing as brightness decreases. However, there is currently no clear pattern comprehensively describing this effect, and the correlation between the tunnel lighting environment and the frequency distribution of action points is weak. This may be because the rate of change in brightness is the primary factor affecting driver behavior in this interval. The "white hole" effect reduces drivers' ability to recognize different lighting conditions at the entrance.

3.2.2 Tunnel Interior Zone

This section primarily analyzes the position interval [500 m, 4500 m] within the tunnel. In the following state, the driver's behavior is influenced by the lead vehicle: participants tend to approach it and then maintain a small and fluctuating distance. This study divides subsequent action points into two groups based on the proximity to the leading vehicle, using the action point corresponding to the first minimum of the relative distance as the boundary point. Figure 5 shows the distribution of the magnitude of action points for relative velocity versus distance (Δv-Δx). for the two types. The intensity of the color reflects the absolute value of the magnitude.

As shown in Figure 5 (a), when the driver initially enters the following state, the action points (APs) on the Δv-Δx plane tend to be biased towards $\Delta v < 0$, with all deceleration points and most acceleration points located in this quadrant. The distribution along the Δx axis is more uniform, indicating that during this phase, acceleration and deceleration behaviors are more closely related to Δv and exhibit distinct characteristics.

In contrast, in Figure 5 (b), when the driver approaches the leading vehicle, the action points cluster and become more symmetric around Δv=0. Nevertheless, they are highly unevenly distributed along the Δx axis, with most points concentrated in a smaller Δx region. This suggests that the distance from the leading vehicle becomes the primary factor influencing driver behavior during this phase. Therefore, we separately analyzed the different characteristics of the two types of subsequent action points under various lighting environments.

3.3 Impact of Tunnel Lighting on Traffic Carbon Emissions

3.3.1 Tunnel Traffic Flow Fitting based on the IDM

According to Section 1.2, in comparison with traditional models such as the Gazis-Herman-Rothery (GHR) model, the Gipps model, Full Velocity Difference

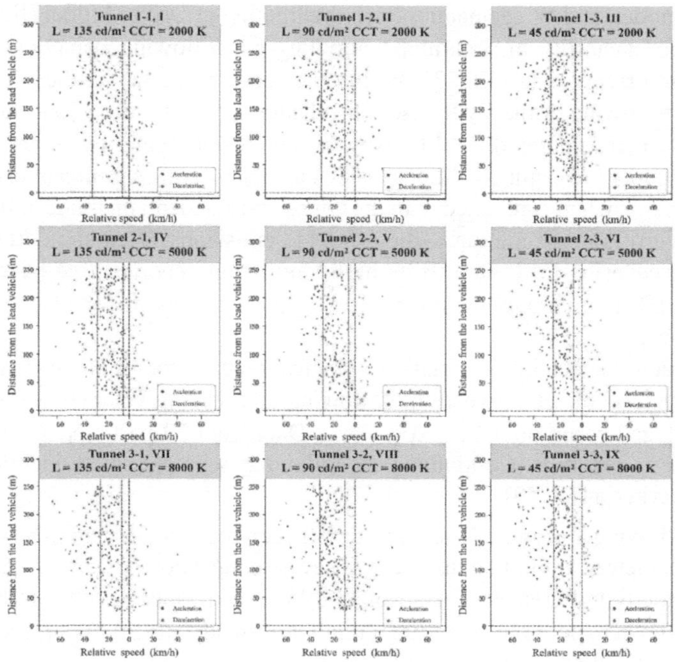

(a)

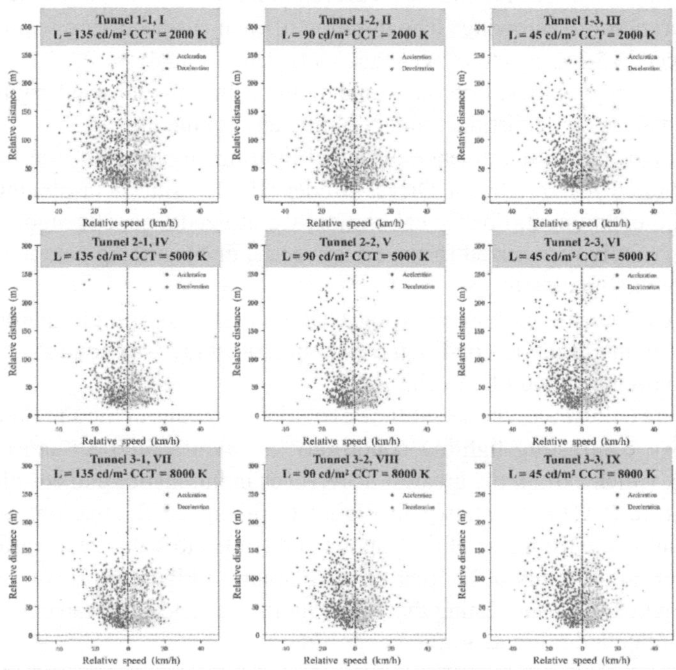

(b)

Fig. 5 Distribution of the magnitudes of the following action points.

(FVD) model, and Wiedemann model, the Intelligent Driver Model (IDM) shows enhanced capability in accurately modeling car-following behaviors based on empirical driving data [22, 53]. Within the scope of tunnel studies, research on vehicle-following primarily focuses on enhancing driving safety [54–56]. Yu et al. (2023) further integrated IDM to explore vehicle-following dynamics and traffic patterns under low-light conditions inside tunnels, offering significant insights[57]. Additionally, when contrasted with the Gipps and Wiedemann models, IDM yields reduced errors in estimating carbon dioxide emissions when simulating convoy driving scenarios [58]. Due to its high simulation accuracy, we have opted to employ IDM for modeling traffic flow within tunnels under varying lighting conditions.

Velocity, acceleration, relative velocity, relative longitudinal distance, and relative lateral offset between the following and leading vehicles were collected and analyzed based on a VR experiment. Data from 32 participants yielded 288 valid driving records. Continuous car-following segments were identified using specific criteria [53, 59]:

1. Following Distance: The distance between the vehicles had to be within 7 to 150 meters, excluding instances of severe congestion or free-flow conditions.
2. Velocity Difference: The absolute difference in speed needed to be under 9 kilometers per hour to ensure a relatively steady car-following scenario.
3. Lateral Offset: The absolute lateral offset was required to be no more than 3.5 meters, ensuring both vehicles remained within the same lane.
4. Duration of Following: The duration of each car-following segment had to exceed 15 seconds to provide adequate data for analysis.

After applying these criteria, a total of 405 valid car-following segments were extracted across nine distinct tunnel lighting environments.

To investigate the influence of tunnel lighting environments on traffic flow, we calibrated the Intelligent Driver Model (IDM) using data obtained from a VR-based experiment. For the calibration, we utilized a genetic algorithm (GA), which is widely recognized for its effectiveness in optimizing parameters within car-following models[60, 61].

To explore the carbon emission characteristics of vehicle platoons under different tunnel lighting conditions, a simulation study was conducted using the IDM, involving a total of 25 vehicles.

Figure 6 illustrates the speed-time evolution diagrams of the simulated platoon under two contrasting lighting conditions: (1 cd/m^2, 2000 K) and (10 cd/m^2, 8000 K), which exhibit the greatest differences in luminance and correlated color temperature (CCT). As shown in Figure 6, the speed fluctuations were more significant in the front part of the platoon and tended to stabilize towards the rear. Moreover, the stability at the rear of the platoon in Figure 6 (a) was superior to that in Figure 6 (b), suggesting that lower luminance and CCT can enhance traffic flow stability and mitigate traffic congestion.

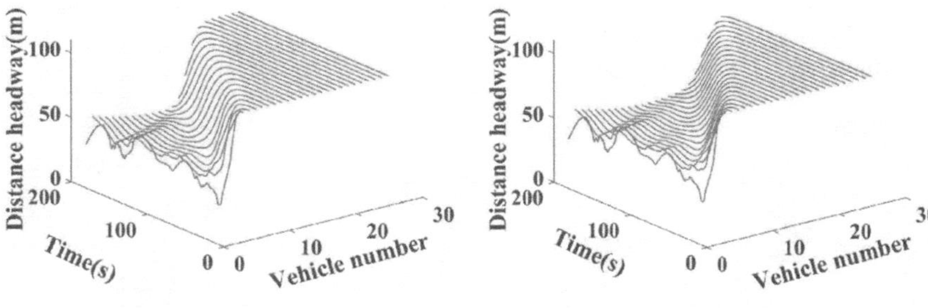

(a) 1 cd/m², 2000 K (b) 10 cd/m², 8000 K

Fig. 6 The space-time evolution of velocity under different tunnel lighting environments.

3.3.2 Calculation of Tunnel Traffic Carbon Emissions based on the MOVES Model

In MOVES, the Vehicle Specific Power (VSP) is determined by a function that considers instantaneous velocity and acceleration, with the assumption that the road grade is zero. The formula for calculating VSP for light-duty vehicles is specified as follows [62]:

$$\text{VSP} = v[1.1a + 0.132] + 0.000302v^3 \qquad ...(4.2)$$

where VSP represents the vehicle-specific power, kw/ton; v is the vehicle's instantaneous speed, m/s; a is the vehicle's instantaneous acceleration, m/s^2.

MOVES categorizes its operational modes into 23 speed-VSP bins[63]. The emission rate for each bin is derived from extensive empirical data. As a result, the carbon emissions of a single vehicle are calculated on a second-by-second basis, depending on the distribution across these bins. The default parameters in MOVES are derived from vehicle tests conducted in the United States. To adapt these calculations to reflect China's vehicle carbon emission characteristics, it is essential to localize the internal parameters of MOVES[64].

Figure 7 illustrates the total carbon emissions and carbon emissions per unit distance for vehicle platoons under nine different tunnel lighting conditions[65]. From this, it can be concluded that the lowest carbon emission levels occur under the lighting conditions of (1 cd/m^2, 5000 K) and (3 cd/m^2, 2000 K). Specifically, the total carbon emissions in the (3 cd/m^2, 2000 K) environment are approximately 20% lower than those in the (3 cd/m^2, 8000 K) environment, which has the highest emissions. Therefore, these two tunnel lighting configurations are most conducive to promoting eco-friendly driving behaviors and have positive implications for achieving low-carbon traffic in tunnels.

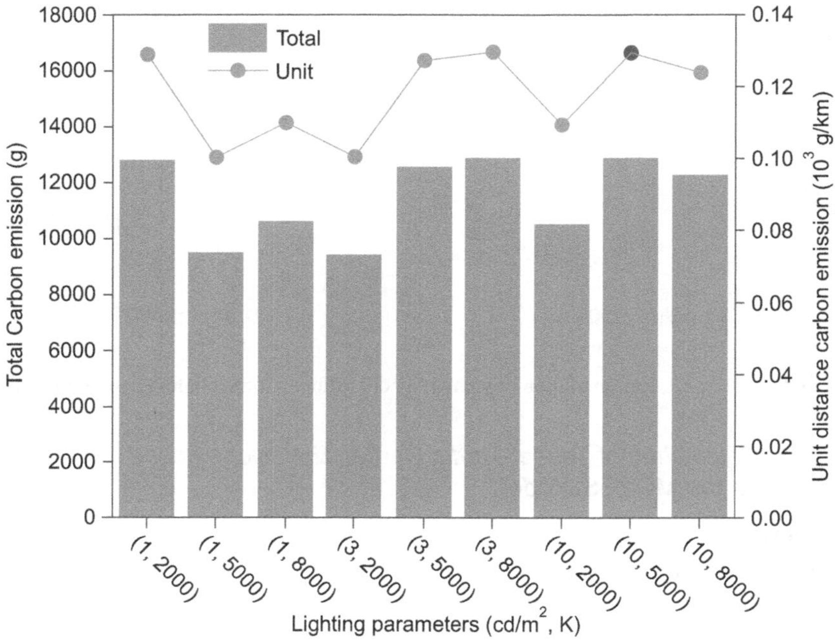

Fig. 7 Total carbon emissions and the carbon emissions per unit distance for the platoons in nine tunnel lighting environments.

4. Conclusion

This chapter reviews the impact of tunnel lighting on driving behavior, traffic safety, and energy conservation and carbon reduction, and explores key factors in tunnel environment design through virtual reality (VR) technology. First, research on tunnel lighting has mainly focused on optimizing drivers' visual comfort and reducing fatigue by adjusting brightness and correlated color temperature (CCT), while also improving driving safety. Studies have shown that under different brightness and CCT conditions, drivers' reaction times, attention levels, and visual performance vary significantly.

In addition, microscopic traffic models are crucial for simulating interactions between vehicles in tunnels, particularly studies on car-following behavior, which reveal drivers' decision-making patterns under specific lighting conditions. Additionally, experiments using VR technology show that variations in brightness at tunnel entrances significantly influence driver behavior, such as slowing down and speeding up. In terms of energy conservation, the adoption of green lighting, intelligent control systems, and other solutions can effectively reduce energy consumption and carbon emissions. For instance, an optimized tunnel lighting environment can save approximately 12.03% of energy consumption.

Additionally, the VR-based experimental platform developed in the case study offers a safe, controlled, and highly replicable environment for simulating various tunnel lighting conditions and driving behaviors. In car-following scenarios, driving behavior is influenced by the combined effects of the leading vehicle and the lighting environment. Near tunnel entrances, drivers typically decelerate within the 200 meters before entering and then accelerate over the subsequent 200 meters. This behavior suggests an adaptation to the changing lighting conditions. Inside the tunnel, when following another vehicle, drivers exhibit more frequent and smaller adjustments in speed, especially when closer to the leading vehicle. The study also reveals that lower luminance and CCT (e.g., 1 cd/m^2, 5000 K) conditions provide better driving stability and reduced frequency of speed adjustments, highlighting the importance of optimizing lighting parameters to enhance driver comfort and safety. Traffic carbon emissions are also influenced by tunnel lighting conditions. The study finds that lower luminance (1 cd/m^2) and moderate CCT (5000 K) produce the lowest carbon emissions, with total emissions approximately 20% lower than those under higher luminance (10 cd/m^2) and CCT (8000 K) conditions. This indicates that optimizing lighting parameters not only enhances traffic flow but also contributes to environmental sustainability by reducing carbon emissions. The findings suggest that tunnel lighting design should prioritize lower luminance and moderate CCT to achieve both traffic and environmental benefits.

In conclusion, the VR-based driving experiments allow for detailed analysis of driver responses under different scenarios, providing a cost-effective alternative to traditional field experiments. The use of VR technology ensures that the experimental conditions can be precisely controlled and repeated, enhancing the reliability and validity of the findings.

5. Future Prospects for VR in Tunnel Driving Research

The rapid advancement of VR technology offers unprecedented opportunities for tunnel driving research. Future studies can leverage the immersive and interactive nature of VR to explore several critical areas:

1. Enhanced realism and scenario complexity: Future VR platforms can incorporate more realistic environmental simulations, including dynamic weather conditions, varying traffic densities, and complex tunnel geometries. This will allow researchers to investigate how drivers adapt to multifaceted scenarios, providing a more comprehensive understanding of driving behavior in tunnels.
2. Multi-sensory integration: Current VR studies primarily focus on visual and auditory stimuli. Future research can integrate additional sensory inputs, such as haptic feedback through advanced steering wheels and seat vibrations, to create a more holistic driving experience. This multi-sensory approach will enhance the realism of the simulation and potentially reveal new insights into driver fatigue and comfort.
3. Personalized driving profiles: VR platforms can be used to develop personalized driving profiles by incorporating individual physiological and psychological data. By integrating EEG, heart rate variability, and eye-

tracking data, researchers can tailor tunnel lighting and environmental conditions to optimize driving performance and safety for different driver demographics.

4. Public awareness and training: VR simulations can be used to educate drivers about the challenges and safety measures associated with tunnel driving. By exposing drivers to various scenarios in a controlled environment, VR can improve public awareness and enhance driver preparedness for real-world tunnel driving conditions.

Reference

[1] Du, F., Weng, J., Hu, Y.K., et al., Study of road tunnel threshold zone lighting reduction coefficient, *Journal of Central South University* 25(9) (2018) 2040–2048.

[2] Kang, C., Lin, G.L., Ye, L., et al., Influence of luminance and luminance longitudinal uniformity of tunnel lighting on drivers' comprehensive visual performance, Tunn. Undergr. Space Technol. 140 (2023).

[3] Liu, Y.Y., Peng, L., Lin, L.Y., et al., The impact of LED spectrum and correlated color temperature on driving safety in long tunnel lighting, Tunn. Undergr. Space Technol. 112 (2021).

[4] Shao, X.J., Chen, F., X.X. Ma, et al., The impact of lighting and longitudinal slope on driver behaviour in underwater tunnels: A simulator study, Tunn. Undergr. Space Technol. 122 (2022).

[5] Chen, X.F., Feng, S.Z., Li, J., et al., Effects of Sidewall Brightness on LED Lighting Environment and Visual Performance in Road Tunnels, Applied Sciences-Basel 12(10) (2022).

[6] He, S.Y., Liang, B., Pan, G.B., et al., Influence of dynamic highway tunnel lighting environment on driving safety based on eye movement parameters of the driver, Tunn. Undergr. Space Technol. 67 (2017) 52–60.

[7] Qin, P.C., Wang, M.N., Chen, Z.W., et al., Characteristics of driver fatigue and fatigue-relieving effect of special light belt in extra-long highway tunnel: A real-road driving study, Tunn. Undergr. Space Technol. 114 (2021).

[8] Liang, B., He, S.Y., Tähkämö, L., et al., Lighting for road tunnels: The influence of CCT of light sources on reaction time, Displays 61 (2020).

[9] Niu, J.A., Liang, He, S.Y. et al., Assessment and optimization of tunnel lighting quality based on drivers' visual comfort: From methodology to application, Tunn. Undergr. Space Technol. 143 (2024).

[10] He, S.Y., Tähkämö, L., Maksimainen, M., et al., Effects of transient adaptation on drivers' visual performance in road tunnel lighting, Tunn. Undergr. Space Technol. 70 (2017) 42–54.

[11] Kircher, K. and Ahlstrom, C. The impact of tunnel design and lighting on the performance of attentive and visually distracted drivers, Accid. Anal. Prev. 47 (2012) 153–161.

[12] Li, X.J., Ling, J.X., Shen, Y., et al., The impact of CCT on driving safety in the normal and accident situation: A VR-based experimental study, Adv. Eng. Inform. 50 (2021).

[13] van Wageningen-Kessels, F., van Lint, H., Vuik, K., et al., Genealogy of traffic flow models, *EURO Journal on Transportation and Logistics. 4* (2015) 445–473.

[14] Pipes, L.A. An Operational Analysis of Traffic Dynamics, Journal of Applied Physics 24(3) (1953) 274–281.

[15] Gipps, P.G. A behavioural car-following model for computer simulation, Transportation Research Part B: Methodological 15(2) (1981) 105–111.

[16] Newell, G.F. Nonlinear Effects in the Dynamics of Car Following, Operations Research 9(2) (1961) 209–229.

[17] Newell, G.F. A simplified car-following theory: a lower order model, Transportation Research Part B: Methodological 36(3) (2002) 195–205.

[18] Saifuzzaman, M. and Zheng, M. Incorporating human-factors in car-following models: A review of recent developments and research needs, Transportation Research Part C: Emerging Technologies 48 (2014) 379–403.

[19] Gazis, D.C., Herman, R. and Rothery, R.W. Nonlinear Follow-the-Leader Models of Traffic Flow, Operations Research 9(4) (1961) 545–567.

[20] Bando, M., Hasebe, K., Nakayama, A., et al., Dynamical model of traffic congestion and numerical simulation, Physical Review E 51(2) (1995) 1035–1042.

[21] Treiber, M., Hennecke, D., Helbing, A Fau. and Helbing, D. Congested traffic states in empirical observations and microscopic simulations, Physical review E 62(1063–651X) (2000).

[22] Zhu, M., Wang, X., Tarko, A., et al., Modeling car-following behavior on urban expressways in Shanghai: A naturalistic driving study, Transp. Res. Pt. C-Emerg. Technol. 93 (2018) 425–445.

[23] Barlovic, R., Santen, L., Schadschneider, A. et al., Metastable states in cellular automata for traffic flow, The European Physical Journal B - Condensed Matter and Complex Systems 5(3) (1998) 793–800.

[24] Schadschneider, A. and Schreckenberg, M. Traffic flow models with 'slow-to-start' rules, Annalen der Physik 509(7) (1997) 541–551.

[25] Simon, C.B., Neil, F.J. and Hui, P.M. Cellular automata models of traffic flow along a highway containing a junction, Journal of Physics A: Mathematical and General 29(12) (1996) 3119.

[26] Ö. Yeldan, A. Colorni, A. Luè, et al., A Stochastic Continuous Cellular Automata Traffic Flow Model with a Multi-agent Fuzzy System, Procedia - Social and Behavioral Sciences 54 (2012) 1350–1359.

[27] Wuhong, W., Wei, Z., Dehui, L.I., et al., Improved action point model in traffic flow based on driver's cognitive mechanism, IEEE intelligent vehicles symposium (Parma 2004-06-14), IEEE, Piscataway NJ, 2004, pp. 447–452.

[28] Wagner, P. A time-discrete harmonic oscillator model of human car-following, The European Physical Journal B 84(4) (2011) 713–718.

[29] Pariota, L. and Bifulco, G.N. Experimental evidence supporting simpler Action Point paradigms for car-following, Transportation Research Part F: Traffic Psychology and Behaviour 35 (2015) 1–15.

[30] Bifulco, G.N., Pariota, L. and Brackstione, M., et al., Driving behaviour models enabling the simulation of Advanced Driving Assistance Systems: revisiting the Action Point paradigm, Transportation Research Part C: Emerging Technologies 36 (2013) 352–366.

[31] Michaels, R.M. Perceptual Factors in Car-Following, 1963.

[32] Olstam, J.J. and Tapani, A., Comparison of car-following models, 2004.

[33] Bowman, D.A. and McMahan, R.P. Virtual reality: How much immersion is enough?, *Computer 40*(7) (2007)36–43.

[34] Boboc, R.G., Butila, E.V. and Butnariu, S. Leveraging Wearable Sensors in Virtual Reality Driving Simulators: A Review of Techniques and Applications, Sensors 24(13) (2024).

[35] Xia, Z., Lv, Y., Pan, X., et al., Research on Design Pattern of City Tunnel Side Wall Based on the Driver Visual Effect, International Conference on Human Factors in Transportation, Fl, 2016, pp. 689–701.

[36] Zhao, X., Ju, Y., Li, H., et al., Safety of Raised Pavement Markers in Freeway Tunnels Based on Driving Behavior, Accid. Anal. Prev. 145 (2020).

[37] Manser, M.P. and Hancock, A. The influence of perceptual speed regulation on speed perception, choice, and control: Tunnel wall characteristics and influences, Accid. Anal. Prev. 39(1) (2007) 69–78.

[38] Qin, X., Zhang, N., Zhang, W., et al., How does tunnel interior color environment influence driving behavior? Quantitative analysis and assessment experiment, Tunn. Undergr. Space Technol. 98 (2020).

[39] Wan, H . Du, Z., Ran, B., et al., Speed Control Method for Highway Tunnel Safety Based on Visual Illusion, Transp. Res. Rec (2485) (2015) 1–7.

[40] Lim, J., Park, H., Oh, T., et al., The impact of information delivery systems in tunnels depending on lighting intensity and speed limit, Tunn. Undergr. Space Technol. 157 (2025).

[41] Shen, Y., Ling, Y., Li, X.J., et al., Holistic digital-twin-based framework to improve tunnel lighting environment: From methodology to application, Building and Environment 224 (2022).

[42] Yan, Y., Zhang, Y.,, Yuan, H.Z., et al., Safety effect of tunnel environment self-explaining design based on situation awareness, Tunn. Undergr. Space Technol. 143 (2024).

[43] Domenichinia, L., La Torre, F., Vangi, D., et al., Influence of the lighting system on the driver's behavior in road tunnels: A driving simulator study, Journal of Transportation Safety & Security 9(2) (2017) 216–238.

[44] M. Chen, Q. Wang, J. Zhen, et al., A review of researches on lighting energy saving technology methods for highway tunnels, 4th International Workshop on Renewable Energy and Development (IWRED), Electr Network, 2020.

[45] Song, Y., Zhu, H., Shen, Y., et al., Green tunnel lighting environment: A systematic review on energy saving, visual comfort and low carbon, Tunn. Undergr. Space Technol. 144 (2024).

[46] Yu, Y., Gu, H., Liang, B., et al., Evaluation and Optimization of Adjacent Tunnel Light Environment Scheme to Low Carbon, Energy Science & Engineering (2025).

[47] Zhang, X., Liu, T., Shen, Y., et al., Analysis of the Effect of Lighting on Carbon Emission in the Full Life Cycle of Highway Tunnels, Modern Tunnelling Technology 60(6) (2023) 40–47.

[48] Zhu, H., Deng, Y., Shen, Y., et al., Application Analysis of Life-cycle Green Index of Highway Tunnel Light Environment, China Journal of Highway and Transport 35(1) (2022) 13–22.

[49] Xu, L., Zhu, H., Shen, Y., et al., Lighting environment design evaluation with digital workflow: A life cycle framework for integrating carbon and cost reduction strategies in tunnels, Tunn. Undergr. Space Technol. 157 (2025).

[50] Zhang, C., Kang, C., Ye, L., et al., The influence of dimming-induced luminance change on driving safety in tunnels, Heliyon 9(1) (2023).

[51] Pena-Garcia, A., Salata, F. and Golasi, I. Decrease of the Maximum Speed in Highway Tunnels as a Measure to Foster Energy Savings and Sustainability, Energies 12(4) (2019).

[52] Xia, Y.J., Liao, C.L., Chen, X.Q. et al., Future reductions of China's transport emissions impacted by changing driving behaviour, Nature Sustainability 6(10) (2023) 1228–1236.

[53] Wang, X., Sun, P., Zhang, X., et al., Calibrating Car-following Models on Freeway Based on Naturalistic Driving Data, China Journal of Highway and Transport 33(5) (2020) 132–142.

[54] Shang, T., Lu, J.X., Luo, Y., et al., Understanding the traffic flow in different types of freeway tunnels based on car-following behaviors analysis, Tunn. Undergr. Space Technol. 143 (2024).

[55] Yeung, J.S. and Wong, Y.D. The effect of road tunnel environment on car following behaviour, Accid. Anal. Prev. 70 (2014) 100–109.

[56] Zhang, T., Chen, F., Huang, Y.D. et al., Analysis of Car-Following Behaviors under Different Conditions on the Entrance Section of Cross-River and Cross-Sea Tunnels: A Case Study of Shanghai Yangtze River Tunnel, International Journal of Environmental Research and Public Health 19(19) (2022).

[57] Yu, S., Zhao, C. and Song, L., et al., Understanding traffic bottlenecks of long freeway tunnels based on a novel location-dependent lighting-related car-following model, Tunn. Undergr. Space Technol. 136 (2023).

[58] Meng, D.-l., Song, G.-h. and Lu, Y. et al., Impact of stochasticity on platoon stability of car-following models for emissions estimation, Journal of Central South University 30(8) (2023) 2772–2789.

[59] LeBlanc, D.S., Bao, S., Sayer, S.R., et al., Longitudinal Driving Behavior with Integrated Crash-Warning System Evaluation from Naturalistic Driving Data, Transp. Res. Rec (2365) (2013) 17–21.

[60] Kesting, A., Treiber, M., Calibrating Car-Following Models by Using Trajectory Data Methodological Study, Transp. Res. Rec (2088) (2008) 148–156.

[61] Wang, T., Cheng, R. and Wu, Y Stability analysis of heterogeneous traffic flow influenced by memory feedback control signal, Applied Mathematical Modelling 109 (2022) 693–708.

[62] Qu, L., Li, M., Chen, M., et al., Multivariate analysis between driving condition and vehicle emission for light duty gasoline vehicles during rush hours, Atmospheric Environment 110 (2015) 103–110.

[63] EPA, Exhaust emission rates for light-duty onroad vehicles in MOVES4, 2023. https://nepis.epa.gov/Exe/ZyPDF.cgi?Dockey=P101872R.pdf. (Accessed May 30 2024).

[64] X. Jia, X. Qin, J. Zhu, et al., Carbon Emission Pattern of Driving Car on Vertical Curves of Highway, Sustainability 15(8) (2023).

[65] Huang, X., Rui, Y., Dou, S., et al., Impact of tunnel lighting on traffic emissions based on VR experiment and car-following method, Displays (2025).

9 VR Model Based on Tunnel Inspection and Maintenance, AR

Zeyu Dai[1], Peinan Li[2*] and Yishu Wang[3]

With the gradual improvement of urban transportation networks, tunnels have become an important safeguard for maintaining fast traffic between regions. The structural safety of tunnels is key to ensuring smooth transportation. Once damage occurs, the tunnel structure typically undergoes rapid changes, and severe cases can lead to significant property damage and loss of life. Currently, regular inspections and maintenance, as required by regulations, remain the primary method to ensure tunnel structural safety. Emerging technologies and methods have developed new solutions for tunnel inspection and maintenance. Among these, VR and AR technologies provide more comprehensive and intuitive means for operations and maintenance engineers. VR technology creates immersive virtual environments that allow inspectors to collaborate remotely and train, improving data visualization. AR technology merges virtual information with the real world, providing on-site engineers with real-time data and operational guidance, reducing human error and increasing inspection efficiency. The benefits of these technologies in tunnel maintenance and inspection are demonstrated through case studies. These benefits include increased accuracy and efficiency of inspections, support for remote training and cooperation, greater teamwork, and increased data-sharing efficiency. This chapter outlines the regulatory requirements for tunnel inspection and maintenance, intelligent methods involved, the progress of AR and VR applications, and discusses their prospects.

1. Introduction

With the acceleration of urbanization and expansion of transportation networks, tunnel inspection and maintenance have become key issues regarding public safety and economic benefits. Tunnel inspection and maintenance span the construction and operation phases of tunnels, serving as the main means of monitoring tunnel

[1,3] Department of Geotechnical Engineering, College of Civil Engineering, Tongji University, Shanghai 200092, China.
[2] College of Environmental Science and Engineering, Donghua University, Shanghai 201620, China.
Email: daizeyu@tongji.edu.cn[1]; wangyishu@tongji.edu.cn[3]
* Corresponding author: lipeinan@dhu.edu.cn

DOI: 10.1201/9781003514480-9

conditions. Various tunnel diseases are the primary targets of inspection and maintenance tasks. These diseases can lead to issues such as water leakage, sand leakage, etc., and severe conditions may cause tunnel damage. During the operational phase of tunnels, inspection and maintenance are periodic tasks crucial for ensuring tunnel safety. Specifically, when the tunnel's operational state is poor, external disturbances can exacerbate the deterioration of tunnel diseases. Timely protection of tunnel structures before diseases deteriorate is a critical component of tunnel maintenance. The frequent collapses during the construction of the Heathrow Airport Express Tunnel in London in 1994 are one of the most severe civil engineering disasters in the UK caused by tunnel damage (Figure 1 (a)). Additionally, in 2020, a major convergence deformation occurred in a subway operational tunnel in Shanghai (Figure 1 (b)), with some areas deforming by more than 60 mm, seriously threatening the safety of subway operations [2]. Tunnel structural damage results from a variety of factors. Some scholars have proposed the Swiss cheese model to describe the relationship between tunnel structural failures [3]. If each hole in the cheese represents a vulnerability, a disaster will occur when all these holes connect. Efficient inspection methods and scientific maintenance measures are effective ways to detect and mitigate these vulnerabilities, preventing most accidents.

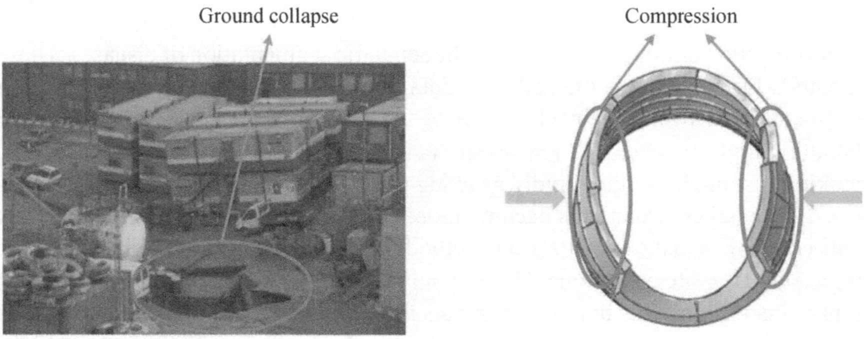

(a) (b)

Fig. 1 Tunnel accidents: (a) collapse accidents [1]; (b) tunnel convergence deformation [2].

Currently, the common tunnel inspection methods in engineering can be divided into manual methods and automated methods. The former requires inspectors to enter closed, dark tunnel environments, relying on visual inspection and handheld devices to collect data, such as tunnel convergence, tunnel disease images, and water leakage. The latter typically involves various sensors or automated equipment, such as industrial cameras, laser scanning systems, and infrared thermal imagers [4], [5]. The former can flexibly adjust the inspection range and focus based on on-site conditions, providing more intuitive detection for complex or special tunnel structures, while the latter is typically customized for specific data collection tasks. These customized services are usually deployed in sensitive areas that require frequent data collection to make up for the low frequency of manual methods. Although the data collection methods of the two approaches differ, the data collected are ultimately analyzed by professional engineers.

Many companies have developed different types of tunnel health inspection devices (e.g., tunnel inspection vehicles, tunnel inspection robots). However, the high research and development costs, maintenance costs, and limited usage conditions still hinder their application in real-world projects. The image or point cloud data collected by these automated inspection devices is crucial for tunnel inspection and maintenance. On one hand, these data provide the most direct indication of the severity of tunnel diseases. On the other hand, compared to the few automated monitoring sensors arranged in point form, images provide more detailed information, such as spatial location, type, and development trends. Thus, the collection of these data allows engineers to overcome the limitations of time and space, especially when the tunnel is severely damaged and poses a safety threat. Using mobile carriers equipped with cameras permits engineers to view the real situation of the site from a safe area.

In recent years, the latest artificial intelligence (AI) technology has also been applied to improve the safety of tunnels and their surrounding environments [6], [7]. The main method is to apply AI to the analysis of collected data, such as the rapid detection of tunnel diseases and the semantic segmentation of diseases. These methods further process the collected data to mine key features of tunnel disease characteristics from the inspection data. AI technology has the advantages of handling high-dimensional non-linear data and rapid real-time dynamic analysis, making it suitable for the rapidly growing demand for tunnel inspection [8–10].

As tunnel environments become more complex, higher precision 3D models and convenient data visualization methods are needed to help engineers better understand the site conditions. This not only sets higher technical requirements for tunnel inspection tasks but also promotes the development of inspection methods toward a combination of manual and automated approaches. For example, high-precision 3D laser scanning technology and industrial cameras can be used to obtain high-density point cloud and image data from the tunnel interior, which are then merged and rendered into realistic 3D models. This helps engineers understand the types, locations, and deformation of tunnel diseases in an immersive way. Meanwhile, combining digital twin technology, engineers can simulate and analyze the development of tunnel diseases in a virtual environment. Digital twin models deeply integrate real-time inspection data with virtual models, enabling dynamic display of multidimensional information, such as tunnel structural health and external geological changes. This integration of virtual and real worlds not only improves the efficiency and visualization of information transfer but also helps engineers make quick decisions and respond promptly to potential issues.

In recent years, with the rapid development of virtual reality (VR) and augmented reality (AR) technologies [11], [12], their applications in tunnel inspection and maintenance have gradually attracted attention. VR/AR technologies can create highly realistic virtual environments, combining the advantages of manual and automated methods, allowing inspectors to observe and analyze tunnels comprehensively in an immersive way, thus discovering potential problems more effectively and developing corresponding maintenance strategies. This chapter will summarize the VR model-based tunnel inspection and maintenance technologies, including their principles, current applications, and future development directions.

2. Tunnel Inspection and Maintenance Work

Tunnels, over time, face challenges to their structural safety and stability due to natural environmental erosion, vehicle load impacts, and material aging. Tunnel inspection and maintenance are the solutions to address these issues. The primary goal is to detect potential structural problems in the tunnel through regular inspections. After completing the inspection tasks, effective maintenance measures must be taken based on the inspection results to prevent tunnel problems from escalating into major disasters, ensuring safe and smooth transportation throughout the tunnel's design life.

This chapter will systematically explore the current state of tunnel inspection and maintenance in two main aspects. For tunnel inspection, it will review regulatory specifications, analyze traditional inspection methods, and discuss their pros and cons. For tunnel maintenance, the chapter focuses on the current strategies, including preventive and corrective maintenance, as well as the challenges faced during maintenance decision-making.

2.1 Basic Regulatory Work for Tunnel Inspection

Since the late 19th century, with the increasing construction of railway and highway tunnels, people's attention to tunnel safety and stability has grown. Most inspection work is regulatory. Since different countries have varying tunnel construction periods and operation times, each country has set corresponding regulations for its tunnels. The inspection periods and times for tunnels in China, Japan, South Korea, and the UK are summarized in Table 1 (Figure 2 shows the regulations for these four countries) [13].

Traditional tunnel inspection projects mainly include appearance inspection, structure inspection, and installation and maintenance of the monitoring system. Among them, the visual inspection of the visual area of the inner surface of the tunnel, such as cracks, water leakage, damage, etc. (as shown in Figure 3). This method is simple and efficient, but is influenced by detection accuracy and integrity. Structure testing uses professional testing equipment to inspect the tunnel structure, such as concrete strength tests, lining thickness measurements, etc. These test methods require sophisticated equipment and professional engineers, and are complex and costly to operate. A number of sensor networks (such as displacement, stress, strain, temperature, humidity, etc.) that can track the tunnel's

Fig. 2 Details inspection cycles for China, Japan, South Korea, and the UK.

structural condition in real time are progressively forming throughout the tunnel as a result of advancements in Internet of Things technology. However, in order to guarantee data accuracy, the enormous quantity of sensor equipment necessitates maintenance and calibration for numerous specialists. At the same time, the amount of monitoring data is huge, which requires complex data processing and analysis.

After the completion of the tunnel detection task, the health status of the tunnel needs to be evaluated according to the test results. The health status evaluation process of the tunnel is shown in Figure 4. Each sub-item of the tunnel was evaluated separately by expert and normative indicators. The safety evaluation

Table 1 Test cycles and time in different countries [14-17].

		China	*Japan*	*Korea*	*UK*
Initial inspection		Before the operation	Before the operation	Before the operation	Before the operation
Daily inspection		1 Time/month	1 time / 2 years	1 / 3 months	1 Time/month
Regular inspection		1 Time/year	1 time / 10 to 20 years	1 time / 2 years	1 Time/year
Special inspection	**Construction in the protected area**	1~2 Times/week	(–)	(–)	(–)
	Extreme disaster event (fire, earthquake, flood, explosion, car accident)	After the disaster	After the disaster	After the disaster	After the disaster
Special examination		≥2 times/month	irregularly scheduled	1 time / 5 years	(–)
Disease review		2 times/month	(–)	(–)	(–)

Fig. 3 Type of tunnel disease: (a) water leakage; (b) damage; (c) crack.

of the general tunnel will be divided into 5 levels, which can be assigned 0~4 points. At the same time, the different structures in the tunnel will give different coefficients according to their importance. After obtaining the single score, the final score of the tunnel can be obtained through the calculation format of the tunnel comprehensive evaluation, as shown in formula (1). As the results of expert evaluation will be affected by human factors, the status of the tunnel can usually be more accurately evaluated by the five-level tunnel service rating [18], [19], two-dimensional cloud model [20], and traditional machine learning methods.

$$S = 100\left[1 - \frac{1}{4}\sum_{i=1}^{n}(S^{i} \times \omega_{i})\right]$$

Among them, S is the final comprehensive evaluation result, ω is the weight of each sub-item of tunnel structure, and S^i is the evaluation structure of each sub-item of tunnel structure.

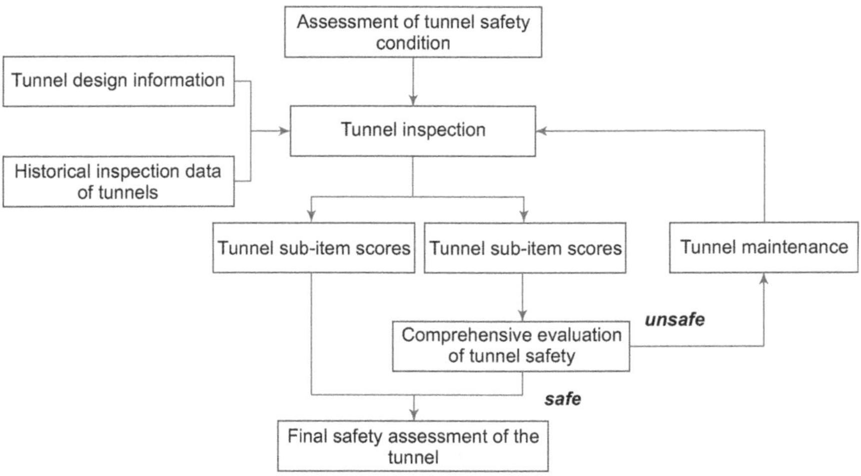

Fig. 4 Tunnel health state assessment process.

2.2 Basic Regulatory Work for Tunnel Maintenance

As the years of service of tunnels increase, maintenance work has gradually gained more attention. Tunnel maintenance can be divided into preventive maintenance and corrective maintenance. Preventive maintenance mainly targets tunnels with good overall structural performance, minor diseases, or small risks, but external disturbances such as construction impacts may be present. Corrective maintenance primarily addresses tunnels with significant disease risks or those that can no longer provide safe and stable transportation services. Table 2 shows the main detection methods and maintenance techniques for tunnel diseases [21].

Table 2 Major Tunnel Disease Detection and Maintenance Methods.

Detection Method	*Maintenance Method*	*Maintenance Method (Minor)*	*Maintenance Method (Severe)*
Uneven Settlement	Total Station, Leveling Instrument, Monitoring Equipment	Grouting, MJS	Grouting, Surface unloading and filling, and stop operation
Convergence Deformation	Total Station, Leveling Instrument, 3D Laser Scanner, Monitoring Equipment	Grouting, MJS	Steel ring reinforcement, MJS
Leakage/ Sand Leakage	Infrared Thermography, Ultrasound, Fiber Optic Sensors, Camera, 3D Laser Scanner, Visual Inspection	Grouting, Enhanced Sealing Measures	Grouting, connecting to the diversion channel, and adding sealing measures
Damage	Ultrasound, Fiber Optic Sensors, Camera, Visual Inspection	Repair	Steel ring reinforcement and repair
Cracks	Ultrasound, Fiber Optic Sensors, Camera, Visual Inspection	Surface Sealing	Surface closure and repair
Step Displacement	Ultrasound, Fiber Optic Sensors, Camera, Visual Inspection	Repair	Steel tie strip or steel ring reinforcement

Preventive maintenance aims to prevent external disturbances from damaging the structural performance of the tunnel. It involves proactively carrying out maintenance operations in weak or hazardous areas. Typically, construction around tunnel protection zones is regulated by government departments, and safety inspections are required before the construction phase can begin. Many countries and regions implement this regulatory service and apply preventive maintenance to projects that significantly impact operational tunnels. Due to different regional geology and tunnel characteristics, preventive maintenance plans vary by region. However, in general, the process consists of three main steps:

1. **Structural Inspection:** Use modern inspection equipment and technology to assess the structure, including stability, horizontal and vertical deformation, groundwater levels, and leakage issues.
2. **Comprehensive and Three-Dimensional Assessment:** Conduct a comprehensive and three-dimensional evaluation of the tunnel based on the inspection results.
3. **Analysis of External Disturbances:** Analyze the impact of external disturbances on the tunnel, and use the health evaluation structure to provide the basis for maintenance and reinforcement plans. In terms of maintenance and reinforcement, techniques such as surface repair, reinforcement treatments, and enhanced waterproofing and moisture-proofing measures are typically used.

Corrective maintenance is needed to perform functional repairs, structural repairs, or periodic maintenance on tunnels with serious diseases. Different levels of corrective maintenance are applied based on the severity of the disease. For example, for damage or cracks in general tunnel segment areas, methods such as filling materials and spray coatings can be used for repairs. For special tunnel structure parts, such as arches, high-strength materials like reinforced concrete or carbon fiber are needed for reinforcement to improve load-bearing capacity and seismic performance. If multiple diseases occur simultaneously, the deterioration of the diseases must be controlled in stages, with each one addressed individually.

In the tunnel maintenance decision-making process. First, the technical problem of underground engineering operation and maintenance is one of the main challenges. For example, how to accurately assess the degree of disease and remaining life of tunnel structures, and how to select appropriate repair materials and techniques, are all issues to be addressed. Second, economic factors are also an important challenge. Tunnel maintenance requires a significant investment, including testing, repair, and preventive maintenance. How to allocate resources reasonably within the limited budget and ensure the safe and stable operation of the tunnel is an important issue to be considered in the maintenance decision. Moreover, environmental factors also have an impact on tunnel maintenance decisions. For example, environmental factors such as the geological conditions, groundwater level, temperature, and humidity of the tunnel will affect the performance and maintenance requirements of the tunnel structure. Finally, the management factors should not be ignored. Tunnel maintenance involves multiple departments and professional fields. How to coordinate the resources of all parties to ensure the smooth progress of maintenance work is a problem to be solved in maintenance decisions.

3. Intelligent Methods for Tunnel Inspection and Maintenance

The inspection and maintenance of tunnels generate large amounts of data. Improving data processing speed, disease detection accuracy, and feedback quality is a major challenge in the inspection field. With the development of artificial intelligence (AI), Internet of Things (IoT), and digital twin technologies, tunnel inspection and maintenance methods are becoming increasingly intelligent. Compared to traditional data processing methods, intelligent approaches offer advantages such as more efficient data processing, stronger predictive capabilities, better visualization, enhanced collaboration, higher maintenance efficiency, and lower costs. In the face of increasing inspection and maintenance demands, these technologies significantly improve the safety, reliability, and operational efficiency of tunnels, providing strong support for tunnel intelligent management.

In terms of tunnel disease detection from collected image data, computer vision-based technologies are popular research topics. For example, Xue et al. [22] proposed a convolutional neural network-based tunnel disease detection algorithm. Specifically, for tunnel leakage, an optimized Fast R-CNN method was established for the semantic segmentation of leakage regions. Zhou et al. [23] developed an enhanced YOLO algorithm for detecting tunnel diseases such as damage, cracks, and exposed rebar. In addition to image data, laser point cloud data is also commonly used for tunnel disease detection. Laser point clouds provide more spatial distance information, making them more suitable for 3D reconstruction of tunnels. This section will explore this further. Some researchers have also developed methods for detecting tunnel diseases based on laser point cloud data. For example, after segmenting and calibrating point cloud data, Li et al. [24] proposed a method for extracting tunnel leakage locations and areas from 3D laser scan point clouds. Wang et al. [25] researched an automatically optimized Bayesian k-nearest neighbors method for detecting leaks and classifying multiple objects from 3D point cloud data [26].

In addition to AI methods, digital twin technology is also applied to tunnel inspection and maintenance. By creating a virtual model of the tunnel, digital twins reflect the actual status of the tunnel in real-time. Combining sensor data and AI algorithms, digital twin models can be used for predictive maintenance and to optimize tunnel operational performance. For example, N.S. Diren & S. Althen introduced the concept of "Managed Digital Twin" by integrating BIM [27], asset management, traffic control, and simulation technologies, providing comprehensive digital management of tunnels[28-30]. The Altona Tunnel project in Germany demonstrated the potential of managed digital twins in improving operational efficiency, optimizing traffic flow, and enhancing safety.

Using robots and drones, combined with automated inspection algorithms, is gradually replacing traditional tunnel inspection vehicles. Compared to traditional inspection vehicles, drones and robots are smaller, more flexible, versatile, and mobile. They can quickly reach and inspect high, narrow, and hazardous areas of the tunnel, greatly improving inspection efficiency and safety. Additionally, by using automatic path planning algorithms, they can identify obstacles, ensuring effective automated inspection[31].

VR and AR technologies serve as the ultimate perception and interaction platform, providing a perfect display for the aforementioned methods (Figure 5). The collected data and processed results are integrated effectively, offering an immersive experience to managers and maintenance personnel, allowing them to intuitively understand the tunnel's actual condition.

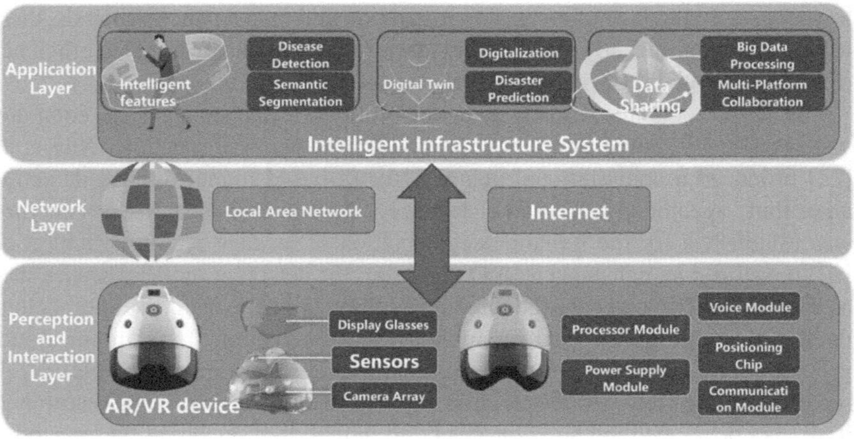

Fig. 5 Refers to AR/VR device applications and framework.

4. VR Technology in Tunnel Inspection and Maintenance

The development of artificial intelligence has improved the efficiency of data collection and analysis for tunnel inspection tasks, solving many tunnel inspection needs in real-world projects. However, most methods focus on single data sources and often overlook comprehensive decision-making behavior in inspection tasks. With the continuous advancement of technology, the advent of VR technology has opened new possibilities for tunnel inspection. On the one hand, VR technology compensates for the shortcomings of traditional methods, enabling inspectors to overcome the constraints of the physical world. On the other hand, it provides more realistic visual effects, allowing for more efficient and comprehensive inspections.

4.1 VR Technology Overview

Virtual reality (VR) technology is a method that uses computers to generate highly realistic virtual environments, allowing users to interact with the virtual world in an immersive way[32], [33]. Its core principle is to use computer graphics, sensor technology, and optical imaging techniques to create a highly realistic 3D virtual world. As shown in Figure 6, users can enter and operate the virtual environment through devices such as head-mounted displays (HMDs), controllers, and gloves.

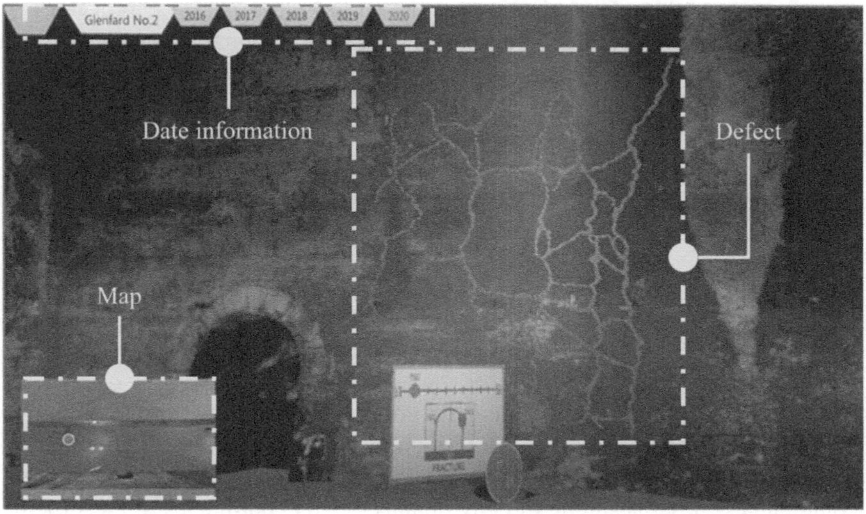

Fig. 6 Refers to VR scenarios of tunnel diseases [34].

VR technology can be traced back to the 1990s when it was proposed by Jaron Lanier [35], primarily for military and aerospace applications. In recent years, with advancements in computer technology, hardware, and software, VR technology has been increasingly applied in industrial fields. In tunnel inspection, researchers have gradually recognized the advantages of VR technology, specifically:

1. **Providing an Immersive Experience:** VR can create highly realistic virtual environments, making inspectors feel as though they are inside the tunnel. This immersive experience not only improves inspector concentration but also reduces misjudgments caused by environmental factors.
2. **Supporting Remote Collaboration and Training:** Experts from different locations can connect to the same virtual environment to collaboratively inspect and analyze the tunnel. Particularly, by sharing perspectives or screens, experts can focus on the same issues.
3. **Improving Data Visualization:** VR can map multi-source data onto models, presenting complex inspection data in an intuitive manner. Virtual reality tools allow data to be viewed in 3D virtual tunnel models, with animations showing data over time and space to support interpretation.

4.2 Construction of VR Models

The construction of VR models is key to providing tunnel inspection services with VR technology. Compared to traditional inspection methods, VR model-based inspection technology is a deep processing approach to the collected data. Specifically, the process of constructing a VR model generally includes four parts: data collection and processing, model creation, model optimization and visualization, and model evaluation (as shown in Figure 7). These processes are carried out through a series of algorithms, whereas traditional inspection methods are typically performed by manual inspections or simple statistical algorithms.

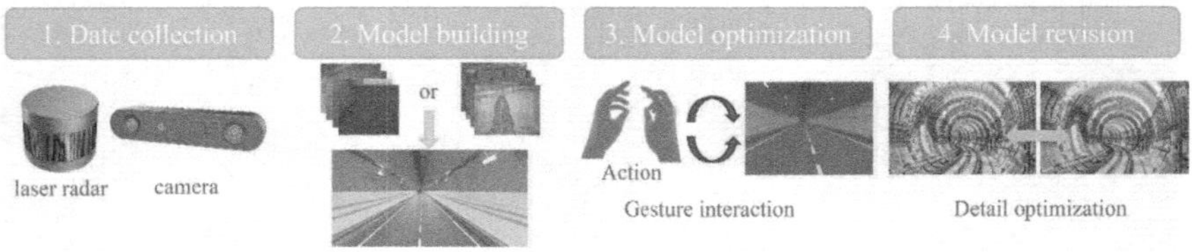

Fig. 7 VR Model Construction Process.

Data collection forms the foundation of constructing a VR model. Common data collection methods include laser scanning and industrial camera photography. Both methods have their respective advantages. In terms of data accuracy and completeness, laser scanning directly obtains the 3D coordinates of the structure and attributes such as reflectivity, while camera-captured data needs to be processed using SfM (Structure from Motion) algorithms to infer the position and orientation of the camera and thus obtain point cloud information [36], [37]. Laser scanning also does not require external light sources to capture the interior surface images of the tunnel. Particularly, fixed laser radar equipment can conduct high-precision spatial scanning even in environments with weak GPS signals and generate realistic color and texture data [38]. However, laser scanning generates a huge amount of point cloud data, which is often unstructured and requires extensive post-processing work (such as matching, denoising, and simplifying) [39]. At the same time, high equipment costs, maintenance costs, and the low detection accuracy for small diseases hinder its use in tunnel data collection. With the gradual improvement of SfM and MVS (Multi-View Stereo) algorithms, capturing tunnel interior image sequences using cameras has become a popular method for data collection in tunnel inspection tasks. Compared to laser scanning, camera-based image collection provides higher resolution, allowing a clear representation of various tunnel diseases.

Based on data collection, the data must be cleaned and matched. For example, laser scanning point cloud data requires the use of filtering algorithms or threshold methods to remove noise and outliers [40]. Feature matching algorithms can convert point cloud data obtained from different coordinate systems into a unified coordinate system. After image filtering and the application of SfM algorithms, image data obtained from cameras can yield camera intrinsic and extrinsic parameters and sparse point clouds. These are then converted into dense point clouds using MVS algorithms. At this point, the camera's collected data are successfully transformed into the same data structure as the laser scanning data. The point cloud data obtained must also be converted into continuous network models through interpolation or fitting operations for subsequent realistic VR model construction [41]. To ensure the model's authenticity, original image textures are usually used to map the grid, generating high-quality 3D models. Some studies have also integrated local disease results from tunnel inspections with BIM (Building Information Modeling) models [42], [43]. This method effectively reduces the data volume processed but sacrifices some precision.

To improve the performance and user interaction experience of VR models, further optimization and visualization of the generated tunnel interior surface models are necessary. VR engines such as Unity3D and Unreal Engine can provide excellent optimization services, adding features like scene rotation, zoom, and more to allow inspectors to freely observe the site. The data presentation areas of the models also need to integrate corresponding data sources for real-time display of key information.

After these operations, the constructed VR model needs to be compared with real engineering data to correct any distorted data. Additionally, the VR model should be adapted based on the user's VR equipment to balance loading speeds and rendering results, ensuring a smooth user experience. Through this series of VR model construction processes, the collected inspection data can be transformed into a high-fidelity VR model.

4.3 Applications of VR Models in Inspection and Maintenance

Based on detailed VR models, a large number of application functions have gradually been developed for various use cases. In terms of tunnel structural health status and damage assessment, some researchers have proposed integrating document materials and structural health monitoring data into VR models. Through real-time data transmission, experts can quickly assess the health status of a detected tunnel by combining the data, documents, and VR model using HTC Vive headsets and controllers.

For tunnel disease detection tasks, combining data collected by drones or robots and establishing VR models for disease monitoring is gradually being explored. Table 3 compares the pros and cons of drones and robots in terms of collection methods, accuracy, and operational efficiency. By designing flight paths, flight operations, and aerial image collection, detailed disease data can be obtained. Based on this, Abdurahman et al. [45] created damage-enhanced VR models by combining the detected disease data with 3D models, significantly reducing the need for on-site inspections. Pahwa et al. [46] utilized geometric prior knowledge and improved beam adjustment technology to generate fully dense 3D scenes for accurate tunnel disease detection and modeling. In complex cable tunnel inspections, Zhou et al. [31] optimized the structure of quadruped robots. Equipped with infrared imaging and visual sensors, these robots can automatically plan inspection paths and recognize obstacles to ensure effective automatic inspection.

Table 3 Comparison of Data Collection Methods for VR Models.

Drones	*Unmanned aerial vehicle*	*Robots*
Flexibility	High flexibility, capable of conducting multi-angle, full-range inspections in the air	Can only travel on land, and some climbing robots can crawl on the walls
Data Collection Range	Suitable for long-distance, large-scale tunnel inspection tasks; able to inspect long tunnels in one go	Suitable for the detection of short distances and local areas, and it is suitable for the detailed inspection of specific parts
Data Collection Accuracy	Can acquire high-resolution images and point cloud data, but accuracy may be influenced by flight posture in complex environments	Can carry a variety of high precision sensors for close, high precision data collection, but for long distance may be lacking
Efficiency	Fast flight speeds, capable of completing large-scale data collection in a short time	Moves slowly and requires more data acquisition time
Cost	Relatively low acquisition cost, but maintenance costs may be higher (e.g., battery replacement, sensor calibration)	The purchase cost of the robot is high, but the maintenance cost is relatively low, and the service life is long

The precise modeling of VR models also provides more simulation operation space for maintenance scenarios, covering the entire maintenance process. Before maintenance operations, engineers can simulate structural maintenance and fault troubleshooting in the virtual environment, enabling them to develop more effective maintenance plans. After completing maintenance tasks, the VR model records the location and key operations for subsequent maintenance inspections, preserving original data. Additionally, VR technology can be used for training new maintenance personnel by simulating various maintenance scenarios to improve their skills and emergency response abilities. In predictive maintenance, combining AI and big data analysis enables real-time monitoring and forecasting of tunnel health status. After mapping the predictive results onto the VR model, maintenance personnel can view the damage to vulnerable areas in advance within the virtual environment.

5. Augmented Reality (AR) Technology in Tunnel Inspection and Maintenance

During inspections, tunnel conditions and environments may change in real-time. To reduce the need for inspectors to frequently refer to handheld devices or paper materials, AR technology has been applied to tunnel inspection. Unlike VR technology, which immerses users in a fully virtual environment, AR technology is primarily used by on-site engineers, making its functions slightly different. AR systems can instantly update and display the latest data, ensuring that inspectors always have the most accurate information available to make timely decisions.

5.1 Overview of AR Technology

Augmented reality (AR) is a technology that blends virtual information with the real world [44]. Through cameras and other sensors, it captures the surrounding environment, then superimposes the required data (such as images, text, video, etc.) onto the real-world scene. The final image is displayed to the user through devices such as AR glasses, smartphones, or tablets. The user can perceive the real world while receiving additional supportive information and guidance. As shown in Figure 8, this integration preserves the user's perception of the physical environment while significantly enhancing the efficiency of information acquisition and the accuracy of decision-making.

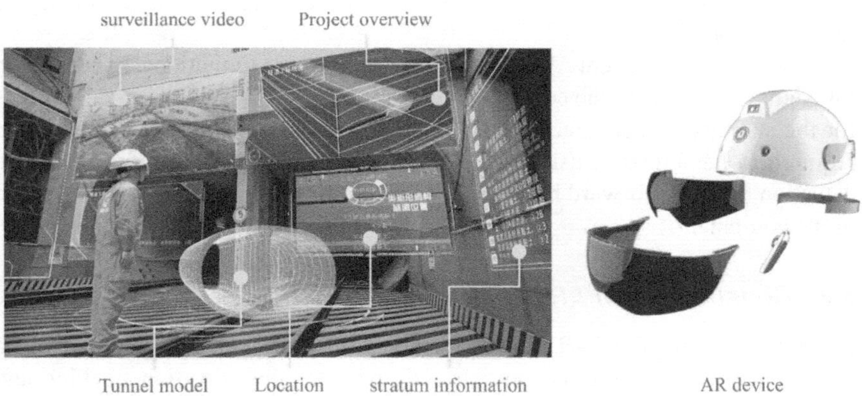

Fig. 8 AR Scenario in Tunnel Inspection.

Currently, AR models are widely applied in various industries such as industrial maintenance [47], medical diagnosis [48], education [49], and gaming [50]. Table 4 compares the differences between AR and VR technologies in terms of equipment requirements, immersion, and other factors. From the table, it is evident that VR technology requires higher hardware specifications for graphics rendering, while AR technology requires more hardware for data collection.

AR technology has also gradually shown its key advantages in the field of tunnel detection. First, the inspectors can instantly access more refined real-world critical data, such as structural health status, crack location, etc., in the actual tunnel. Secondly, the AR model is able to show the potentially dangerous areas and reduce the disaster accidents caused by human negligence. Thirdly, the AR model provides the information access ability of the boundary, which can reduce the time consumption of data access. Finally, through real-time data sharing,

Table 4 Comparison of AR and VR Technologies.

Feature	*VR Technology*	*AR Technology*
Immersion	Complete immersion in a virtual environment	The user can still perceive the real environment
Hardware	Head-mounted display, controllers, high-performance computer, or standalone VR device	Display devices, sensors, and cameras
Software	Graphics rendering, real-time simulation, and virtual modeling technologies	Image processing, recognition, and positioning technologies
Platform	Simulated inspection process, data analysis module	Real-time data transmission, historical database

experts are able to provide remote guidance, which improves the efficiency of team collaboration.

Of course, the current AR technology still has some limitations. Especially in low light environment, the complex tunnel environment puts higher requirements for the stability and accuracy of AR models. For users, professional training is required to be skilled in using AR equipment. At the same time, the long-time detection task puts forward higher requirements for the endurance and endurance of the equipment.

5.2 Construction of AR Models

The process of building AR models for tunnel inspections mainly involves data collection, environmental perception and understanding, position tracking and alignment, optimization, and testing, as shown in Figure 9. The primary difference between AR model construction and VR model construction is that AR models require real-time data collection, and therefore, the display content needs to be aligned with the real world.

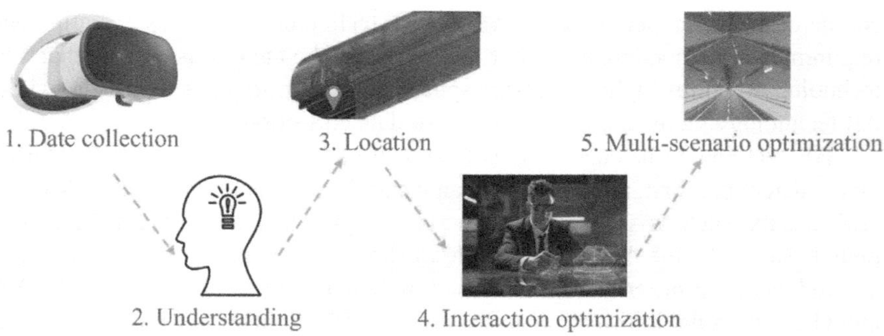

Fig. 9 AR Model Construction Process.

Data collection for AR models consists of two types: real-world environmental data and virtual object data. Typically, real-world environmental data inside the tunnel is captured using cameras or other sensors, which can generate point clouds, mesh models, or depth maps of the environment. These data are then used to understand the environment and precisely overlay virtual objects. Virtual objects, such as 3D models, are typically created using 3D modeling software like Blender, 3ds Max, or Maya [51]. Developers must ensure that the number of models, textures, and resolutions align with the surrounding environment as closely as possible to balance rendering quality and performance needs.

Environmental perception and understanding are a crucial step in constructing AR models. This involves extracting useful information from the collected data to understand the structure and content of the real world. Computer vision technologies play an important role in this process, employing methods like feature extraction, object recognition, and scene analysis to achieve comprehensive environmental perception [52]. For example, semantic segmentation algorithms can be used to categorize objects within the environment and identify their relationships, enhancing adaptability to complex environments. These perceptive and understanding capabilities are essential for accurately positioning and interacting with virtual elements.

Position tracking and alignment are key to ensuring that virtual information integrates seamlessly with the real-world space. This is particularly important in confined environments like tunnels. Position tracking is carried out by capturing user location changes during the inspection process. Common position tracking methods include vision-based tracking, sensor-based tracking, and hybrid tracking that combines both vision and sensors [53], [54]. Vision-based tracking uses images captured by cameras to match feature points and estimate motion, one example being SLAM (Simultaneous Localization and Mapping) [55], [56]. Sensor-based tracking relies on devices like IMUs (Inertial Measurement Units) and uses filtering algorithms like Kalman filters to estimate position and orientation. To enhance robustness and accuracy, a combination of methods is typically used to achieve high-precision, low-latency position tracking. Position alignment refers to the process of aligning virtual elements with the real world. During modeling, factors like scene depth and lighting conditions need to be considered to enhance the realism and immersion of virtual elements.

Similar to VR, optimization and testing are essential phases in constructing AR models. These steps ensure that the system performs well and provides a high-quality user experience. Optimization primarily involves improving interactivity and visual effects. The testing environment for AR models, however, needs to account for multiple real-world scenarios, including various lighting conditions, complex environments, and multi-user interactions, ensuring that the AR model functions stably under all circumstances. Through iterative testing and continuous optimization, the AR system's performance and user experience can be gradually improved, leading to high-quality augmented reality applications.

5.3 Applications of AR Models in Inspection and Maintenance

Due to limitations in device weight and battery capacity, AR technology is less commonly used in tunnel environments compared to VR. However, as hardware and software optimization continue, the usage of AR technology in tunnel inspection and maintenance tasks is increasing, and related research is gradually gaining more attention.

In tunnel inspections, AR technology can superimpose virtual digital information onto the real tunnel environment in real-time, providing inspectors with more intuitive and comprehensive inspection data [57]. Through AR devices, inspectors can view location data, material information, dimensions, and other external details of the areas they are inspecting. Notably, unlike traditional image recognition or semantic segmentation, this information can be presented in the form of 3D models for better visual effects [58]. Through accurate coordinate transformations and image matching, AR can also provide more precise monitoring results. For example, in tunnel segment displacement detection, AR systems can precisely align the virtual model with the actual tunnel segment to measure displacement accurately. Recording work during inspections can also be handled by AR technology. Studies have shown that using AR for displacement detection can reduce inspection time by 25% [59]. Similar to VR, AR technology can also enable multiple users to participate in the inspection process simultaneously, effectively improving team efficiency.

AR technology also plays an important role in tunnel maintenance tasks [60]. First, it can provide maintenance personnel with detailed guidance and operational support. Through AR devices, maintenance workers can access real-time maintenance plans, operation steps, and relevant precautions. Additionally, key information such as equipment operation data, maintenance records, and fault information can be closely monitored during maintenance operations. For instance, in a tunnel project in South Korea, Zaher et al. [61] used AR technology to quickly access facility information, thereby improving maintenance efficiency and safety. Finally, AR technology can be used for safety training, simulating real work environments, and enhancing training effectiveness.

In addition to the aforementioned applications, integrated solutions combining AR with other technologies are also being introduced. After combining AR with BIM, GIS, or IoT technologies, more comprehensive tunnel maintenance solutions can be provided. Moreover, hybrid solutions that combine AR and VR are gaining innovation in the tunnel inspection field. Since both technologies share many similar steps in model construction, such integration allows users to seamlessly switch between virtual and real environments. For example, in tunnel lining inspections, inspectors can simulate the inspection process in a virtual environment and then use AR devices in the real world to collect data and assess the lining conditions, enabling quick decision-making.

6. Conclusion

Tunnel inspection and maintenance tasks are key to minimizing the operational impact of tunnel structural failure. Particularly, when tunnel structures experience damage, it is usually rapid, irreversible, and significantly damaging. This chapter summarizes the application of AR and VR models in tunnel inspection and maintenance work. For regulatory tunnel inspection and maintenance tasks, the chapter introduced the main methods and regulatory processes currently used in tunnel inspection and maintenance. Driven by emerging technologies, it discussed the image-based methods, digital twin technologies, and automated equipment used in current tunnel inspections. Through the analysis of existing technologies and practical application case studies, we revealed the main construction methods, application scenarios, and advantages of VR and AR models in tunnel inspection and maintenance tasks. Both methods provide engineers with comprehensive observation and analysis of tunnels, improving inspection efficiency and accuracy. They also support remote collaboration and training, enhancing team collaboration and data-sharing efficiency. Based on the research and discussions in this chapter, it is clear that the future of VR and AR models in tunnel inspection lies in integrating various technologies, including different data sources, formats, and processing methods. These integrated data sources will provide more comprehensive tunnel status information, offering stronger support for inspection and maintenance decisions. The integration of multiple data formats will enhance visualization and spatial-temporal information, though this requires innovative data processing approaches for unified, corroborated data across time and space. Through continued technological innovation and practical exploration, VR and AR technologies can further support tunnel safety and stable operations.

References

[1] A tunnel collapse at Heathrow nearly ruined the airport and a Tube line. MyLondon, (2023, October 11). https://www.mylondon.news/news/nostalgia/london-underground-heathrow-tunnel-collapse-17579206.

[2] Wang, R., Yuan, Q., Liang, F.Y. and Wang, L.J. (2023). Case and remediation technology of deformation in soft soil subway shield tunnel caused by road fill soil. *Chinese Journal of Geotechnical Engineering*, 45(1), 12.

[3] Reason, J. (1990). *Human error*. Cambridge University Press.

[4] Ding, Y.F., Chen, X.H., Zhang, Y.X. and Huang, X.Y. (2025). Smart building evacuation by tracking multi-camera network and explainable Re-identification model. *Engineering Applications of Artificial Intelligence*.

[5] Zhang, Y.X., Ding, Y.F., Chen, X.H. and Huang, X.Y. (2025). Multi-scale analysis of fire and evacuation drill in a multi-functional university high-rise building. *Developments in the Built Environment*, 100626.

[6] Wang, X., Yin, Z.Y., Wu, W. and Zhu, H.H. (2025). Differentiable finite element method with Galerkin discretization for fast and accurate inverse analysis of multidimensional heterogeneous engineering structures. *Computer Methods in Applied Mechanics and Engineering*, 473, 117755.

[7] Wang, X. and Yin, Z.Y. (2024). Interpretable physics-encoded finite element network to handle concentration features and multi-material heterogeneity in hyperelasticity. *Computer Methods in Applied Mechanics and Engineering*, 431, 117268.

[8] Dai, Z., Li, P., Zhu, M., Zhu, H., Liu, J. and Zhai, Y., et al. (2023). Dynamic prediction for attitude and position of shield machine in tunneling: a hybrid deep learning method considering dual attention. Advanced Engineering Informatics, 57.

[9] Li, P., Dai, Z.*, Rui, Y., Ling, J., Liu, J., Zhai, Y. and Fan, J. (2024). Generative adversarial network for optimization of operational parameters based on shield posture requirements. *Automation in Construction*, 105553.

[10] Dai, Z., Li, P., Liu, J., Liu, X., Rui, Y. and Zhai, Y. (2023). Data-driven prediction for curved pipe jacking performance during underwater excavation of ancient shipwreck using an attention - based graph convolutional network approach. *Expert Systems With Applications*, 121393.

[11] Melnyk, O., Huymajer, M., Fenzl, D., Huemer, C., Wenighofer, R. and Mazak-Huemer, A. (2024). Augmented reality for enhanced documentation and anchor inspection reporting in conventional tunnelling. Tunnelling and Underground Space Technology, 153, 106040.

[12] Negi, P., Singh, R., Gehlot, A., Kathuria, S., Thakur, A.K. and Gupta, L.R., et al. (2024). Specific soft computing strategies for the digitalization of infrastructure and its sustainability: a comprehensive analysis. Archives of Computational Methods in Engineering, 31(3).

[13] Wang, Y., An, Z.L., Ma, W.B., Zheng, Z.F. and Guo, X.X. (2022). Current status and development trends of railway tunnel detection technology in China, Japan, and South Korea. *Tunnel Construction (Chinese & English)*, 42(7), 11.

[14] GB/T 39559.3-2020. (2020). *Specifications for operational monitoring of urban rail transit facilities—Part 3: Tunnels*. National Standards of the People's Republic of China.

[15] Japanese Ministry of Land, Infrastructure, Transport and Tourism. (2007). *Maintenance and management standards for railway structures and related explanations (Structures section)*. Japan Society of Civil Engineers.

[16] Ministry of Land, Infrastructure and Transport. (2023). *Tunnel Standard Specifications (KCS 27 00 00)* [Revision Notice No. 2023-512]. Republic of Korea.

[17] Highways England. (2020). Design Manual for Roads and Bridges: CS 452 Inspection and records for road tunnel systems.

[18] Yuan, Y., Bai, Y. and Liu, J.H. (2012). Assessment of service state of tunnel structure. *Tunnelling and Underground Space Technology Incorporating Trenchless Technology Research*, 27(1), 72–85.

[19] Li, X.J., Lin, X.D., Zhu, H.H., Wang, X. and Liu, Z. (2017). Condition assessment of shield tunnel using a new indicator: The tunnel serviceability index. *Tunnelling and Underground Space Technology*, 67, 98–106.

[20] Zhu, M., Zhu, H., Guo, F., Chen, X. and Ju, J.W. (2020). Tunnel condition assessment via cloud model-based random forests and self-training approach. Computer-Aided Civil and Infrastructure Engineering(2).

[21] Shandong Provincial Market Supervision and Administration Bureau. (2024). *Guidelines for maintenance and management of highway tunnel (civil engineering)* (DB37/T 4774-2024). Shandong Provincial Market Supervision and Administration Bureau.

[22] Xue, Y. and Li, Y. (2018). A fast detection method via region-based fully convolutional neural networks for shield tunnel lining defects. Computer-Aided Civil and Infrastructure Engineering, 33(4).

[23] Zhou, Z., Zhang, J. and Gong, C.. (2022). Automatic detection method of tunnel lining multi-defects via an enhanced you only look once network. Computer-aided civil and infrastructure engineering(6), 37.

[24] Li P, Wang Q, Li J., et al. Automated extraction of tunnel leakage location and area from 3D laser scanning point clouds[J]. Optics and Lasers in Engineering, 2024(Jul.):178.DOI:10.1016/j.optlaseng.2024.108217.

[25] Wang, K., Zhang, Z., Wu, X. and Zhang, L. (2022). Multi-class object detection in tunnels from 3d point clouds: an auto-optimized lazy learning approach. Advanced Engineering Informatics, 52, 101543.

[26] Azari, H., Nazarian, S. and Yuan, D. (2014). A case study on tunnel lining assessment using stress wave-based nondestructive testing methods. In *Geo-Congress 2014 Technical Papers* (GSP 234, pp. 439-452). American Society of Civil Engineers.

[27] Diren, N.S. and Althen, S. (2023). Tunnel 4.0: Managed digital twin for tunnel operations. In E. Anagnostou, A. Benardos and P. Marinos (Eds.), *Expanding Underground: Knowledge and Passion to Make a Positive Impact on the World* (pp. 2635-2642). CRC Press.

[28] Sadhu, A., Peplinski, J.E., Mohammadkhorasani, A. and Moreu, F. (2023). A review of data management and visualization techniques for structural health monitoring using bim and virtual or augmented reality. *Journal of Structural Engineering*, 149(1), 18.

[29] Marzouk, M. and Hisham, M. (2012). Bridge information modeling in sustainable bridge management. In Proceedings of the International Conference on Sustainable Design and Construction (ICSDC) (pp. 457–466). American Society of Civil Engineers.

[30] Delgado, J.M.D., Butler, L.J., Gibbons, N., Brilakis, I., Elshafie, M. and Middleton, C. (2016). Management of structural monitoring data of bridges using BIM. ICE Proceedings, 170(3), 204–218.

[31] Zhou, Y., Chen, Y., Wang, C., Wu, C., Wang, X., Jiang, X. and Yuan, J. (2021). Automatic inspection method of cable tunnel in complex environment based on quadruped robot. In *2021 IEEE 3rd International Conference on Frontiers Technology of Information and Computer (ICFTIC)* (pp. 599-603).

[32] Zhang and Yan, C. (2012). The use of massively multiplayer online games to augment early-stage design process in construction. Loughborough University.

[33] Yang, Z., Chen, X. and Jia, G.. (2011). Modeling and Visualizing Dynamic Component for Virtual Educational Experiments. ICVRV2011. IEEE.

[34] Insa-Iglesias, M., Jenkins, M.D. and Morison, G. 3d visual inspection system framework for structural condition monitoring and analysis. Automation in Construction, 128.

[35] Rheingold, H. (1991). *Virtual Reality*. Summit Books.

[36] Kendall, A., Martirosyan, H., Dasgupta, S., Henry, P., Wang, D., Tosa, D. and Cohen, J. (2017). End-to-end depth and camera pose estimation for driving. In *Proceedings of the IEEE International Conference on Computer Vision (ICCV)* (pp. 194-202).

[37] Ummenhofer, B., Zhou, H., Sun, J., Jampani, V., Liu, M.Y., Kautz, J. and Garg, R. (2020). Demo: Learning structure and motion from a single RGB camera in real-time. In Proceedings of the IEEE/CVF Conference on Computer Vision and Pattern Recognition (CVPR) (pp. 1354-1363).

[38] Kim, M.-k., Hwang, D. and Park, D. (2023). Analysis of maintenance techniques for a three-dimensional digital twin-based railway facility with tunnels. *Platforms*, 1, 5–17.

[39] Chen, D.L., Lu, Y.Y. and Jia, D.Z. (2018). Automatic 3D reconstruction of highway tunnel using terrestrial laser scanning technology. Fresenius Environmental Bulletin, 27(11), 7677–7684.

[40] Feng, X. and Wan, A. (2024). Research on digital twin system for complex environment subway tunnels. *Software Engineering and Applications*, 13(1), 125–132.

[41] Niina, Y. (2020). Self-organized model fitting method for railway structures monitoring using lidar point cloud. Remote Sensing, 12.

[42] Lee, M.J.L., Wen, W.S. and Au, S.L.M. (2023). Integrating Building Information Modeling and Panoramic Structure-from-Motion for Accurate Camera Pose Estimation. Fresenius Environmental Bulletin, 27(11), 7677–7684.

[43] Wang, C., Wang, J.Y., Xu, K.L., Liu, C.Q. and Wang, Z.Q. (2023). Research on a BIM Model and Real-Scene 3D Model Integration Method. Science of Surveying and Mapping, 48(10), 159–168.

[44] Zheng, Liu, Yehuda and Kleiner. (2013). State of the art review of inspection technologies for condition assessment of water pipes. Measurement.

[45] Yiit, A.Y. and Uysal, M. Virtual reality visualisation of automatic crack detection for bridge inspection from 3d digital twin generated by uav photogrammetry. Measurement, 242.

[46] Pahwa, R., Chan, K.Y., Bai, J. and Foong, S. (2019). Dense 3d reconstruction for visual tunnel inspection using unmanned aerial vehicle.

[47] Hollmann, M. and Hildebrandt, T. (2019). Augmented Reality in Industry 4.0: A Systematic Literature Review. Procedia CIRP, 85, 1132–1137.

[48] Santanam, A. and Saba, M. (2020). Augmented Reality in Medicine: A Review. *Journal of Medical Systems*, 44(4), 1–16.

[49] Zyda, M.J. (2005). From Visual Simulation to Virtual Reality to Games. *Computer*, 38(9), 25–32.

[50] Billinghurst, M. and Thomas, B. (2001). Human-Computer Interaction with Augmented Reality. Annual Review of Information Technology and the Social Sciences, 1(1), 1–31.

[51] Ning-Ning, ZhouYu-Long and Deng. (2009). Virtual reality: a state-of-the-art survey. *International Journal of Automation&Computing*.

[52] Wu, X., Shi, Z. and Zhong, Y. (2010). A Review of UAV Vision Navigation. Journal of System Simulation.

[53] Siltanen, S. (2012). Theory and applications of marker based augmented reality.

[54] Kelly, J. and Sukhatme, G.S. (2011). Visual-inertial sensor fusion: localization, mapping and sensor-to-sensor self-calibration. *International Journal of Robotics Research*, 30(1), 56–79.

[55] Kohlbrecher, S., Stryk, O.V., Meyer, J. and Klingauf, U. (2011). A flexible and scalable slam system with full 3d motion estimation. IEEE.

[56] Chen, C., Wang, B., Lu, C.X., Trigoni, N. and Markham, A. (2020). A survey on deep learning for localization and mapping: towards the age of spatial machine intelligence.

[57] Zhou, Q.Y., Park, J. and Koltun, V. (2018). Open3d: a modern library for 3d data processing.

[58] Campos, C., Elvira, R., Rodriguez, J.J.G., Montiel, J.M.M. and Tardos, J.D. (2021). Orb-slam3: an accurate open-source library for visual, visual–inertial, and multimap slam. IEEE Transactions on Robotics: A publication of the IEEE Robotics and Automation Society(6), 37.

[59] Zhou, Y., Luo, H. and Yang, Y. (2017). Implementation of augmented reality for segment displacement inspection during tunneling construction. Automation in Construction, 82(oct.), 112–121.

[60] Wang, M. and Yin, X. (2022). Construction and maintenance of urban underground infrastructure with digital technologies. Automation in construction.

[61] Zaher, M., Greenwood, D. and Marzouk, M. (2018). Mobile augmented reality applications for construction projects. *Construction Innovation*, 18(2), 152–166.

10 Virtual Reality Studies on Choice Guidance in Road Tunnel Fire and Evacuation

Yifei Ding[1], Max Kinateder[2], Yuxin Zhang[3*] and Xinyan Huang[4*]

Road tunnel evacuation incidents are uncommon, but when they do happen, they present several safety and human factors-related difficulties. Empirical observations from real-world road tunnel evacuations are rare. To address this gap, Virtual Reality (VR) has emerged as a tool that promises to provide useful insights in this area. Compared with traditional evacuation studies such as field experiments and real-life drills, VR offers a flexible, cost-effective, and controlled alternative for simulating emergency conditions and collecting detailed behavioral data. However, the limited ecological validity of virtual environments and indirect locomotion techniques remain challenges. This chapter summarizes how VR can be used to simulate and study occupant behaviors during tunnel fire evacuations, for example, to evaluate the design of evacuation guidance systems and the social dynamics during emergencies. The experimental results emphasize the importance of designing tunnel guidance systems that rely more on timely information and easy-to-understand cues to optimize evacuation safety. The results also show the feasibility, validity, and flexibility of studying human behavior in virtual environments, especially simulating complex scenarios. The chapter also discusses the limitations of VR experiments on scenario realism and participant interaction. Additionally, the proposed Fire Dynamic Simulator (FDS)-enhanced smoke environment would be a promising solution to optimize VR experiments.

1. Introduction

In recent decades, the number of road tunnels has grown as traffic volumes have increased, especially in urban areas [1]. At the same time, modern tunnel designs have also become larger, more complex, and multilayered [2]. Functionally, tunnels act as sociotechnical systems, involving vehicles, technical infrastructure, and human users, among others. During emergencies, these systems must enable safe evacuation of individuals, either out of the tunnel, to a neighboring bore, or into refuge areas [3]. Among various tunnel emergencies, fires have drawn significant

[1,4] Department of Building Environment and Energy Engineering, The Hong Kong Polytechnic University, Hong Kong.
[2] National Research Council, Canada.
[3] Department of Urban Management Kyoto University, Kyoto, Japan.
* Corresponding author: yx.zhang@polyu.edu.hk; xy.huang@polyu.edu.hk

DOI: 10.1201/9781003514480-10

attention due to the hazards they pose to human life. Although major tunnel fires are relatively rare, their consequences can be catastrophic [4].

Adverse conditions, including high temperatures, low visibility, and toxic combustion products (e.g., smoke), can pose serious threats to occupants during a tunnel fire [5]. Evacuation must occur before these conditions become untenable. Compared to typical building fires, tunnel fires are often more hazardous due to the confined space, where toxic gases are difficult to dissipate, and lighting conditions are poor, increasing the evacuation difficulty [6]. As a result, occupants must respond quickly and initiate self-evacuation upon recognizing the danger or receiving an evacuation order. However, most tunnel occupants may only be familiar with road tunnels as drivers or passengers in a vehicle, but not with the egress routes that may need to be used for pedestrian evacuation, e.g., foot paths that lead to a place of safety. Studies have shown that while many occupants are aware of the risks posed by tunnel fires, they often lack the knowledge of how to respond and evacuate appropriately [7].

Therefore, it is critical to develop effective guidance, including signage and audio broadcasting systems that support occupant evacuation effectively when needed. It is also important to understand how other sources of information, such as social (i.e., the behavior of other people) or physical (e.g., smoke) cues, impact evacuee behavior and the effectiveness of guidance systems [8]. Numerous previous studies have examined individual or combined forms of different external guidance, demonstrating their positive impacts in fire evacuation guidance [9–13].

Before VR technology became an established research tool, researchers relied on field observations to study evacuation from tunnels. For example, Nilsson et al. [14] investigated the impact of lighting in evacuation and found that flashing lights and strobe lights were more helpful compared to standard emergency exit design, and green lights performed better than blue lights or red lights. Some studies related to technical installations strongly indicate that pre-recorded alarms effectively prompt occupants to leave their vehicles in a tunnel [15]. In a full-scale evacuation study in a road tunnel, Fridolf et al. [16] evaluated the effectiveness of different emergency exit portal designs. Zhang et al. [17] studied the role of guidance from technical installations on evacuation performance.

However, traditional experimental approaches often rely on real-world simulations and surveys, which are costly, hard to control, and yield limited repeatable experimental data [8]. In contrast, VR tests allow researchers to compare experimental conditions systematically, more accurately manipulate experimental variables, and also maintain strong control over the tests. In addition, VR supports rapid prototyping and evaluation of fire evacuation training and experimental designs before relatively higher-cost, higher-risk physical testing. Thus, VR experiments [18] have emerged as a promising complementary method and have seen increasing applications in recent years. A VR experiment constitutes a controlled yet immersive environment that is well-suited for investigating evacuation behavior under external guidance, as it allows for the presentation of complex, realistic scenarios, facilitates the manipulation of various conditions with specific cues, and enables the collection of detailed behavioral data. To this end, this chapter summarizes and discusses the current research on VR-based experiments on choice guidance for route and exit choice in road tunnel fire evacuation.

2. Display Format of VR and its Application in Fire Safety Research

The evolution of VR technologies and corresponding hardware devices has spanned decades, from the stereoscope invented by Wheatstone in the 19th century [19] to nowadays' consumer-grade VR headsets [20]. The timeline of major milestones of VR development is shown in Fig. 1(a). Currently, the most common display formats of the immersive environment depend on head-mounted displays (HMD) [21] and projection-based displays (PBD) [22], examples of which are respectively shown in Fig. 1(a) and Fig. 1(b). An HMD requires it to be worn on the head and is unrestricted by the room size. With the rise of consumer-grade HMDs in recent years, more and more researchers have started to use HMDs for experiments due to their lower cost, higher portability, and good immersion. PBD has been widely used since the 1990s as an immersive VR device. It provides a large, shared visualization with satisfactory immersion, and small group collaboration depending on the large screen, but the fixed installation of the screen or projection limits the users' mobility and causes a higher cost. Cave Automatic Virtual Environment (CAVE) [23] (shown as Fig. 1c) systems are a subset of PBD and facilitate a very high immersive experience with surround projection on four walls, ceiling, and floor. CAVEs also allow multiple users to share the same immersive environment, allowing for collaborative immersion, but their design of a multi-wall projection room causes higher cost and complexity.

The application of VR in fire science research and fire protection engineering has developed rapidly over the past decade. For example, the potential of using VR for simulating fire and emergency evacuation scenarios to study human behaviors and conduct firefighting and egress training was systematically reviewed and highlighted [18, 24]. To improve the immersive experience, multisensory simulation has been used in prototype applications incorporating heat to enhance realism during fire escape training [25]. In addition, VR-based fire extinguisher training has also been shown to have advantages in safety, efficiency, and repeatability compared to traditional firefighting training using real fire, although the challenges in achieving higher levels of visual and environmental fidelity still remain [26]. To enhance interaction among different users in a simulated scenario, researchers attempted to develop multi-user immersive VR applications that support teamwork and improve fire safety training performance [27]. More recent work has explored the integration of VR, augmented reality (AR), and VR, aiming to balance visual immersion with physical task performance. AR enhances the reality view by overlaying the virtual objects generated by computer vision technology. For example, one study compared AR and video in fire safety training

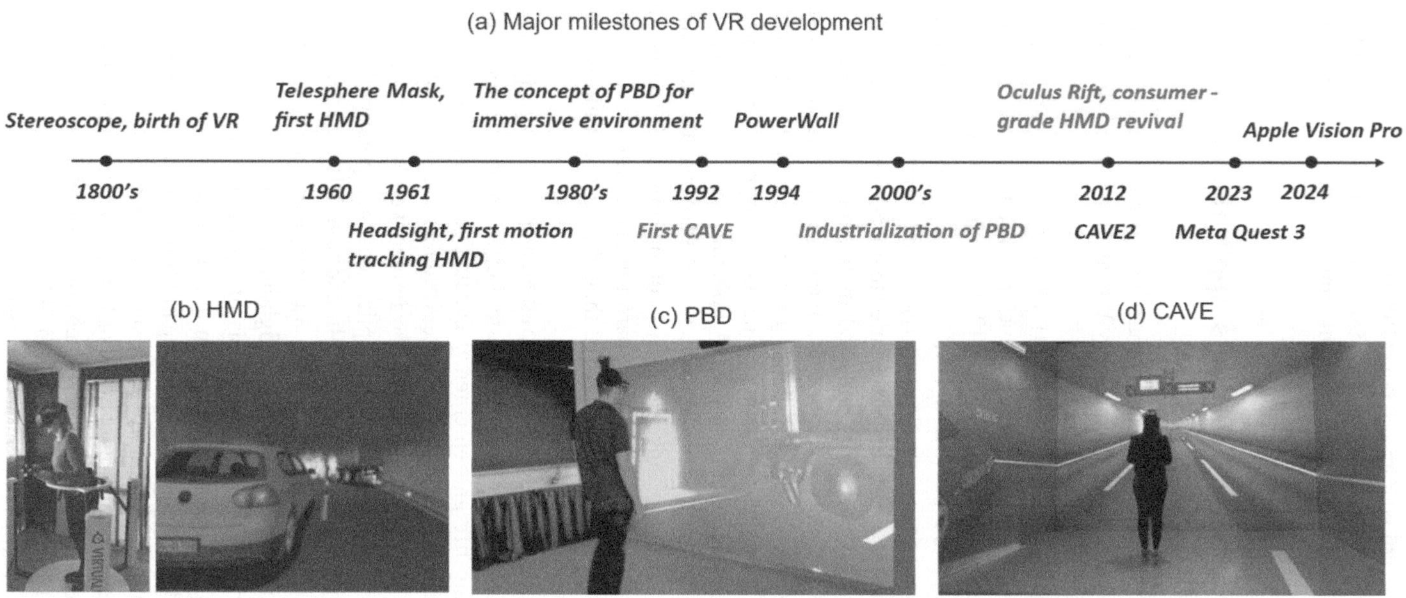

Fig. 1 (a) Timeline of major milestone of VR development, and different VR display technologies, (b) Head-Mounted Displays (The original images are from Ref [21] under a creative commons 4.0.), (c) Projection-Based Displays (The original images are from Ref [22]. Copyright © 2014 Elsevier Ltd.), and (d) Cave Automatic Virtual Environment (The original images are from Ref [23]. Copyright © 2018 Elsevier Ltd.).

and found that see-through AR improved operational efficiency compared to VR-only approaches [28]. Most recently, there has been a growing trend that integrates all kinds of extend reality (XR) technologies, i.e., VR, AR, mixed reality (MR) that allows digital objects to appear anchored to and interact with reality, and other advanced simulation systems for the more immersive and multimodal training ecosystems serving for firefighting training and fire safety research [29].

3. VR Experiments for Choice Guidance Simulation in Road Tunnel Evacuation

The following section summarizes selected studies to illustrate how VR can be used to study road tunnel evacuation.

3.1 Experiment on Social Influence on Route Choice

Social influence (SI) has been identified as a crucial factor in human behavior during emergencies. Kinateder et al. [30] investigated how SI impacts route choice during a tunnel fire evacuation using a VR experiment (shown as Fig. 2). Using a CAVE system, participants were placed in a simulated tunnel fire scenario and observed whether the presence of a virtual agent (VA) running towards an emergency exit influenced their evacuation decisions. The scenario included a two-lane road tunnel with emergency exits and emergency phones, as well as a burning heavy goods vehicle emitting smoke. Participants were randomly assigned to two experimental groups: half were alone in the scenario (control group), while the other half saw a VA running towards the emergency exit (SI group). The study analyzed evacuation decisions based on destination choice (emergency exit vs. emergency phone) and route selection (direct vs. indirect path).

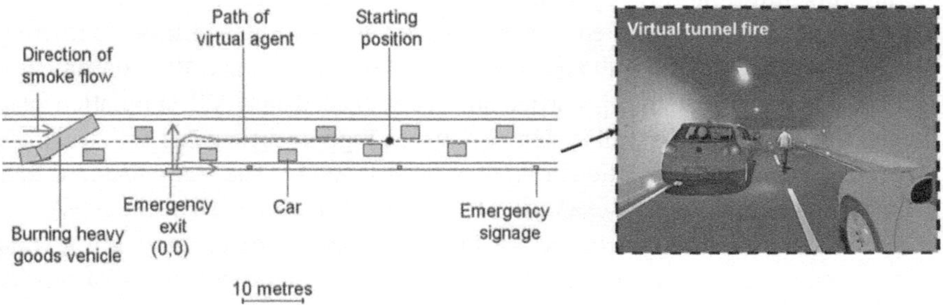

Fig. 2 Floor plan of the experimental virtual evacuation scenario and the virtual tunnel fire situation in the CAVE system, where a VA was running to the emergency exit (The original images are from Ref [30]. Copyright © 2014 Elsevier Ltd.).

The results showed that most participants (33 out of 42) chose the emergency exit as their destination, while fewer opted for emergency phones or walked away from the exit. However, participants in the Social Influence group were more likely to follow the path taken by the VA, while those in the control group often took longer routes along the tunnel walls. This behavior aligned with real-world observations where individuals tend to mimic others in unfamiliar or high-stress environments [31]. Trajectories of participants who went to the emergency exit and the corresponding heatmap (Fig. 3) show that participants in the Social Influence group occupied specific areas of the tunnel more frequently, following the trajectory of the VA. These findings suggested that the behavior of others can influence how people navigate within a tunnel.

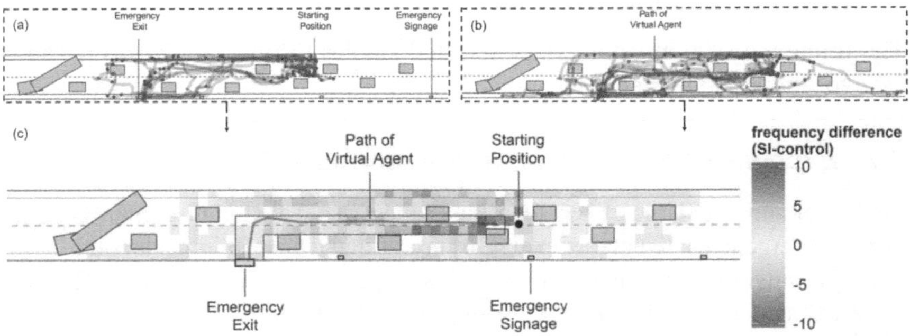

Fig. 3 Movement trajectories of participants in (a) the control group and (b) the social influence group, and (c) heatmap (The original images are from Ref [30]. Copyright © 2014 Elsevier Ltd.).

3.2 Experiment on Conflicting Information on Route Choice

In a tunnel fire scenario filled with thick smoke, it can be very hard for occupants to look for the optimal evacuation path, for example, when visibility is poor. Occupants may need to rely on signage to find an emergency exit. However, conflicting information may occur from social behavior and the emergency signage. To this end, Kinateder et al. also investigated if and how conflicting social information might influence evacuation behavior using a VR experiment.

In this experiment [22], participants were placed in a VR simulation of a tunnel fire using a powerwall system. Shown as Fig. 4, participants had to choose between two possible directions to move towards safety. The study featured four experimental conditions to examine SI: (1) a control condition without VA, (2) a no-conflict condition where the VA moved toward an emergency exit, (3) an active conflict condition where the VA moved away from the exit, and (4) a passive conflict condition where the VA remained motionless.

This study analyzed two primary phases of evacuation behavior: the pre-movement phase (time taken before initiating movement) and the movement phase (route selection and time taken to reach safety). Results indicated that SI significantly affected both phases. Participants in the passive conflict condition exhibited the longest pre-movement delays, hesitating significantly before deciding on an evacuation path. In contrast, those in the active conflict condition, where the VA moved away from the exit, were more likely to follow the incorrect path

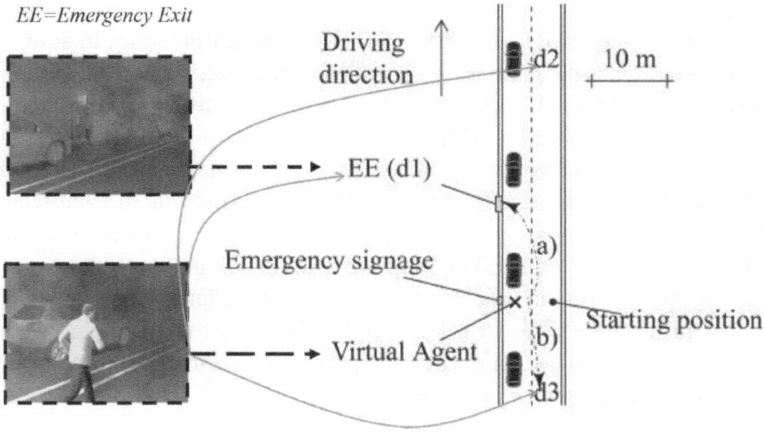

Fig. 4 Experimental setup with virtual tunnel model and simulated SI by the VA. At the beginning of each trial, participants were situated at the starting position. Orange lines indicate the possible destination choices of participants: d1 = EE; d2 = past EE into tunnel; d3 = away from EE into tunnel (The original images are from Ref [22]. Copyright © 2014 Elsevier Ltd.).

rather than the designated emergency route. This finding aligned with the classic diffusion of responsibility hypothesis [32], suggesting that passive bystanders can inhibit proactive decision-making in emergencies.

Destination choices analysis revealed that while most participants moved toward the emergency exit, their route selection was heavily influenced by social cues. In the no-conflict condition, 85% of participants followed the VA toward the emergency exit, compared to only 61% in the active conflict condition. Interestingly, in the passive conflict conditions, many participants first moved toward the VA before continuing to an exit, suggesting a reliance on social cues even when they provided no clear guidance. Statistical modeling confirmed that the presence of conflicting social information decreased the likelihood of selecting the correct evacuation route.

3.3 Experiment on Visual and Auditory Competing Guidance for Egress

To help occupants navigate exits in tunnel fire, most tunnels are equipped with highly visible exit signs to guide occupants to safe points of egress. Additionally, the public audio systems installed in some road tunnels provide auditory emergency instructions to guide occupants. Some distracting information is provided, such as warning signs or physical barriers on the path to dangerous regions, to keep residents from entering those areas. Similar to what was discussed in Section 3.2, this kind of contradictory information of contracting and detracting guidance may cause confusion and negative effects for evacuees' decision-making.

To study the competing information from visual and auditory guidance, one study [33] conducted a VR experiment tested two common forms of attracting cues (attractors), i.e., "EXIT" signs and audio instructions, and two forms of detracting cues (repellers), i.e., "DO NOT ENTER" signs and traffic cones to study which cue combination has the greatest influence on evacuee exit choices in a tunnel fire scenario. Participants completed a series of trials in a virtual tunnel environment (Fig. 5 d) while wearing an HMD. The VR simulation featured a bidirectional two-lane tunnel with two emergency exit doors. Participants were required to select one of the exits based on the guidance cues present in each trial. This study was divided into three VR experiments. The first experiment examined the effectiveness of attractors ("EXIT" signs and audio instructions), shown as Fig. 5 a. The second experiment assessed the impact of repellers ("DO NOT ENTER" signs and traffic cones), shown as Fig. 5 b. The third experiment explored conflicts between attractors and repellers, where participants encountered contradicting guidance cues at the same exit, shown as Fig. 5 c.

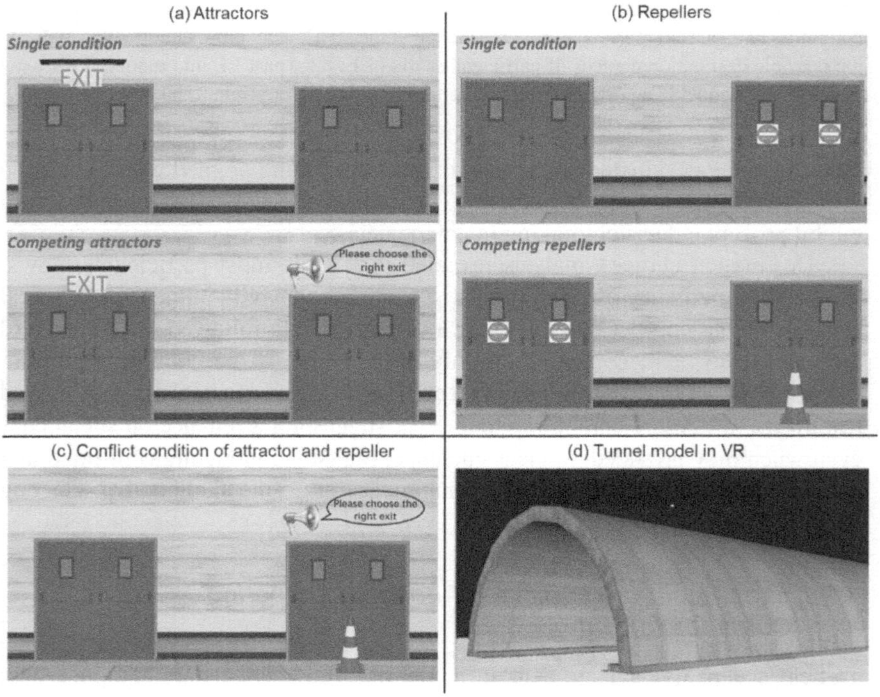

Fig. 5 Experiment configurations, (a) competing attractors ("EXIT" sign: left door, and audio instructing the right door), (b) competing repellers ("DO NOT ENTER" signs: left door, and traffic cone: right door), (c) A traffic cone (repeller) was placed in front of the right door, and the audio instruction said "Please choose the right exit" (attractor). (d) The schematic diagram of the virtual road tunnel (The original images are from Ref [33]. Copyright © 2025 Springer Nature).

Results revealed that audio instructions were the most effective attractor, with nearly all participants following spoken guidance. In contrast, "EXIT" signs were frequently ignored, especially when other cues were present. Among repellers, traffic cones were more effective deterrents than "DO NOT ENTER" signs, suggesting that physical barriers have a stronger impact on decision-making. When conflicting cues were presented, participants prioritized audio instructions over visual cues, demonstrating a preference for real-time, dynamic guidance over static signage. The statistical analysis ranked the effectiveness of guidance cues in the following order: (1) audio instructions, (2) traffic cones, (3) "DO NOT ENTER" signs, and (4) "EXIT" signs. These findings suggested that temporary and adaptive guidance systems are perceived as more credible and useful than permanent signage. Additionally, post-experiment surveys revealed that approximately 30% of participants did not notice "EXIT" signs in some trials, further supporting the conclusion that static signage alone is insufficient for effective tunnel evacuation.

4. Challenges and Perspectives

4.1 Limitations of VR Experiments on External Guidance Effects

The above sctions summarize several typical VR experiments on external influences on humans' evacuation behaviors in tunnel fire. The experiments showed the flexibility of VR scenarios and informative results about exit and path choices. However, there are limitations to the present experimental research. In Kinateder et al.'s experiments (Section 3.1 and Section 3.2) [22, 30], VAs played the role of providing external guidance or influence on participants in the tunnel fire. The issue is that only one VA was present per trial, whereas real emergencies typically involve multiple people. Therefore, potential group dynamics were not considered in the experimental design, which may create stronger or more ambiguous SI. Moreover, VA adopted a fixed movement speed in the whole egress process, but it was higher than typical speeds in low-visibility tunnel evacuations, potentially skewing timing data. Additionally, participants couldn't communicate with the VA, limiting realism and potential SI, especially in scenarios suggesting interaction.

4.2 Limitations of VR Scenarios in Tunnel Fire Simulations

VR has become an increasingly popular tool for simulating fire scenarios in training, safety analysis, and behavioral studies, which benefits from the high internal validity of VR experiments. Validity broadly concerns whether a study measures what it claims to measure. Internal validity refers to research that can ensure the observed results are caused by the variables manipulated by the researchers rather than other irrelevant factors. In VR-based evacuation research, they can provide strong internal validity because the to-be-studied environmental and human factors can be easily controlled in a virtual scenario [34]. However, several limitations hinder its effectiveness, especially regarding immersive experience, realism, and user interaction [35].

Most importantly, current VR fire simulations often fail to deliver a truly immersive experience, which refers to the issue of ecological validity. Ecological validity evaluates whether the experimental setups accurately reflect real-life environments, ensuring participants' behaviors can represent those in reality. It is critical in fire evacuation studies, especially those involving risk and danger receptions, because real-world conditions, i.e., smoke, heat, pressure, and other SI, have a powerful impact on behavior [36, 37]. Although participants can see visual representations of flames and smoke, other critical sensory cues, such as heat, smell, noise, and physical resistance, are typically absent or inadequately simulated. These missing elements significantly limit the ecological validity as well as the degree to which findings can be expected to hold up in real emergencies. For example, the visual and physical accuracy of VR environments remains limited. Fire dynamics, material degradation, smoke movement, and lighting effects are often simplified to reduce computational load, which can make the scenario appear artificial. This lack of environmental realism may prevent users from fully engaging with the simulation or responding exactly as they would in an emergency. Overall, the weakness of external validity induced by immersive issues is still a technical limitation in the current stage. When the research focuses on the impacts of heat and smoke odors on human behavior in VR, this weakness could be the biggest challenge to the authenticity of the results [38]. However, if the research is to explore the design options or compare relative differences between conditions in hypothetical scenarios, for example, whether SI and auditory instruction can determine the evacuees' exit choices, internal validity can be more important than external validity.

Another key limitation is the lack of efficient and realistic interaction among multiple users within the same VR scenario. In many VR platforms, collaboration between users is limited, uncoordinated, or unnatural due to technical constraints or poor interface design. This impedes the simulation of group behavior during evacuation or firefighting efforts, where communication, coordination, and shared decision-making are crucial. The inability to mimic social dynamics and interpersonal responses undermines realism and limits the scope of research or training based on group behavior in emergencies.

4.3 Perspective of FDS-enhanced VR Scenarios

To enhance the physical realism and immersive experience of VR scenarios, a promising solution is to insert computational fluid dynamics (CFD) [39] data into the VR devices to generate more realistic smoke environments. Shown as Fig. 6, it begins with simulating a fire scenario using the Fire Dynamic Simulator (FDS) [40] to calculate fire-driven fluid flow, simulating the behavior of smoke, heat, and toxic gases within a defined space. In this process, it generates a set of output files including smoke density, temperature fields, and visibility data. To prepare this data for VR use, it must be exported in a way that captures the evolution of the smoke plume over time and space. SmokeView [41] can be used to convert

FDS output into 3D datasets or sequences, such as voxel grids or volume textures, which represent smoke concentration at different time steps. These datasets are then processed into formats compatible with 3D creation engines, e.g., Unity [42] and Unreal Engine(UE) [43], often using scripts or custom tools to transform raw simulation data into volumetric representations suitable for real-time rendering. Once the visual effects are optimized for real-time performance, the scene is prepared for deployment to VR platforms. In this final stage, user interaction elements such as movement, scenario playback controls, and evacuation guidance are added to enhance VR experiments.

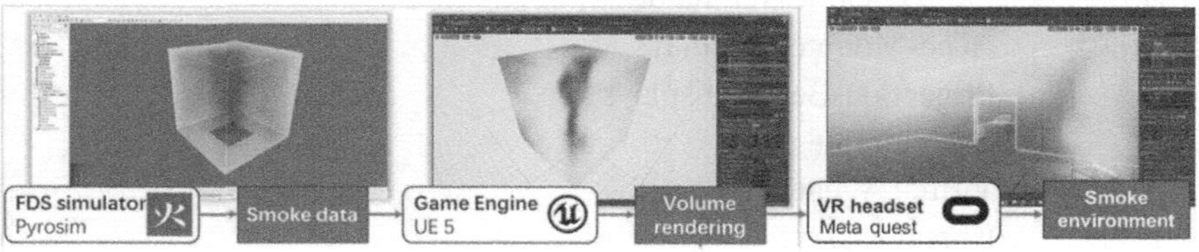

Fig. 6 Pipeline of FDS-enhanced smoke VR scenario.

5. Summary

Tunnels pose significant challenges during fire emergencies due to intense heat, low visibility, and the rapid spread and accumulation of toxic gases. While traditional evacuation studies relied on field experiments and real-life drills, VR offers a flexible, cost-effective, and controlled alternative for simulating emergency conditions and collecting detailed behavioral data. This chapter summarizes how VR can be used to simulate and study occupant behaviors during road tunnel fire evacuations, with a particular focus on external guidance like signage, SI, and auditory instructions.

Through studies on SI, conflicting information, and competing guidance cues, VR has demonstrated its value as a research method for isolating the effects of specific environmental and social variables on evacuees' egress behavior, thereby improving the internal validity of the results. At the same time, the chapter also discusses technical limitations, such as restricted sensory experience, simplified fire dynamics, and limited multi-user interaction, that still critically challenge the ecological validity of current VR systems. These findings also raise several research directions for future work, for example, the exploration of the impacts of realistic degree of VR scenarios on the validity of road tunnel safety instruction to evacuees, the comparison of the importance of internal validity and ecological validity on different VR tests, and the new technologies to enhance immersive experiences of VR e.g. heat, smell, and smoke. VR is a potential methodological option for studying human behaviors in limited spaces affected by several variables that need to be controlled. Using VR scenarios requires researchers to balance validity, feasibility, and ethics to support safer road tunnel evacuation instruction design and more effective emergency management.

Abbreviations

AR	Augmented Reality
VR	Virtual Reality
MR	Mixed Reality
XR	Extended Reality
CAVE	Cave Automatic Virtual Environment
CFD	Computational Fluid Dynamic
EE	Emergency Exit
FDS	Fire Dynamic Simulator
HMD	Head-Mounted Displays
PBD	Projection-Based Displays
SI	Social Influence
VA	Virtual Agent
UE	Unreal Engine

References

[1] Dong, J., Yan, Z., Zhang, Y. and Zhu, H. Research methods in tunnel fire distinguishing from buildings in evacuation, in: Proc. GeoShanghai 2018 Int. Conf. Tunn. Undergr. Constr., Springer, 2018: pp. 666–673.

[2] J. Ji, C.G. Fan, W. Zhong, X.B. Shen, J.H. Sun, Experimental investigation on influence of different transverse fire locations on maximum smoke temperature under the tunnel ceiling, Int. J. Heat Mass Transf. 55 (2012) 4817–4826.

[3] Y. Zhang, X. Huang, A Review of Tunnel Fire Evacuation Strategies and State-of-the-Art Research in China, Fire Technol. (2022) 1–34. https://doi.org/10.1007/S10694-022-01357-5/FIGURES/17.

[4] E. Ronchi, P. Colonna, J. Capote, D. Alvear, N. Berloco, A. Cuesta, The evaluation of different evacuation models for assessing road tunnel safety analysis, Tunn. Undergr. Sp. Technol. 30 (2012) 74–84.

[5] L.C. Boer, D.W. Veldhuijzen van Zanten, Behaviour on tunnel fire, in: Pedestr. Evacuation Dyn. 2005, Springer, 2005: pp. 91–98.

[6] A. Kallianiotis, D. Papakonstantinou, I.C. Tolias, A. Benardos, Evaluation of fire smoke control in underground space, Undergr. Sp. 7 (2022) 295–310.

[7] D. Nilsson, M. Johansson, H. Frantzich, Evacuation experiment in a road tunnel: A study of human behaviour and technical installations, Fire Saf. J. 44 (2009) 458–468.

[8] Y. Zhang, M. Kinateder, X. Huang, W.H. Warren, Modeling competing guidance on evacuation choices under time pressure using virtual reality and machine learning, Expert Syst. Appl. 262 (2025) 125582. https://doi.org/https://doi.org/10.1016/j.eswa.2024.125582.

[9] M. Kinateder, W.H. Warren, K.B. Schloss, What color are emergency exit signs? Egress behavior differs from verbal report, Appl. Ergon. 75 (2019) 155–160.

[10] C. Kim, M. Hur, Y. Oh, J. Choi, J. Jeong, The Effect of the Running-Man Emergency Exit Sign and Its Installed Location on Human Directional Choice, Appl. Cogn. Psychol. 30 (2016) 1014–1019.

[11] G. Cosma, E. Ronchi, D. Nilsson, Way-finding lighting systems for rail tunnel evacuation: A virtual reality experiment with Oculus Rift®, J. Transp. Saf. Secur. 8 (2016) 101–117.

[12] J. Olander, E. Ronchi, R. Lovreglio, D. Nilsson, Dissuasive exit signage for building fire evacuation, Appl. Ergon. 59 (2017) 84–93.

[13] H. Omori, E.D. Kuligowski, S.M. V Gwynne, K.M. Butler, Human response to emergency communication: a review of guidance on alerts and warning messages for emergencies in buildings, Fire Technol. 53 (2017) 1641–1668.

[14] D. Nilsson, Exit choice in fire emergencies-Influencing choice of exit with flashing lights, Lund University, 2009.

[15] L.K. Siebeneck, T.J. Cova, Spatial and temporal variation in evacuee risk perception throughout the evacuation and return-entry process, Risk Anal. An Int. J. 32 (2012) 1468–1480.

[16] K. Fridolf, E. Ronchi, D. Nilsson, H. Frantzich, The representation of evacuation movement in smoke-filled underground transportation systems, Tunn. Undergr. Sp. Technol. 90 (2019) 28–41. https://doi.org/https://doi.org/10.1016/j.tust.2019.04.016.

[17] Y. Zhang, H. Zhu, Q. Guo, R. Carvel, Z. Yan, The effect of technical installations on evacuation performance in urban road tunnel fires, Tunn. Undergr. Sp. Technol. 107 (2021) 103608. https://doi.org/10.1016/J.TUST.2020.103608.

[18] M. Kinateder, E. Ronchi, D. Nilsson, M. Kobes, M. Müller, P. Pauli, A. Mühlberger, Virtual reality for fire evacuation research, in: 2014 Fed. Conf. Comput. Sci. Inf. Syst., IEEE, 2014: pp. 313–321.

[19] N.J. Wade, Wheatstone and the origins of moving stereoscopic images, Perception 41 (2012) 901–924.

[20] G.A.M. Vasiljevic, L.C. De Miranda, Brain–computer interface games based on consumer-grade EEG Devices: A systematic literature review, Int. J. Human–Computer Interact. 36 (2020) 105–142.

[21] J. Skjermo, C. Moscoso, D. Nilsson, H. Frantzich, Å.S. Hoem, P. Arnesen, G.D. Jenssen, Analysis of visual and acoustic measures for self-evacuations in road tunnels using virtual reality, Fire Saf. J. 148 (2024) 104224. https://doi.org/https://doi.org/10.1016/j.firesaf.2024.104224.

[22] M. Kinateder, M. Müller, M. Jost, A. Mühlberger, P. Pauli, Social influence in a virtual tunnel fire - Influence of conflicting information on evacuation behavior, Appl. Ergon. 45 (2014) 1649–1659. https://doi.org/10.1016/j.apergo.2014.05.014.

[23] D. Nilsson, H. Frantzich, E. Ronchi, K. Fridolf, A. Lindgren Walter, H. Modig, Integrating evacuation research in large infrastructure tunnel projects - Experiences from the Stockholm Bypass Project, Fire Saf. J. 97 (2018) 119–125. https://doi.org/https://doi.org/10.1016/j.firesaf.2017.07.001.

[24] Z. Feng, V.A. González, R. Amor, R. Lovreglio, G. Cabrera-Guerrero, Immersive virtual reality serious games for evacuation training and research: A systematic literature review, Comput. Educ. 127 (2018) 252–266.

[25] G. Lawson, E. Shaw, T. Roper, T. Nilsson, L. Bajorunaite, A. Batool, Immersive virtual worlds: Multi-sensory virtual environments for health and safety training, ArXiv Prepr. ArXiv1910.04697 (2019).

[26] M. Saghafian, K. Laumann, R.S. Akhtar, M.R. Skogstad, The evaluation of virtual reality fire extinguisher training, Front. Psychol. 11 (2020) 593466.

[27] I. Capasso, C. Bassano, F. Bracco, F. Solari, E. Viola, M. Chessa, A VR Multiplayer Application for Fire Fighting Training Simulations, in: Ext. Real. First Int. Conf. XR Salento 2022, Lecce, Italy, July 6–8, 2022, Proceedings, Part I, Springer-Verlag, Berlin, Heidelberg, 2022: pp. 130–138. https://doi.org/10.1007/978-3-031-15546-8_11.

[28] L.I. Domgue K, D. Paes, Z. Feng, S. Mander, S. Datoussaid, T. Descamps, A. Rahouti, R. Lovreglio, Video see-through augmented reality fire safety training: A comparison with virtual reality and video training, Saf. Sci. 184 (2025) 106714. https://doi.org/https://doi.org/10.1016/j.ssci.2024.106714.

[29] D. Hancko, A. Majlingova, D. Kačíková, Integrating Virtual Reality, Augmented Reality, Mixed Reality, Extended Reality, and Simulation-Based Systems into Fire and Rescue Service Training: Current Practices and Future Directions, Fire 8 (2025) 228.

[30] M. Kinateder, E. Ronchi, D. Gromer, M. Müller, M. Jost, M. Nehfischer, A. Mühlberger, P. Pauli, Social influence on route choice in a virtual reality tunnel fire, Transp. Res. Part F Traffic Psychol. Behav. 26 (2014) 116–125. https://doi.org/10.1016/j.trf.2014.06.003.

[31] N. Walz, A. Mühlberger, P. Pauli, A Human Open Field Test Reveals Thigmotaxis Related to Agoraphobic Fear, Biol. Psychiatry 80 (2016) 390–397. https://doi.org/https://doi.org/10.1016/j.biopsych.2015.12.016.

[32] C. Mynatt, S.J. Sherman, Responsibility attribution in groups and individuals: A direct test of the diffusion of responsibility hypothesis., J. Pers. Soc. Psychol. 32 (1975) 1111.

[33] Y. Zhang, M. Kinateder, W. Warren, A. Templeton, A Virtual Reality Experiment on Visual and Auditory Guidance for Egress in Road Tunnel Fires, Fire Technol. (2025). 10.1007/s10694-024-01668-9.

[34] M. Kinateder, W.H. Warren, Social influence on evacuation behavior in real and virtual environments, Front. Robot. AI 3 (2016) 43.

[35] Checa, D. and Bustillo, A. Advantages and limits of virtual reality in learning processes: Briviesca in the fifteenth century, Virtual Real. 24 (2020) 151–161.

[36] Arias, S., Ronchi, E., Wahlqvist, J., Eriksson, J. and Nilsson, D. ForensicVR: Investigating human behaviour in fire with Virtual Reality, LUTVDG/TVBB (2018).

[37] Haghani, M. Empirical methods in pedestrian, crowd and evacuation dynamics: Part I. Experimental methods and emerging topics, Saf. Sci. 129 (2020) 104743. https://doi.org/https://doi.org/10.1016/j.ssci.2020.104743.

[38] Gagliardi, E., Bernardini, G., Quagliarini, E., Schumacher, M. and Calvaresi, D. Characterization and future perspectives of Virtual Reality Evacuation Drills for safe built environments: A Systematic Literature Review, Saf. Sci. 163 (2023) 106141. https://doi.org/https://doi.org/10.1016/j.ssci.2023.106141.

[39] Merci, B. Introduction to fluid mechanics, in: SFPE Handb. Fire Prot. Eng., Springer, 2016: pp. 1–24.

[40] Shen, T.-S., Huang, Y.-H. and Chien, S.-W Using fire dynamic simulation (FDS) to reconstruct an arson fire scene, Build. Environ. 43 (2008) 1036–1045.

[41] Forney, G.P. and McGrattan, K.B. User's Guide for Smokeview Version 4: A Tool for Visualizing Fire Dynamics Simulation Data, US Department of Commerce, National Institute of Standards and Technology …, 2004.

[42] J.K. Haas, A history of the unity game engine, (2014).

[43] Romero, M. and Sewell, B., Blueprints Visual Scripting for Unreal Engine 5: Unleash the true power of Blueprints to create impressive games and applications in UE5, Packt Publishing Ltd, 2022.

11 Using Virtual Reality to Improve Tunnel and Underground Station Fire Safety

Axel Mossberg[1*]

Tunnel and underground station safety is a critical issue. This is highlighted by several fire incidents, such as the Baku metro fire (1995), the Mont Blanc tunnel fire (1999), and the Kings Cross fire (1987). These events highlight the challenges in evacuation from such environments. Despite advancements in evacuation systems like escalators and occupant evacuation elevators (OEEs), the increasing number of metro systems worldwide calls for innovative solutions to improve evacuation safety. This chapter proposes a structured design method where Virtual Reality (VR) can be incorporated to enhance evacuation safety. The method involves a behavioral analysis to identify critical aspects of evacuation design, followed by VR testing to simulate and refine these designs. VR allows for the realistic testing of environments and systems not yet constructed, which could help improve the usability of unfamiliar evacuation routes and/or understanding navigation through smoke, to give some examples. The effectiveness of VR in evacuation design is demonstrated through the development of the Stockholm Metro's Sofia station, where VR testing significantly improved the evacuation system design. While VR offers numerous advantages, including cost-efficiency and reduced physiological risks, it also has limitations that must be addressed. Ultimately, VR serves as a powerful tool for both research and practical applications in improving tunnel and underground station fire safety.

1. Introduction

Tunnel and underground station safety is a complex issue. Several fire incidents, such as the Baku metro fire in Azerbaijan in 1995 [1], the Mont Blanc tunnel fire in France in 1999 [2], and the Kings Cross fire in England in 1987 [3], witnessed the difficulties in evacuation from both tunnels and underground stations. Since these incidents, several studies have investigated various aspects of evacuation, including movement through smoke [4] and the identification of evacuation routes [5]. Furthermore, the question of egressability has received increased attention within the design and research communities [6], spurring the development of new

[1] Bengt Dahlgren, Fire Research, Sweden.
* Corresponding author: axel.mossberg@bengtdahlgren.se

DOI: 10.1201/9781003514480-11

evacuation solutions, such as escalators [7] and occupant evacuation elevators (OEEs) [8].

While these challenges still exist, the number of tunnels, underground metro systems, and underground stations is increasing across the globe. Since the year 2000, the number of cities with metro systems in the world has increased by 70%. Furthermore, the projection for the coming years is that new or extended metro systems will keep increasing at an even higher rate [9]. As mentioned, these structures are associated with several challenges to the evacuation procedure. The most important aspects that need to be considered in the design are deemed to be:

1. The structures may be located at considerable depths, meaning that vertical evacuation, sometimes over vast distances, can be needed to reach the outside.
2. Both the general premises and the evacuation routes in tunnels and underground stations may be unfamiliar to the people intended to use them, meaning that both identification and, in certain cases, proper use of evacuation routes may be challenging to achieve.
3. Evacuation may, in some cases, take place in a smoke-filled environment, which further challenges proper identification and use of evacuation routes.

There could be several different solutions for these issues, e.g., OEEs to overcome the difficulties of vertical evacuation, or new evacuation designs or guidance systems to improve the usability of unfamiliar evacuation routes and/or navigation when evacuating through smoke.

2. Design Method

To successfully incorporate new evacuation systems and/or new guidance system designs, it is imperative to adopt a structured design method. Using a systematic approach, where the behavioral aspects and challenges of the design/system are thoroughly assessed, increases the probability of achieving the intended behavior from the evacuees in a real situation. This is especially important in complex environments where the correct use of the evacuation systems is imperative for a successful evacuation.

The general principles of the design method are based on previous work where similar methods have been proposed for research on evacuation guidance systems [10], [11], for elevator evacuation signage [12], and for evacuation designs incorporating elevator evacuation in general [13]. In principle, the design method can be applied to any complex environment where evacuation behavior is a critical issue. The method is illustrated in Figure 1. Depending on the situation and uncertainties associated with the design situation, all or some of the steps of the method can be applied. However, the first step of the method, i.e., the behavioral analysis, can be a useful step to perform regardless of design complexity.

Each step of the method is briefly discussed below.

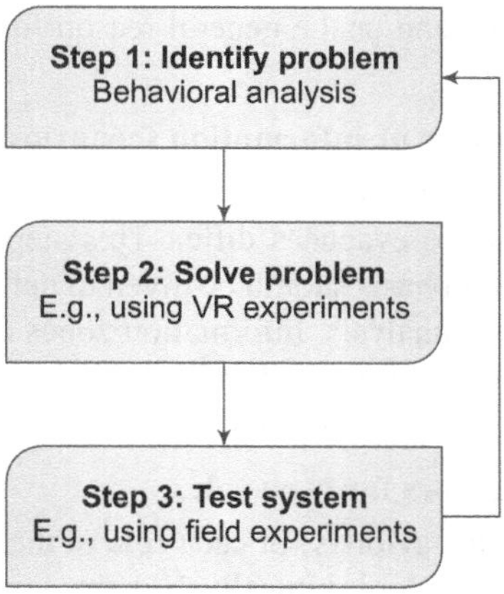

Fig. 1 Illustration of the design method (adopted from [10], [12], [13]).

3. Behavioral Analysis

A behavioral analysis is the initial step to discover important features of the suggested design. The crucial elements could be decision nodes or other parts of the suggested design where the evacuees' ability to make the correct decisions is crucial for a successful evacuation process. This step is proposed to follow a systematic framework, and the Situational Awareness Requirement (SARA) analysis has previously been proposed as a possible such framework [13]. Situational Awareness is a concept where the designer analyzes the intended behavior of a person in a given situation, what information that person would need to behave in the intended way, and how that information can be given to the person in the given situation. The concept has previously been applied to several different situations, such as for pilots in military aircraft [14], air traffic control [14], or evacuation elevators [15].

With a basis in the SARA approach, a framework for behavioral analysis of evacuation design/systems has previously been developed [13]. This framework was developed to give a more practical approach to the behavioral analysis step in the design method. The framework is illustrated in Figure 2, and the steps are shortly described below.

1. Define Scenario

In this step, a qualitative, or semi-quantitative, description of the relevant evacuation scenario should be given. Normally, there are several uncertainties associated with an evacuation scenario, and thus, the description may benefit from being given on a general level. Note that several scenarios probably should be reviewed for each design studied.

2. Define the roles of the evacuees.

The roles of the evacuees may be defined by their initial position in the scenario. However, the role may also be affected by other factors, such as physical ability for evacuation, familiarity, and/or the general reason for the person to be in the premises in the first place.

3. Define information zones or information scenarios.

Within the evacuation scenario, there may be different zones or situations where the information needs of the evacuees differ. This step is a crucial step in the behavioral analysis, and emphasis should be given to defining proper information zones, or scenarios, for the analysis. Information zones could be different rooms or parts of an underground station, and information scenarios could be such as evacuation in a tunnel with or without smoke.

4. Define intended behaviors for each role.

In this step, the intended behavior(-s) of each role of the evacuees is defined and described. The intended behavior is typically the behavior that would give the most efficient and/or safe evacuation of the premises. A intended behavior could be to use a certain evacuation route or means of evacuation, e.g., OEEs, or similar. The intended behavior could vary depending on scenario, role, and information zone.

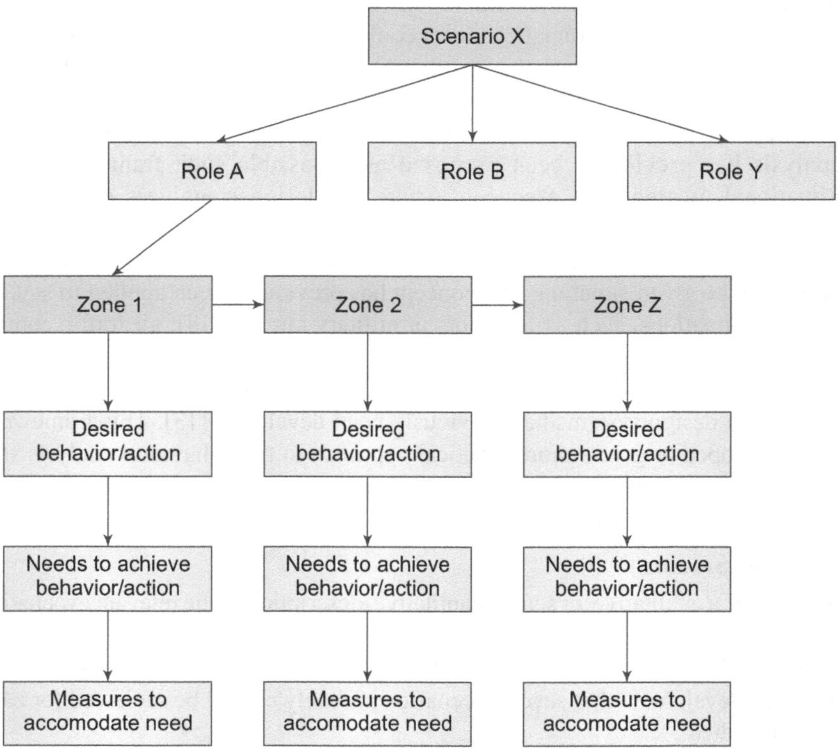

Fig. 2 Illustration of a framework for behavioral analysis (adopted from [13]).

5. Define the needs to achieve the intended behavior.

The needs to achieve an intended behavior varies depending on the situation and scenario. This could consist of informational needs, physical needs, or other needs. In situations where the evacuees are unfamiliar with the premises or use an unfamiliar route in the evacuation scenario, the needs will probably be greater than if familiar routes are supposed to be used.

6. Define measures to accommodate the needs.

In this step, different measures to accommodate the needs defined in the previous step should be analyzed and discussed. This could include guidance systems, actions from personnel, or other measures.

The general process is outlined in Figure 2.

When the behavioral analysis has been performed, the design/system should be tested to make sure that it performs in the intended way. This is described further below.

4. Testing the Design

Once a design proposal has been developed using the behavioral analysis described in the previous step, it is important to test if the design performs as intended. The reason for this is to avoid design mistakes, e.g., where the designer assumes that a design/system will be perceived in a certain way but in reality the evacuees perceive it differently, and/or to refine the proposed design to perform better.

Many evacuation concepts cannot be evaluated in a real-world situation beforehand. This is particularly true for intricate or novel evacuation solutions that could be required to accomplish a safe evacuation in underground or tunnel environments. Thus, Virtual Reality (VR) is a tool that can be very powerful for this type of testing. With VR, environments that are not yet constructed can be tested, as well as systems that are not yet fully developed.

Fig. 3 A person with a Head-Mounted Display (HMD) VR equipment is immersed in the VR environment.

One example of behavioral analysis and testing of the design has been performed in the development of the Stockholm Metro system. Within this development, one of the stations, Station Sofia, is planned to have a platform depth of more than 100 meters below ground level [8]. For this station, elevators will be used to enter and leave the station, both in normal use and if there is an evacuation. As this is a novel evacuation route for underground metro stations, and the station is not yet constructed, the testing and development of the evacuation system have partly been performed in VR [16].

A virtual version of the station was developed, and different guidance systems, with the purpose of increasing OEE use, were tested. The systems tested were developed using a behavioral analysis, as described in the previous step of the method. Systems tested were different evacuation alarms, signage, count-down timers showing the time until an elevator arrives, as well as guidance through a smartphone. The blueprint of the VR environment is shown in Figure 4, and an example of the view the participants see is shown in Figure 5.

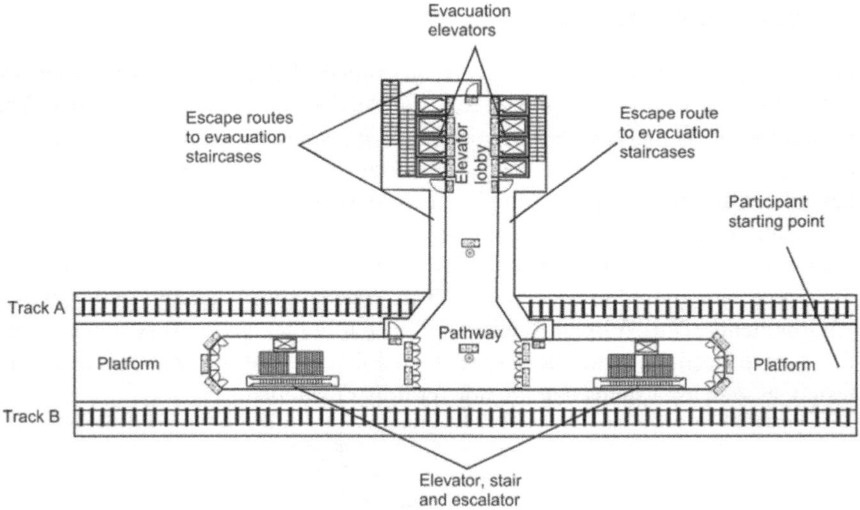

Fig. 4 An overview of the station layout. The small signs above the doors show where the exit signs were located in the model. The signs with dots that have a circle around them show signage that was reinforced with green flashing lights in two of the scenarios (enhanced and timer). (adopted from [16]).

The VR experiments showed that the design that had been developed (named "Enhanced" in the trials) performed significantly better than the baseline scenario, which represented a scenario with a design coherent with the requirements in the building code. The Enhanced design included a more informative voice alarm, i.e., one that stated that the elevators could be used for evacuation, enforced signage at key decision points, i.e., signs with green flashing lights, and information signs

Fig. 5 Example from the VR environment.

Fig. 6 Examples of the information/direction given by the smartphone during the evacuation. The arrow was pointing in the evacuation direction. Furthermore, the smartphone gave different information messages during the evacuation, such as that they should evacuate. The messages were given in Swedish, which all participants were fluent in.

about the station depths (~100 m) and that the elevators were safe to use adjacent to the evacuation routes. The OEE usage, which was the intended evacuation behavior, is shown in Figure 7. The usage increased from 3% in the baseline scenario to 68% in the enhanced scenario.

With smartphone guidance, which was also tested, the elevator usage increased to 30%. Figure 6 illustrates how the advice was displayed. During the process, several design faults were found in this design solution. For example, the guidance on the smartphone changed at different waypoints along the evacuation route.

However, the participants did not always notice this, as there was no audio or vibration signal from the phone when the information changed. Thus, much of the information was not registered by many of the participants. It is probable that the system could perform even better if these flaws were addressed properly [16].

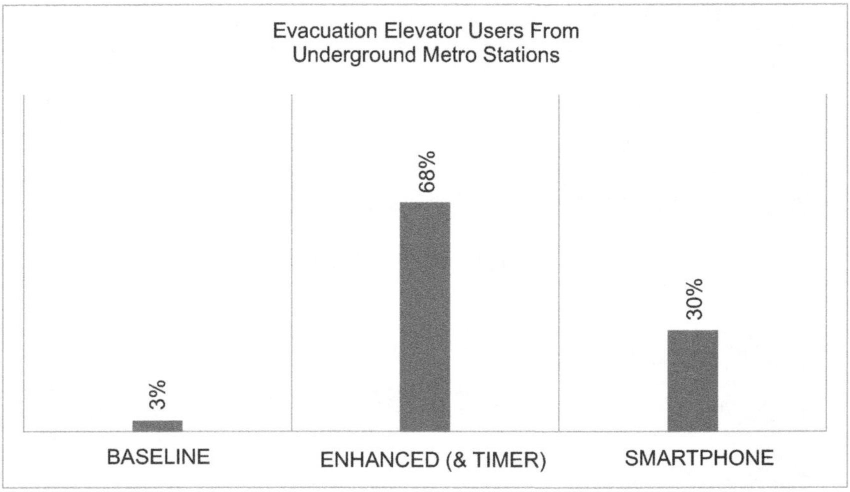

Fig. 7 OEE usage in different scenarios in the VR experiments [16].

As the metro station studied is still under construction, the final step of the design method has not yet been performed. However, with the results from the VR experiments, the design could be further developed and refined. Several design improvements could be proposed based on the results from the experiments.

Even though the design has been tested in VR, it is still imperative to validate the results from the VR testing with field tests to ensure that the results are coherent across the experimental methods. This step is discussed further below.

5. Advantages and Limitations of VR

Due to several reasons, VR has become a popular tool, both in design situations and for evacuation behavior research. In recent years, a large number of studies have been performed on evacuation behavior using VR, such as the design of evacuation signs [17], way-finding during evacuation [18], [19], and the behaviors that are recorded during evacuation [20]. VR has also been used to replicate real fires for forensic purposes [21]. For tunnel situations, some examples of VR studies performed are on the design of road portals [22], the design of evacuation systems [23], how dangerous goods vehicles affect the perceived threat in a fire emergency [24], and how different lighting systems can improve way-finding in rail tunnels [25]. Furthermore, both movement speed [26] and driving behavior in smoke [27] have been studied.

The benefits of using VR for evacuation behavior research and for design purposes are several. For design purposes, the most obvious advantage is probably that environments that are not yet constructed can be analyzed realistically. Furthermore, the environments and systems that are relevant to assess can easily be changed in a highly cost-efficient way. Another important aspect is that the physiological risks and stress factors are lower for participants in VR experiments, making ethical considerations easier to adhere to.

Even though the advantages are many, there are still considerations that need to be considered when performing VR experiments, both for research and design purposes. One such consideration is the VR equipment used. The person in Figure 3 wears a HMD, which is a common VR piece of equipment . However, other equipment, such as the Cave Automatic Virtual Environment (CAVE). In a CAVE simulation, the person stands in a room, and the virtual environment is projected on screens covering the floor, walls, and sometimes even the roof of the room. With different equipment, the feeling of immersion will differ. Similarly, locomotion in VR can be performed using different strategies and equipment, and this factor can also affect the feeling of immersion. Apart from the feeling of immersion, different equipment and locomotion strategies can be assumed to affect the behavioral realism of the scenario studied [28]. Thus, it is important to assess that the equipment and locomotion strategy chosen are suitable for studying the behavior that the experiments are trying to capture or evaluate.

6. Validating the Results

Whether it is for research or design purposes, it is always important to assess the validity of the chosen methods for the study at hand. This is important regardless of the method chosen. An example of how different methods can affect the results has previously been studied for high-rise buildings [29].

In Figure 8, results from different studies on the willingness to use OEEs for evacuation are presented. The results show the willingness recorded on floor 16 for a high-rise building. Three studies were performed as hypothetical scenario studies, also called studies of behavioral intent, which are the three stacks in red to the left [30], [31], [32]. This type of study is performed by asking the participants to imagine a certain scenario and then answer a questionnaire on how they think they would have acted in such a scenario. Furthermore, for the case of the high-rise building, the participants were asked to imagine their position in the building as well as the fire scenario in two of the studies [31], [32]. The research findings range from 15% to 45% for elevator users. It should be mentioned that these studies were conducted in different countries, and cultural variations may account for some of these variations.

The yellow stacks in the middle of Figure 8 show VR experiments performed with different equipment. The two yellow stacks to the left show the results from a CAVE experiment performed [33], and the two yellow stacks to the right show the results from a replica experiment where HMD equipment was used [34].

Both experiments were designed to mimic a real hotel building where OEEs are installed. As the figure shows, the results for the same scenario vary, with 60% elevator users in the CAVE experiment and 94% elevator users in the HMD experiment. Furthermore, the CAVE experiment shows a significant increase in elevator usage if green flashing lights are mounted on the evacuation sign pointing towards the elevator lobby.

The green stacks to the right in Figure 8 show the results from a field experiment performed in the actual building that the VR experiments mimicked [35]. This experiment was performed as an unannounced evacuation experiment in a hotel building. The participants were given eye-tracking equipment and told that they were participating in a hotel design study. They were then told to go to a room on the 16th floor and act as if checking in at the hotel, and that the researchers would come to collect their eye-tracking equipment after a while. Once the participants had reached the hotel room, the evacuation alarm was activated, and the evacuation behavior was recorded. As the results show, the elevator usage was 91–100% depending on scenario, where the lowest elevator usage was recorded for a scenario where the hotel room that the participants checked into was located adjacent to the evacuation stair, see Figure 9.

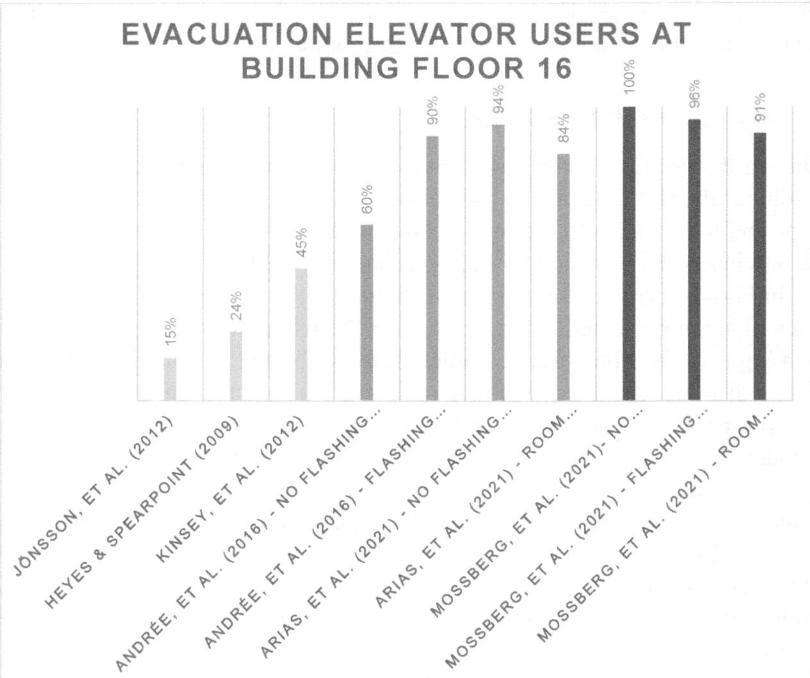

Fig. 8 Elevator usage in experiments performed with different experimental methods. Light purple stacks show results from questionnaire studies, yellow stacks show results from VR studies, and blue stacks show the results from a field experiment. Adopted from [29].

To give context to the differences between the participants' initial position in the experiment, Figure 9 shows the floor plan of the hotel building, where the participants in the scenarios labeled "Room adjacent to stair" started in Room 2. For the rest of the scenarios, the participants started out in Room 1. The flashing lights were mounted on the evacuation signage pointing towards the elevator lobby.

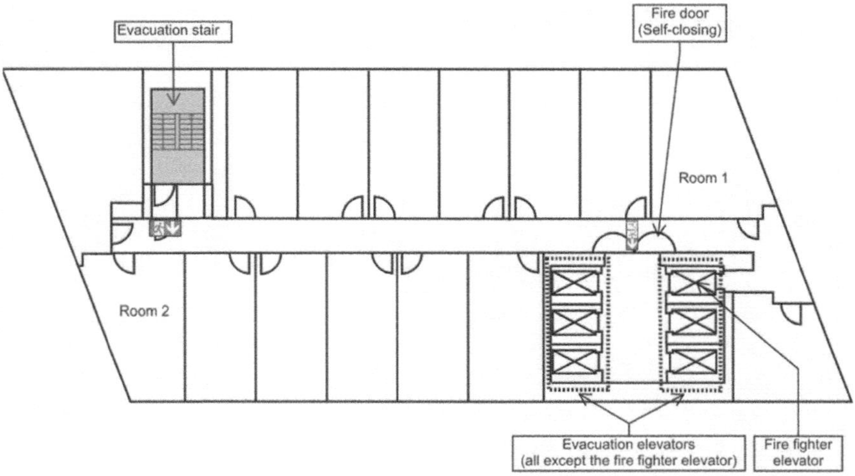

Fig. 9 Floor plan of the hotel building studied in both VR [33], [34] and field experiments [35].

For underground metro stations, the research foundation is not as strong. However, similar comparisons can be made with some performed studies. One such comparison can be made between a questionnaire, or hypothetical scenario study, performed [36] and the VR study of an underground metro station that has been introduced previously [16]. The hypothetical scenario study was performed with two different groups, one with difficulties walking stairs (Group 2) and one without such difficulties (Group 1). Furthermore, two hypothetical depths of the underground metro station were investigated, 8 and 30 meters. However, Group 2 was only investigated for the deeper station.

The results from the studies can be seen in Figure 10. As the results show, there are similarities and differences between the experimental methods for this setting as well. One such similarity is the number of elevator users when comparing the questionnaire results and the VR results for the Baseline scenario for this sort of setting. This can probably be explained by the elevator not being a natural point of entry or exit in most underground metro stations, which is a difference when compared to high-rise buildings. However, a deeper station increases the amount of elevator users, which is consistent with the correlation for high-rise buildings, where elevator usage increases with building floor in the hypothetical scenario studies performed [30], [31], [32].

The results for group 2 indicate that people who are unable to walk the stairs are more likely to use the elevator than people who are able to walk the stairs. This is a reasonable correlation, but as no similar group was included in the VR study, it is not possible to compare the results between experimental methods.

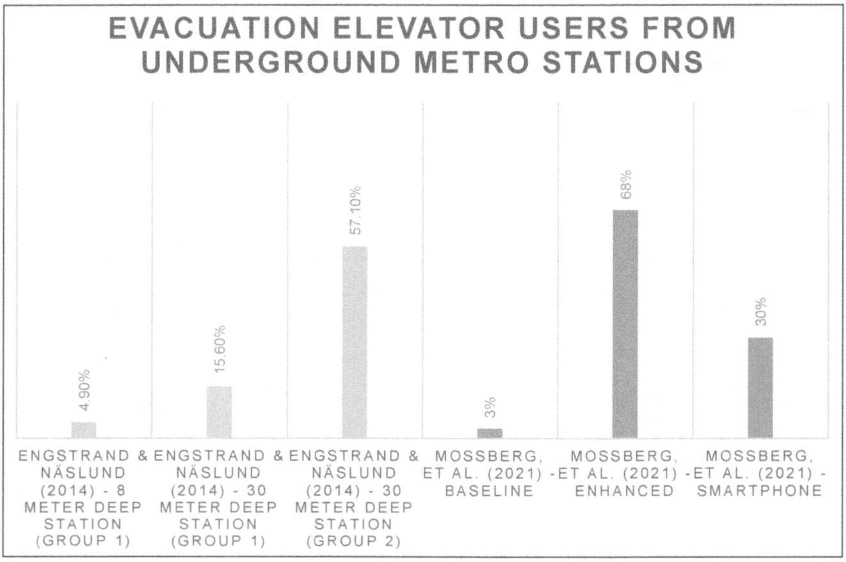

Fig. 10 Comparison between a questionnaire study (stacks in light purple) and a VR study (stacks in yellow) on elevator usage for an underground metro station. Adopted from [29].

As mentioned, to ensure validity in these results for the underground setting, a field study performed in a similar environment to the two studies mentioned above would be needed. However, the results from both settings discussed above highlight the importance of doing structured design work in complex environments and show that VR, used correctly, can be an important tool for such design situations.

7. Improving Realism in VR Environments

The more realistic a VR environment can be designed, the more likely it is that the results from VR studies will be coherent with field experiments or real evacuation behavior. Thus, realism in VR is an important matter, both for the research community and for design situations. In previous work, certain possible areas of improvement for VR experiments have been highlighted, such as interactivity and visual realism [37]. Furthermore, it is challenging to record certain behaviors in VR due to the lack of consequences, both physical and social. Thus, it is important to acknowledge what behaviors, or decision-making, the VR experiments are supposed to record and how the environment is best designed to ensure the best possible foundation for those behaviors or decisions.

It should be noted that there have been, and are ongoing, projects to improve realism in VR by different means. For example, the incorporation of smoke smell in VR environments has been shown to increase the sense of realism [38]. Furthermore, in non-fire-related studies, the effects of realistic avatars have not been shown to significantly improve realism in VR [39]. In the VR study of the underground metro station discussed above, the effects of simulating locomotion, or walking, with a set of "VR shoes" [40]. The shoes were mounted on the participants' feet and had a wheel underneath that mimicked walking when the participants dragged the foot on the ground from a half-seated position. However, only 14% of the participants thought that this way of movement was realistic, highlighting the need for further improvement to locomotion in virtual environments.

8. VR as a Tool to Improve Tunnel and Underground Station safety

This chapter has presented a method to incorporate human behavior in the evacuation design of tunnels and underground stations. The approach starts out with a systematic behavioral analysis, followed by testing the evacuation design in VR and finally testing the design in a field environment as validation. In a design situation, the first two steps, i.e., the behavioral analysis and the VR testing, would be performed in the early design stage of the project, while the field testing would be done once the facility is built, but not yet commissioned. The use of this method is strongly believed to improve the evacuation safety of complex environments, especially when new evacuation strategies, e.g., OEEs, are incorporated into the design.

Furthermore, VR is also regarded as an important tool for research on human behavior aspects in evacuation scenarios. It has been concluded that VR can be a useful tool for studying many different situations, if suitable equipment is chosen and the virtual environment is realistic enough for the situation at hand. However, VR is also associated with several limitations that need to be considered when performing such research. It should be acknowledged that while some of these limitations may be possible to eliminate through further development of equipment or virtual environments, other limitations will probably remain regardless of such development. Thus, further research into the validity of VR for different situations would create benefits for the research community as well as for design situations.

On a final note, the use of elevators for evacuation is a strategy that has the possibility of increasing safety in many different situations, e.g., for tunnels and underground stations as well as in high-rise buildings. The efficiency of elevators has been proven for both high-rise buildings [41], [42] and underground situations [43], but this efficiency is highly dependable of the usage of the elevators. For underground situations, the usage of elevators will rarely be the natural exit route, and correct guidance systems to improve elevator usage will probably be needed. This is a complex situation, and there is a need to use a systematic approach to ensure a proper evacuation design. The use of the method proposed in this chapter gives a foundation for such an approach.

References

[1] Hedefalk, J., Wahlstrom, B. and Rohlen, P. 'Lessons from the Baku subway fire', 1998. Accessed: Mar. 08, 2025. [Online]. Available: https://www.semanticscholar.org/paper/Lessons-from-the-Baku-subway-fire-Hedefalk-Wahlstrom/1e0e7b22787f51c9ec2efe40d02c7146d16e12b5.

[2] Catmur, J., Parsons, M. and Tarada, F. 'A Service Analysis of the Mont Blanc Tunnel Fire', presented at the Safety Critical Systems Symposium, York, 2023.

[3] Moodie, K. 'The King's Cross fire: damage assessment and overview of the technical investigation', *Fire Safety Journal*, vol. 18, no. 1, pp. 13–33, Jan. 1992, doi: 10.1016/0379-7112(92)90045-E.

[4] Fridolf, K., Andrée, K., Nilsson, D. and Frantzich, H. 'The impact of smoke on walking speed', *Fire and Materials*, vol. 38, no. 7, pp. 744–759, 2014, doi: 10.1002/fam.2217.

[5] Nilsson, D., Johansson, M. and Frantzich, H. 'Evacuation experiment in a road tunnel: A study of human behaviour and technical installations', *Fire Safety Journal*, vol. 44, no. 4, pp. 458–468, May 2009, doi: 10.1016/j.firesaf.2008.09.009.

[6] Smedberg, E. 'Egressibility: Applying the concept of accessibility to the self-evacuation of people with functional limitations', Licentiate Thesis, Lund University, 2022.

[7] Zhang, C., Wu, X., Chen, J. and Shen, H. 'Research on escalators used as evacuation stairs under fire scenarios', *Sci Rep*, vol. 13, no. 1, p. 16933, Oct. 2023, doi: 10.1038/s41598-023-43784-6.

[8] Region Stockholm, 'Sofia station – one of a kind', Nya tunnelbanan. Accessed: Aug. 22, 2021. [Online]. Available: https://nyatunnelbanan.se/en/artikel/sofia-station-one-kind

[9] UITP, 'World Metro Figures 2018', the International Association of Public Transport, Brussels, 2018.

[10] Nilsson, D. 'Exit choice in fire emergencies - Influencing choice of exit with flashing lights', Lunds Universitet, Lund, Doktorsavhandling 1040, 2009.

[11] Nilsson, D., Frantzich, H., Ronchi, E., Fridolf, K., Lindgren Walter, A. and Modig, H. 'Integrating evacuation research in large infrastructure tunnel projects - Experiences from the Stockholm Bypass Project', *Fire Safety Journal*, vol. 97, pp. 119–125, Apr. 2018, doi: 10.1016/j.firesaf.2017.07.001.

[12] Nilsson D. and Jönsson, H. 'Design of Evacuation Systems for Elevator Evacuation in High-Rise Buildings', *Journal of Disaster Research*, vol. 6, no. 6, Art. no. 6, 2011.

[13] Mossberg, A., Nilsson, D. and Frantzich, H. 'Evaluating new evacuation systems related to human behaviour using a situational awareness approach – A study of the implementation of evacuation elevators in an underground facility', *Fire Safety Journal*, vol. 134, p. 103693, Dec. 2022, doi: 10.1016/j.firesaf.2022.103693.

[14] Endsley, M.R. 'Toward a Theory of Situation Awareness in Dynamic Systems', *Human Factors*, Mar. 1995, doi: 10.1518/001872095779049543.

[15] Groner, N.E. 'A Situation Awareness Requirements Analysis for the Use of Elevators During Fire Emergencies', Cambridge: Interscience Communication, 2009, pp. 61–72.

[16] Mossberg, A., Nilsson, D. and Wahlqvist, J. 'Evacuation elevators in an underground metro station – A Virtual Reality evacuation experiment', *Fire Safety Journal*, 2020, doi: 10.1016/j.firesaf.2020.103091.

[17] Kinateder, M., Warren, M. and Schloss, K.B. 'What color are emergency exit signs? Egress behavior differs from verbal report', *Applied Ergonomics*, vol. 75, pp. 155–160, Feb. 2019, doi: 10.1016/j.apergo.2018.08.010.

[18] Kobes, M., Helsloot, I., de Vries, B., Post, J.G., Oberijé, N. and Groenewegen, K. 'Way finding during fire evacuation; an analysis of unannounced fire drills in a hotel at night', *Building and Environment*, vol. 45, no. 3, pp. 537–548, Mar. 2010, doi: 10.1016/j.buildenv.2009.07.004.

[19] Arias, S., La Mendola, S., Wahlqvist, J., Rios, O., Nilsson, D. and E. Ronchi, 'Virtual Reality Evacuation Experiments on Way-Finding Systems for the Future Circular Collider', *Fire Technology*, vol. 55, no. 6, Art. no. 6, 2019, doi: 10.1007/s10694-019-00868-y.

[20] Arias, S., Nilsson, D. and Wahlqvist, J. 'A virtual reality study of behavioral sequences in residential fires', *Fire Safety Journal*, vol. 120, p. 103067, Mar. 2021, doi: 10.1016/j.firesaf.2020.103067.

[21] Arias, S., Fahy, R., Ronchi, E., Nilsson, D., Frantzich, H. and Wahlqvist, J. 'Forensic virtual reality: Investigating individual behavior in the MGM Grand fire', *Fire Safety Journal*, vol. 109, p. 102861, Oct. 2019, doi: 10.1016/j.firesaf.2019.102861.

[22] Ronchi, E. *et al.*, 'A Virtual Reality Experiment on Flashing Lights at Emergency Exit Portals for Road Tunnel Evacuation', *Fire Technol*, vol. 52, no. 3, pp. 623–647, May 2016, doi: 10.1007/s10694-015-0462-5.

[23] Zhang, Y., Kinateder, M., Templeton, A. and Warren, W.H. 'A Virtual Reality Experiment on Visual and Auditory Guidance for Egress in Road Tunnel Fires', *Fire Technol*, Jan. 2025, doi: 10.1007/s10694-024-01668-9.

[24] Kinateder, M *et al.*, 'The effect of dangerous goods transporters on hazard perception and evacuation behavior – A virtual reality experiment on tunnel emergencies', *Fire Safety Journal*, vol. 78, pp. 24–30, Nov. 2015, doi: 10.1016/j.firesaf.2015.07.002.

[25] Cosma, G., Ronchi, E. and Nilsson, D. 'Way-finding lighting systems for rail tunnel evacuation: A virtual reality experiment with Oculus Rift®', *Journal of Transportation Safety & Security*, vol. 8, no. sup1, pp. 101–117, Jun. 2016, doi: 10.1080/19439962.2015.1046621.

[26] Davis, C., Sole, C., Khan, H. and Nilsson, D. 'Investigating movement through smoke in virtual reality', *Fire Safety Journal*, vol. 140, p. 103890, Oct. 2023, doi: 10.1016/j.firesaf.2023.103890.

[27] Wetterberg, N., Ronchi, E. and Wahlqvist, J. 'Individual Driving Behaviour in Wildfire Smoke', *Fire Technol*, vol. 57, no. 3, pp. 1041–1061, May 2021, doi: 10.1007/s10694-020-01026-5.

[28] Arias, S. 'Application of Virtual Reality in the study of Human Behavior in Fire: Pursuing realistic behavior in evacuation experiments', Doctoral Thesis (compilation), Division of Fire Safety Engineering, Lund University, Lund, 2021.

[29] Mossberg, A. 'Elevator evacuation: Exploring behavioural aspects', Doctoral thesis, Division of Fire Safety Engineering, Lund University, Lund, 2022.

[30] Jönsson, A., Andersson, J. and Nilsson, D. 'A Risk Perception Analysis of Elevator Evacuation in High-Rise Buildings', Cambridge: Interscience Communication, 2012, pp. 398–409.

[31] Heyes E. and Spearpoint, M. 'Lifts for Evacuation - Human Behaviour Considerations', Cambridge: Interscience Communication, 2009, pp. 73–84.

[32] Kinsey, M. J., Galea, E. R. and Lawrence, P. J. 'Human Factors Associated with the Selection of Lifts/Elevators or Stairs in Emergency and Normal Usage Conditions', *Fire Technology*, vol. 48, no. 1, Art. no. 1, 2012, doi: https://doi-org.ludwig.lub.lu.se/10.1007/s10694-010-0176-7.

[33] Andrée, K., Nilsson, D. and Eriksson, J. 'Evacuation experiments in a virtual reailty high-rise building: exit choice and waiting time for evacuation elevators', *Fire and Materials*, vol. 2016, no. 40, Art. no. 40, 2016.

[34] Arias, S., Mossberg, A., Nilsson, D. and Wahlqvist, J. 'A Study on Evacuation Behavior in Physical and Virtual Reality Experiments', *Fire Technology*, vol. 58, no. 2, Art. no. 2, 2022, doi: 10.1007/s10694-021-01172-4.

[35] Mossberg, A., Nilsson, D. and Andrée, K. 'Unannounced Evacuation Experiment in a High-Rise Hotel Building with Evacuation Elevators: A Study of Evacuation Behaviour Using Eye-Tracking', *Fire Technology*, Oct. 2020, doi: 10.1007/s10694-020-01046-1.

[36] Engstrand, A. and Näslund, J. 'Utrymning med hiss från tunnelbanestationer under mark - En studie om förmåga och riskperception [Evacuation with elevator from subway stations under ground – A study on ability and risk perception]', Lunds Universitet, Lund, 5467, 2014.

[37] Arias, S., Wahlqvist, J., Nilsson, J., Ronchi, E. and Frantzich, H. 'Pursuing behavioral realism in Virtual Reality for fire evacuation research', *Fire and Materials*, vol. 45, no. 4, pp. 462–472, Jun. 2021, doi: 10.1002/fam.2922.
[38] Khan, H. and Nilsson, D., 'Smell of Fire Increases Behavioural Realism in Virtual Reality: A Case Study on a Recreated MGM Grand Hotel Fire', presented at the 22nd IEEE International Symposium on Mixed and Augmented Reality, ISMAR 2023, Sidney, 2023.
[39] Fraser, A.D., Branson, I., Hollett, I.C., Speelman, C.P. and Rogers, S.L. 'Do realistic avatars make virtual reality better? Examining human-like avatars for VR social interactions', *Computers in Human Behavior: Artificial Humans*, vol. 2, no. 2, p. 100082, Aug. 2024, doi: 10.1016/j.chbah.2024.100082.
[40] Mossberg, A., Nilsson, D., Frantzich, H. and Wahlqvist, J. 'Utrymningshissar – Vidareutveckling av informationssystem', Stockholm, BSL 2020:01, 2020.
[41] Hammarberg, J., Niva, H. and Mossberg, A. 'Incorporation of elevator evacuation from a specific floor – A numerical study of an office building', *Collective Dynamics*, no. 5, Art. no. 5, 2020.
[42] Ronchi, E. and Nilsson, D. 'Fire evacuation in high-rise buildings: a review of human behaviour and modelling research', *Fire Science Reviews*, vol. 2, no. 7, Art. no. 7, 2013, doi: https://doi.org/10.1186/2193-0414-2-7.
[43] Mårtensson, O. and Johansson, A.-L. 'Modellering av utrymningshissar som utrymningsväg från djupa undermarksanläggningar', *LUTVDG/TVBB*, 2022, Accessed: Mar. 11, 2025. [Online]. Available: http://lup.lub.lu.se/student-papers/record/9073928.

12 Extended Reality (XR) in Tunnel Engineering for Inspection and Maintenance

Saizhe Ding[1], Rong Deng[2], Yuxin Zhang[3*] and Xinyan Huang[4*]

With the rapid development of urbanization and increasing demand for infrastructure construction, tunnel engineering plays a vital role in urban transportation, municipal projects, and energy transmission. However, various challenges such as construction risks, structural health monitoring difficulties, and emergency evacuation inefficiencies persist. Extended Reality (XR) technology, including Augmented Reality (AR), Virtual Reality (VR), and Mixed Reality (MR), provides innovative solutions to these problems. By integrating XR with digital twin models, BIM, GIS, and IoT, tunnel engineers can enhance real-time monitoring, improve safety training and risk awareness, facilitate intelligent emergency evacuation, and optimize inspection and maintenance efficiency. This paper systematically explores the application of XR technology in tunnel engineering, analyzing its advantages, challenges, and future development trends to provide theoretical and technical insights for research and practical implementations.

1. Introduction

In the context of the accelerated global urbanization process and the growing demand for infrastructure construction, tunnel engineering has become an essential part of urban transportation, municipal construction, and energy transmission. Whether they are subway tunnels, highway tunnels, or drainage tunnels in water conservancy projects, tunnels play a critical role in modern society. However, tunnel engineering construction and operation have always faced numerous challenges, such as construction risks posed by complex geological environments, the difficulty of structural health monitoring, insufficient safety training, and challenges in emergency evacuation. With the advancement of science and technology, people have begun to seek more efficient, safe, and intelligent technological solutions to address these challenges in the tunnel engineering field.

[1,2] Research Center for Smart Urban Resilience and Firefighting, The Hong Kong Polytechnic University, Hong Kong.

[3] Department of Urban Management Kyoto University, Kyoto, Japan.

[4] Department of Building Environment and Energy Engineering, The Hong Kong Polytechnic University, Hong Kong.

* Corresponding author: yx.zhang@polyu.edu.hk; xy.huang@polyu.edu.hk

DOI: 10.1201/9781003514480-12

The emergence of Extended Reality (XR) technology has provided innovative technical support for tunnel engineering. XR is a technological concept encompassing Augmented Reality (AR), Virtual Reality (VR), and Mixed Reality (MR). It uses computer graphics and human-computer interaction technologies to allow users to have an immersive experience between the real and virtual environments. The development of XR technology has gone through several stages, from early laboratory explorations to modern engineering applications. With the advancement of technologies such as computer vision, artificial intelligence, and 5G communication, XR applications in engineering have gradually moved beyond traditional visualization modes to more intelligent and interactive solutions.

In tunnel engineering, the application of XR technology is mainly focused on the following aspects: (1) 3D information monitoring—by building digital twin models and integrating BIM (Building Information Modeling), GIS (Geographic Information System), and IoT (Internet of Things) data, real-time monitoring of tunnel structural health is achieved; (2) Safety training and risk awareness—using VR technology to simulate tunnel construction environments and emergencies, allowing engineers to undergo safety training in virtual environments, thereby improving emergency response capabilities; (3) Intelligent emergency evacuation—combining MR technology to enable multi-user collaborative drills, improving evacuation efficiency in the event of tunnel fires or collapses; (4) Robot inspection and smart patrols—using AR technology to enhance the automation of tunnel inspections, improving inspection efficiency and accuracy. These technological applications not only enhance engineering safety but also bring about significant changes in construction efficiency and management intelligence.

This paper will systematically explore the current application of XR technology in tunnel engineering, analyze its technological advantages and challenges, and forecast future development trends. The article first introduces the basic concepts and development trajectory of XR technology, then analyzes its specific application scenarios in tunnel engineering, discusses current challenges and technological bottlenecks, and finally proposes future development directions, providing theoretical support and technical references for research and practice in tunnel engineering.

2. XR Technologies: A Primer

2.1 Definitions and Scope

The first formal discussion on immersive technology appeared in [1], where Sutherland first conceptualized fundamental principles of human-computer interaction in virtual environments. Since then, several taxonomies and modifications have been raised and can be categorized through multiple dimensions:

Primarily based on the degree of integration between real and virtual environments, we distinguish between Augmented Reality (AR) and Augmented Virtuality (AV), which are two complementary concepts along Milgram's Reality-Virtuality Continuum [2], as shown in Fig. 1.

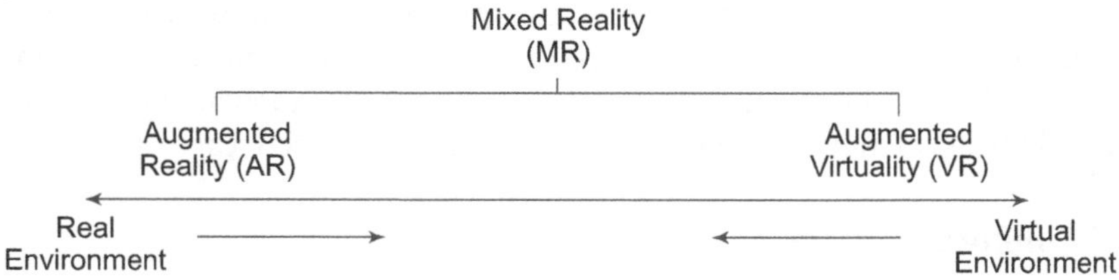

Fig. 1 The Milgram's Reality-Virtuality Continuum.

Secondly, according to different levels of immersion, it can be classified as non-immersive, semi-immersive, and fully immersive [3]. In non-immersive settings, users can manipulate avatars or activities within the virtual environment but maintain awareness of their physical surroundings, which means they are not the focal point of attention [4,5]. Computers, video game consoles, and smartphones are the most common non-immersive devices. By combining visual, aural, and olfactory cues, fully immersive systems give users the impression that they are physically present in the virtual environment. In the meantime, users' body movement can be translated into actions in the virtual world and has interactive effects on digital elements [6,7]. Semi-immersive experiences strike a balance between fully immersive and non-immersive; it gives users a sense of being there while they still have a notion of being in the real world since real physical movements are cut off [8]. Users often use a console, mouse, or keyboard to move in a virtual world through a set of projection screens or wearable devices under semi-immersive settings.

Finally, through technological implementation approaches, mainstream categories emerge as Virtual Reality (VR), Augmented Reality (AR), and Mixed Reality (MR). To be more specific, VR represents an effort to create a totally synthetic environment with visual and immersive elements that provide users with a sensation of reality [9,10]. This immersion can range from fully embodied experiences using head-mounted displays (HMDs) to non-immersive interactions via desktop monitors [11]. In contrast, augmented reality (AR) improves a person's view of virtual items with real-world features by displaying virtual content with residual see-through characteristics. Its see-through capability strengthens the connection between the virtual and real elements while maintaining the flexibility of the virtual space [12]. MR further bridges this spectrum by blending virtual and physical objects into a unified mixed environment, enabling real-time interaction between realms. In MR, virtual elements are not only overlaid onto the real world but can interact with real-world objects and vice versa, creating a more seamless and dynamic experience for users. MR technologies typically use advanced sensors, cameras, and depth mapping to detect physical spaces and objects, integrating these real-world inputs with virtual overlays that behave as if they are part of the physical world. This results in a more immersive and interactive user experience compared to AR, as virtual and real elements coexist and interact in real-time, providing an enhanced sense of presence.

Overall, with the boosted development of software and hardware in immersive technology, increasingly higher levels of immersion and engagement can be achieved. More choices are also provided to developers, allowing them to adopt technological solutions based on various contexts and objectives.

2.2 Immersive Devices

In 1956, American photographer Morton Neiling invented the first immersive simulator called Sensorama [13]. It could project 3D short films with surround sound, vibration effects, and even odor and blowing effects, creating a realistic experience (as shown in Fig. 2). Sensorama was a large 3D simulation display; users had to sit in specific chairs and place their heads inside the device to experience it. However, due to its size and poor user experience, it failed to become a popular entertainment device. In 1968, Ivan Sutherland, the father of computer graphics and Turing Award winner at the Massachusetts Institute of Technology, developed the first head-mounted display, the Sword of Damocles [14] (as shown in Fig. 2(a)). Due to hardware limitations at the time, the design was very complex, and the components were heavy, making it hard to wear. A mechanical arm with a support rod was used to suspend the device for wearing. The creation of Sutherland simplified the interaction between humans and computers, making it a key pioneer in the field of virtual reality and laying a strong foundation for future virtual reality development. Lvan Sutherland is also called the father of virtual reality.

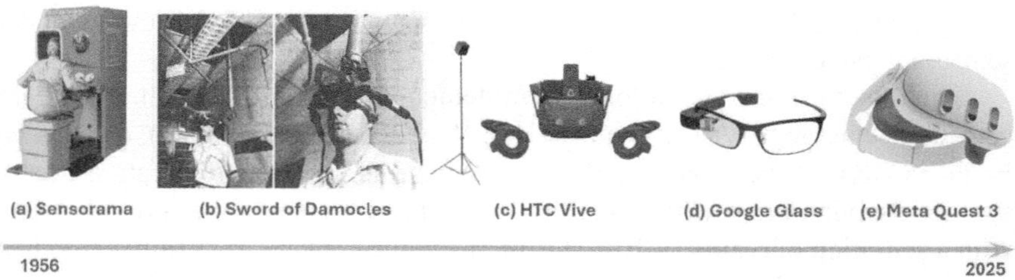

Fig. 2 The Development History of Immersive Devices.

In the 1980s, the theory and concept of virtual reality took shape, and hardware technology research made significant progress. For example, in 1986, Fisher et al. [15] created a three-dimensional virtual Martian surface environment through a virtual display. In 1989, VPL released a commercial virtual reality product called Eyephone [16], which later introduced a data globe for interactive input. Despite the development of several technological products, none were successfully used by social users. The products did not bring breakthroughs but laid a theoretical foundation and basic ideas for future virtual reality technology development. This led to Nintendo releasing the 32-bit gaming console Virtual Boy in 1995 [17]. Due to limitations in computer processing technology, game content could not achieve high 3D effects. The display screen had low resolution and refresh rate, which caused users to feel dizzy and uncomfortable. It failed commercially in less than a year and was named one of the 50 worst inventions in history by *Time* magazine. Despite its short market life, it greatly advanced virtual reality technology and opened the door for VR hardware to enter the consumer market.

In the 21st century, with improvements in computer performance, graphics processing, and motion capture technology, immersive technology saw a breakthrough. In 2012, Palmer Luckey released a developer version of the headset on a crowdfunding platform, which was officially launched in 2016 as the Oculus Rift headset [18]. In 2015, HTC released the HTC VIVE headset [19] (shown in Fig. 2), along with other mobile-specific immersive devices like Google Cardboard and Samsung Gear VR. This marked the entry of immersive technology into the consumer electronics market and sparked interest in the VR industry. At the same time, companies began to explore the potential of AR and MR. For example, Google introduced its Google Glass [20] (shown in Fig. 2(d)) product in 2013. It featured a head-up display (HUD) for functions like music playback, information display, and route navigation. In 2017 and 2019, Google released two enterprise versions of Google Glass for industrial inspections and workforce training. Microsoft also launched the mixed reality headset HoloLens in 2016 and HoloLens 2 in 2019 [21]. These devices included sensors like 3D cameras, depth sensors, gyroscopes, and microphones, enabling features such as environmental and user behavior tracking, as well as eye-tracking. In 2023, Oculus introduced Quest 3 [22] (shown in Fig. 2(e)), which was lightweight and affordable, making immersive devices more popular. With continuous technological updates, it added features like gesture control, full-body tracking, and environmental occlusion depth recognition.

3. Applications for Tunnel Engineering

3.1 3D Information Monitoring

These tunneling monitoring systems using immersive technology emphasize that operators can observe the safety status of tunnels in real time from an immersive perspective without being physically present on-site. They can quickly locate critical problem areas using data-driven navigation functions. At the same time, these technologies enable asynchronous real-time online collaboration and data recording and storage, providing convenience for joint communication and management across multiple departments. By integrating IoT with VR technology, tunnel engineers can not only monitor real-time sensor data but also visualize environmental changes immersively. For instance, VR training can incorporate IoT-driven dynamic simulations, where temperature fluctuations, airflow variations, and smoke dispersion can be rendered in virtual tunnel environments, enhancing training realism. Furthermore, IoT-enabled VR inspections allow engineers to interact with sensor data remotely, detecting anomalies and optimizing maintenance strategies without direct site visits.

For instance, Ma et al. [23] stitched multiple-angle images to create 720-degree panoramic spherical images. By doing this, users can view the tunnel's structural response and design information in a virtual environment. Besides, information related to the tunnel's regular maintenance and inspection is also to be written into the built-in database. Janiszewski et al. [24] used motion digital photogrammetry technology to measure discontinuities of tunnel rock surfaces. The measurement results were imported into a VR digital twin system that supports displaying photorealistic surface textures. The system allows remote visual inspection of the rock surface using specified virtual tools and performs virtual measurements of the discontinuities' direction. Wu et al. [25] proposed a tunnel digital twin application. In this application, a BIM model of the real tunnel structure is built to scale, and some necessary facilities are considered, including the substation, ventilation, shaft, crossing duty booth, and more. At the same time, it also supports the calibration of tunnel surveillance video with the 3D tunnel model, allowing 2D surveillance images to be projected onto the 3D tunnel for viewing. Finally, users can view IoT sensing data in the application, such as temperature, wind speed, humidity, fire alarms, and more. Traxler and Chmelina [26] developed a collaborative tunnel monitoring system on the Geotunnel Viewer platform. Individuals can observe various information (*e.g.*, navigation, radial displacements of the tunnel wall, and tunnel segments) by using a VR headset or traditional desktop devices to actively explore the digital twin model of the tunnel construction scene. At the same time, users can adjust the timeline to observe changes in the data over time. This form of storytelling enhances the interpretability and understandability of complex data analysis. Kim et al. [27] proposed an innovative platform integrating virtual reality (VR), which uses the Leica BLK360 laser radar device to scan six tunnels within the Osong integrated railway test track in South Korea. Combining 2D panoramic images and 3D point cloud data, the Matterport software was used to construct a 3D digital twin platform for virtual-real mapping. The generated 2D panoramic images (4K resolution) support immersive VR workflows, allowing maintenance personnel to inspect tunnel structure details (such as cracks, leaks, etc.) from a panoramic view without being on-site, and to perform precise dimensional analysis using the built-in measurement tools. Besides, the platform allows for digital annotation and dynamic updating of tunnel maintenance information (such as damage markings and maintenance records), creating a traceable time-series database.

3.2 Safety Training

Immersive safety training can be divided into two manifolds. First, some studies focus on the daily use of tunnels, such as emergency response, driving safety, fire safety, and evacuation, etc. For example, Kinateder [28] developed a driving simulator in which participants sit in a specially designed moving PowerRide chair and wear Z800 3D Visor headsets. The car's movement is controlled by a game car steering wheel with a gas and brake pedal, while walking is controlled by a gamepad. During the training, participants must drive the car through a simulated tunnel with accidents, fire, and smoke, and respond to the crises as instructed. In the study by Kirytopoulos et al. [29], they developed a VR driving training tool aimed at improving tunnel driving safety. The tool is based on the concept of serious games and features multiple VR driving scenarios designed from a first-person perspective. Participants simulate the entire process of driving through a tunnel, and the researchers assess the driver's knowledge level based on relevant regulations and driving guidelines, providing real-time guidance when necessary. Cha et al. [30] developed a VR-based fire training simulator that focuses on integrating accurate computational fluid dynamics (CFD) fire simulation data. As shown in Fig. 3(a), by rendering high-temperature areas and the distribution of toxic gases in a 3D environment, the simulator allows trainees to visualize hazards in a virtual tunnel fire scenario and perform basic firefighting training, such as escape, rescue, and fire extinguishing, to enhance emergency response capabilities. Calandra et al. [31] developed a platform that supports multi-user collaborative online training, where trainees can experience multiple roles and collaborate with other trainees. For example, they can play the roles of car and truck occupants. When a fire occurs, they need to maintain a safe distance from the burning car and either follow instructions to find the nearest shelter or attempt to extinguish the fire using fire extinguishers. If they play the role of a firefighter, they need to rescue people trapped in the fire and attempt to connect a fire hose to a fire truck or hydrant to attack the base of the fire. This platform supports various immersive devices, such as HTC Vive or Oculus Rift, and can even be used with conventional screens and keyboards. Subsequently, Yu et al. [32] designed additional roles. Specifically, trainees can play the role of an operator in the control room. As shown in Fig.3(b), when they witness an accident in the surveillance feed, they need to provide timely and correct assistance through remote command. Trainees can also choose to play the role of an emergency response group, helping to evacuate vehicles trapped in the tunnel, locating fire extinguishers, and operating a fire hose.

(a) (b)

Fig. 3 Different role training in the VR tunnel safety drill.

Second, another group of studies focuses on occupational safety training for workers during tunnel construction. For instance, Isleyen et al. [33] developed a training system based on the Vizard game engine to enhance tunnel construction workers' risk assessment and decision-making abilities. During the training, participants wear a headset equipped with motion tracking provided by the Worldviz PPT Precision Position Tracker, which displays a virtual tunnel environment that has just completed a blasting operation. They need to identify potential rockfall risks and then select the appropriate rock bolt length and spacing as support measures. The VR system displays real-time tunnel displacement data and updates the geological stress field and displacement vectors after the support measures are implemented, allowing participants to visually assess the effectiveness of the support. In the work of Tender et al. [34], they used a VR Stereographic Projection Optoma 3D Projector and accompanying 3D glasses to simulate Tunnel Boring Machine (TBM) operations in a tunnel project. During the training process, the actors undergo a series of Occupational Safety and Health training, covering hazard perception, emergency response, and maintenance procedures. Dolsak and Reich et al. [35] focused on a specific task within tunnel construction scenarios, namely pipe umbrella installation. The training system allows users to switch between the roles of basket operator and machine operator to deepen mutual understanding of their respective accountabilities. The training is modular and phased, allowing trainees to progress according to their experience and skill level. It also provides real video demonstrations of tunnel pipe umbrella installation. After each phase, user performance is analyzed, and multiple-choice tests are conducted to assess the skills mastered by the trainees. To enhance the awareness of frontline workers in underground tunnels regarding the hazards of inhalable dust, Ren et al. [36] developed an interactive VR simulation and training tool. In this training platform, data calculated by the computational fluid dynamics model is transformed into visual results and imported into the Oculus Quest 2. Trainees can observe various dust data according to their needs, including dust flow, dust concentration, and dust velocity. They can also switch between multiple pre-set scenarios.

3.3 Emergency Evacuation

On the other hand, by presenting virtual content with residual see-through qualities, augmented reality (AR) enhances a person's perception of virtual objects with real-world aspects. At the same time, it allows for multiple emergency evacuation drills to be conducted at low cost and high efficiency. Furthermore, with the assistance of certain hardware, immersive emergency evacuation can more easily provide real-time data monitoring and feedback. It can collect participants' behavioral data during the simulation, assess the effectiveness of emergency evacuation, and provide a basis for subsequent improvements.

In the work of Kinateder et al. [37], they used duo projectors to project the scene of a two-bore, unidirectional road tunnel filled with smoke onto a Powerwall screen. Participants then wore 3D glasses to experience the virtual effect. In their experiment, they set up a control group without a virtual agent and designed three different movement scenarios for the virtual agent (VA): evacuating in the correct direction (no-conflict condition), evacuating in the opposite direction (active conflict condition), and standing still (passive conflict condition). Through trials, they found that when a virtual agent was present, participants were more likely to wait for the virtual agent to move before taking further action. Moreover, the influence of the virtual agent's decisions was evident: for the no-conflict condition group, 85% evacuated toward the emergency exit, while when the virtual agent fled in the opposite direction of the exit, this proportion dropped sharply to 60%, which proved that individuals' path choices are influenced by those around them. Ronchi et al. [38] created a scene in a 5-sided CAVE system where a burning Heavy Goods Vehicle (HGV) catches fire and is surrounded by smoke in a tunnel. The tunnel consists of two lanes and sidewalks, and the participants were required to stand at a specific starting point and find the only safe exit based on the emergency exit signs in the tunnel. According to the analysis based on functional analysis theory, evacuees do not necessarily follow the shortest theoretical path to the emergency exit. In contrast, they typically choose to evacuate along the wall on the opposite side of the safe exit, with a small proportion walking along the center of the road. Ronchi et al. [39] set up an empty tunnel scene in a CAVE system, with an emergency exit present in the scene. Participants were required to find the emergency exit under different types of lighting conditions for the exit, considering factors such as color, flashing rate, type of light source, and layout (shown in Fig. 4(a)). The final experimental results indicated that, compared to blue lights, green and white flashing lights were more effective, while 1 Hz and 4 Hz flashing frequencies were more effective than 0.25 Hz. A light-emitting diode (LED) light source performed better than single and double strobe lights. The layout of the lights had no impact on the experimental results. To find reasonable and effective different visual and acoustic measures to assist evacuation, Skjermo et al. [40] used a combination of the Cyberith Virtualizer Research & Development Kit walking platform and the HTC Head Mounted Display to create an immersive single-tube tunnel scene. In this scenario, participants were required to experience emergency evacuation under different combinations of exit visual and acoustic settings (the basic combination elements are shown in Fig. 4 (b)).

Through the analysis of wayfinding, route speed, and walking speed, they found that continuous guiding lighting had a positive effect on locating the safe exit, which helped participants accurately identify the door's location without passing it. Participants' evacuation behaviors in a virtual tunnel with various elements, such as VR agents, emergency evacuation signs, and fire extinguishers, were observed by Zhang et al. [41]. At the same time, computational fluid dynamics (CFD) was used to analyze the cumulative distribution of CO, CO2, and temperature along the evacuation path. Based on this, the A* algorithm was used to design the theoretically optimal path.

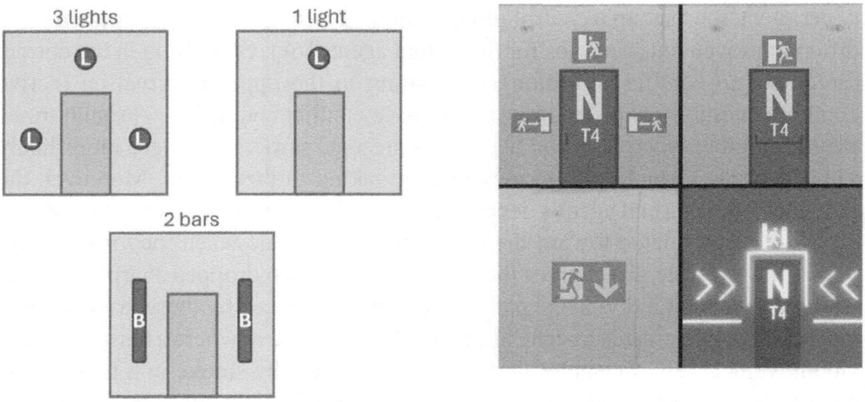

(a) (b)

Fig. 4 Different tunnel safety exits setup considering the lights and sounds effect.

3.4 Inspection and Maintenance

Regular structural and functional inspection and maintenance are essential to monitor the health and reliability of tunnel facilities. Traditional manual inspection methods are labor-intensive and time-consuming, with error-prone and low-accuracy. Some information is hard to acquire by manual inspection, such as tiny defects, structural damage, and functional conflicts. As inspection and maintenance tasks will be performed throughout the life cycle of facilities, XR methods are envisioned to integrate historical data with physical tunnel structures. Inspectors can use portable devices, including cellphones, pads, and headsets, to display and check the structure and function of tunnel facilities by timelines.

In the field of building and industry inspections, AR tools are introduced step-by-step and applied in many construction, inspection, and maintenance practices [42]. As shown in Fig. 5, Marino et al.[43] explored the potential of Augmented Reality (AR) as a key technology in Industry 4.0 to support factory workers during inspection tasks. They introduced an AR tool designed to help operators detect assembly and production errors by superimposing 3D models onto physical products. Papachristos et al. [44] researched the integration of AR and aerial robotics for structural inspection. Human operators received live stereo camera views combined with real-time 3D reconstruction, which allowed intuitive adjustments to the robot's inspection trajectory using head motions. Hsu et al. [45] presented an advanced framework for defect inspection in indoor building components using deep learning object detection and AR. By developing a framework that integrated AR and deep learning, this framework enabled real-time damage detection and localization. Engineers walked through a building, filming the interior to create a 3D environment, where defects can be detected in real-time and marked in the 3D space. Mourtzis et al. [46] discussed the integration of Unmanned Aerial Vehicles (UAVs) into industrial environments using AR for tele-operation. This study focused on designing a tele-operation system that uses AR to improve data collection in factory environments and uses this pipeline for industrial inspection and monitoring.

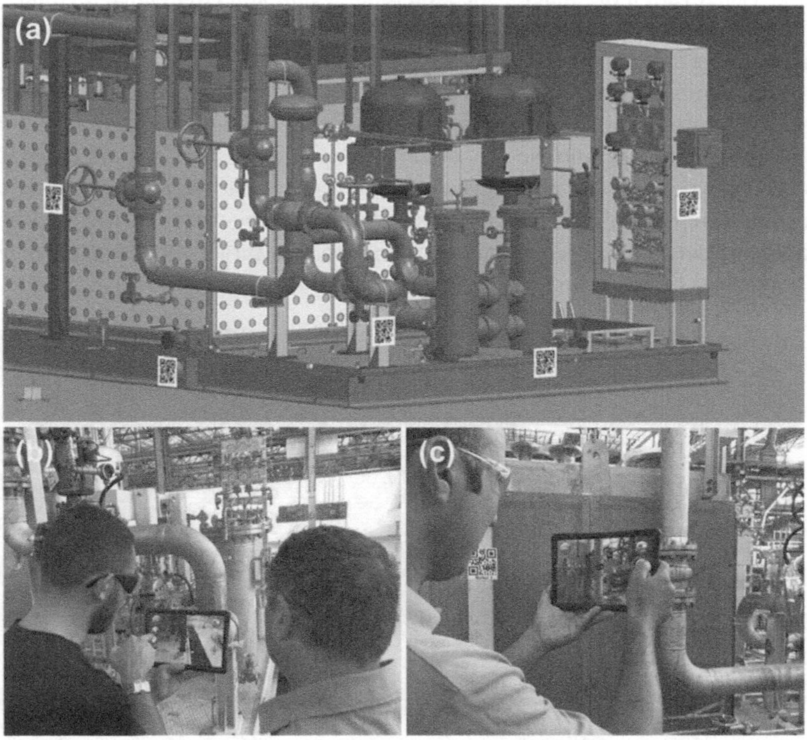

Fig. 5 The AR inspection tool for industrial pipeline systems. (a) 3D AR model of the auxiliary baseplate system for user assessment. (b) The tutor is providing guidance to field workers. (c) The tutor is giving a demo of the AR inspection tool. [43]

For tunnel service inspection and maintenance, XR is a beneficial way to superimpose digital graphics over the real-world view, which achieves a comprehensive diagnosis for the structure and function of tunnel facilities. Zhou et al. [47] investigated the implementation of AR for segment displacement inspection during tunneling construction, achieving automatic evaluation of structural integrity and real-time feedback to improve tunnel stability and prevent accidents. This proposed system was able to overlay a virtual baseline model (created according to quality standards) onto real segment displacement using AR technology, providing a more intuitive and efficient alternative compared to traditional manual inspection. Melnyk et al. [48] studied the application of AR technology for enhanced documentation and anchor inspection reporting in conventional tunnelling (shown in Fig. 6). The AR prototype enhances documentation efficiency and quality management. While input detection issues and hardware limitations still exist, the approach shows promise for real-world applications. AR facilitates real-time data capture, reduces errors, and enables better visualization of tunnel inspection reports compared to manual methods. Yu et al. [49] developed an advanced inspection assistant system for industrial settings that leverages augmented reality (AR) to improve the efficiency, accuracy, and safety of equipment inspections, overlaying virtual information (such as inspection task lists, real-time equipment data, intelligent troubleshooting recommendations, and virtual manuals) onto the actual scene using AR. Through camera calibration and coordinate transformation techniques, the system can register virtual models onto real equipment in real time and ensure that the 3D overlays (such as inspection guides and detailed equipment models) correspond accurately to the actual objects in a dynamic environment.

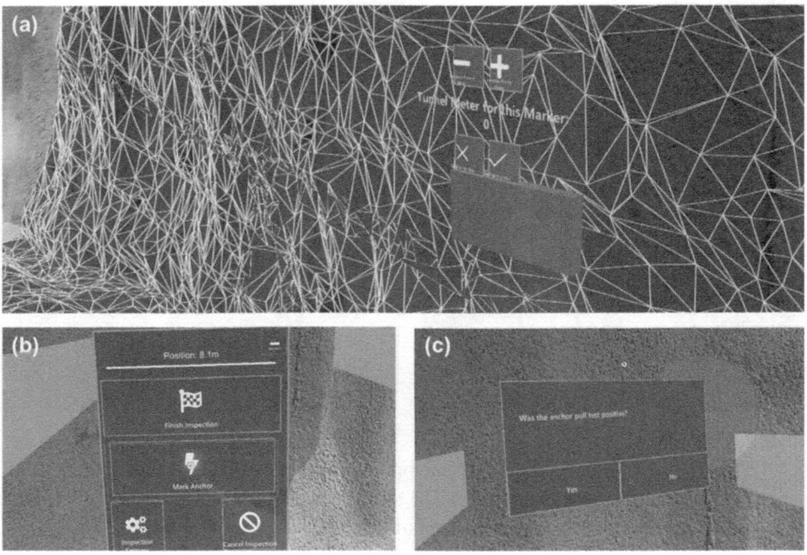

Fig. 6 AR interface of tunnel construction and inspection. (a) Localisation of the inspected tunnel section. (b) Anchor marking dialogue window. (c) Anchor marking placement process. [48]

4. Discussions and Conclusions

In this chapter, we explored the application and potential value of immersive technologies in the field of tunnel engineering. We provided a detailed introduction to the classification (taxonomy) of immersive technologies and related equipment, and further analyzed their specific applications in tunnel engineering, such as information monitoring and safety training.

Sections 3.1-3.4 of the discussion illustrate that immersive technologies have already been applied across various stages of the tunnel engineering lifecycle for different purposes. For example, IoT technology is used to monitor the structural health of tunnels during construction to ensure safety; virtual reality (VR) and augmented reality (AR) are employed to guide and train tunnel workers, improving their understanding and proficiency in their work operations; high-precision sensor technologies enhance the efficiency of routine inspections and maintenance of tunnels; and the dynamics of trapped individuals in emergencies are studied to optimize evacuation strategies. These applications demonstrate the enormous potential of immersive technologies in enhancing the safety and efficiency of tunnel engineering.

Although immersive technologies have already shown significant advantages in tunnel engineering, the development of XR technology still faces technical challenges. At present, the real-time modelling accuracy for complex geological environments needs to be improved, and there are still bottlenecks in integrating multi-source heterogeneous data. Additionally, the comfort of wearing XR devices, their battery life, and the cost of hardware are important factors limiting their large-scale application. At the same time, standardized implementation guidelines for XR technology in the engineering field are still incomplete, and data compatibility between different platforms needs further optimization. In the future, with the deep integration of technologies such as artificial intelligence, big data, and the Internet of Things, the application of XR technology in tunnel engineering will become more intelligent, and is expected to build a more efficient, safe, and collaborative engineering environment. For example, artificial intelligence (AI) can be introduced in the future to achieve high-precision 3D reconstructions of tunnel scenes, reducing the cost and time of manual surveys and increasing the automation level of engineering management. Furthermore, large language models (LLMs) can be integrated to enhance the interactivity and realism of safety training, enabling trainees to receive more targeted guidance in simulated environments, thus improving their ability to handle complex work environments.

In conclusion, immersive technologies are gradually transforming traditional work practices in tunnel engineering, providing innovative solutions for construction monitoring, safety training, and emergency management. In the future, by thoroughly integrating with other emerging technologies like AI, LLMs, and IoT, immersive technologies will play an even bigger role in the tunnel engineering field, further promoting the industry's growth and smart transformation.

CRediT authorship contribution statement

Saizhe Ding: Conceptualization, Methodology, Investigation, Validation, Writing - original draft.

Rong Deng: Resourcing, Writing - original draft.

Yuxin Zhang: Methodology, Supervision, Funding acquisition, Writing - review & editing.

Xinyan Huang: Conceptualization, Supervision, Funding acquisition, Writing - review & editing.

Declaration of Competing Interest

The authors declare there is no conflict of interest.

Acknowledgements

YZ thanks the support from MTR Research Fund (PTU-23005), the National Natural Science Foundation of China (52204232), and PolyU Start-up Fund under the Strategic Hiring Scheme (P0045772). XH thanks the support from MTR Research Fund (PTU-23005) and Start-up Fund under the Strategic Hiring Scheme (P0045772), PolyU Teaching Grant (9BNH). The technical support from PolyU Industrial Centre is acknowledged.

References

[1] Sutherland, I.E. Sketch pad a man-machine graphical communication system, in: Proceedings of the SHARE Design Automation Workshop, ACM Press, New York, USA, 1964: pp. 6.329–6.346. https://doi.org/10.1145/800265.810742.

[2] Milgram, P., Takemura, H., Utsumi, A. and Kishino, F. Augmented reality: A class of displays on the reality-virtuality continuum, in: Telemanipulator and Telepresence Technologies, Spie, 1995: pp. 282–292.

[3] Fernie, K. and Richards, J.D. Creating and Using Virtual Reality: A Guide for the Arts and Humanities, URL: Http://Vads. Ahds. Ac. Uk/Guides/vr Guide/Glossary. Html (n.d.).

[4] Flavián, C., Ibáñez-Sánchez, S. and Orús, C. The impact of virtual, augmented and mixed reality technologies on the customer experience, *J Bus Res* 100 (2019) 547–560. https://doi.org/10.1016/j.jbusres.2018.10.050.

[5] Pleyers, G. and Poncin, I. Non-immersive virtual reality technologies in real estate: How customer experience drives attitudes toward properties and the service provider, *Journal of Retailing and Consumer Services* 57 (2020) 102175. https://doi.org/10.1016/j.jretconser.2020.102175.

[6] Hartmann, C. and Siegrist, M. Virtual reality and immersive approaches to contextual food testing, in: *Context*, Elsevier, 2019: pp. 323–338. https://doi.org/10.1016/B978-0-12-814495-4.00016-7.

[7] Restout, J., Bernache-Assollant, I., Morizio, C., Boujut, A., Angelini, L., Tchalla, A. and Perrochon, A. Fully Immersive Virtual Reality Using 360° Videos to Manage Well-Being in Older Adults: A Scoping Review, *J Am Med Dir Assoc* 24 (2023) 564–572. https://doi.org/10.1016/j.jamda.2022.12.026.

[8] Pastorelli, E. and Herrmann, H. A Small-scale, Low-budget Semi-immersive Virtual Environment for Scientific Visualization and Research, Procedia Comput Sci 25 (2013) 14–22. https://doi.org/10.1016/j.procs.2013.11.003.

[9] Freina, L. and Ott, M. A literature review on immersive virtual reality in education: state of the art and perspectives, in: The International Scientific Conference Elearning and Software for Education, 2015: pp. 10–1007.

[10] Neguţ, A., Matu, S.-A., Sava, F.A. and David, D. Task difficulty of virtual reality-based assessment tools compared to classical paper-and-pencil or computerized measures: A meta-analytic approach, *Comput Human Behav* 54 (2016) 414–424. https://doi.org/10.1016/j.chb.2015.08.029.

[11] Buttussi, F. and Chittaro, L. A Comparison of Procedural Safety Training in Three Conditions: Virtual Reality Headset, Smartphone, and Printed Materials, IEEE Transactions on Learning Technologies 14 (2021) 1–15. https://doi.org/10.1109/TLT.2020.3033766.

[12] Hou, L. Wang, X., Bernold, L. P.E.D. Love, Using Animated Augmented Reality to Cognitively Guide Assembly, *Journal of Computing in Civil Engineering* 27 (2013) 439–451. https://doi.org/10.1061/(ASCE)CP.1943-5487.0000184.

[13] Jones, S. and Dawkins, S. The Sensorama Revisited: Evaluating the Application of Multi-sensory Input on the Sense of Presence in 360-Degree Immersive Film in Virtual Reality, in: T. Jung, M.C. tom Dieck (Eds.), Augmented Reality and Virtual Reality: Empowering Human, Place and Business, Springer International Publishing, Cham, 2018: pp. 183–197. https://doi.org/10.1007/978-3-319-64027-3_13.

[14] Sutherland, I.E. A head-mounted three dimensional display, in: Proceedings of the December 9-11, 1968, Fall Joint Computer Conference, Part I on - AFIPS '68 (Fall, Part I), ACM Press, New York, New York, USA, 1968: pp. 757–764. https://doi.org/10.1145/1476589.1476686.

[15] Fisher, S.S., McGreevy, M., Humphries, J. and Robinett, W. Virtual environment display system, in: Proceedings of the 1986 Workshop on Interactive 3D Graphics - SI3D '86, ACM Press, New York, USA, 1987: pp. 77–87. https://doi.org/10.1145/319120.319127.

[16] Weiss, J., Nolan, J., Hunsinger, J. and Trifonas, P. The international handbook of virtual learning environments, Springer, 2006.

[17] Boyer, S.A Virtual Failure: Evaluating the Success of Nintendo's Virtual Boy, The Velvet Light Trap 64 (2009) 23–33. https://doi.org/10.1353/vlt.0.0039.

[18] Milgram, P., Takemura, H., Utsumi, A. and Kishino, F. Augmented reality: a class of displays on the reality-virtuality continuum, in: Proc.SPIE, 1995: pp. 282–292. https://doi.org/10.1117/12.197321.

[19] Egger, J., Gall, M., Wallner, J., Boechat, P., Hann, A., Li, X., Chen, X. and Schmalstieg, D. HTC Vive MeVisLab integration via OpenVR for medical applications, PLoS One 12 (2017) e0173972. https://doi.org/10.1371/journal.pone.0173972.

[20] Wei, N.J., Dougherty, B., Myers, A. and Badawy, S.M. Using Google Glass in Surgical Settings: Systematic Review, JMIR Mhealth Uhealth 6 (2018) e54. https://doi.org/10.2196/mhealth.9409.

[21] Ungureanu, D., Bogo, F., Galliani, S., Sama, P., Duan, X., Meekhof, C., Stühmer, J., Cashman, T.J., Tekin, B., Schönberger, J.L. Olszta, P. and Pollefeys, M. HoloLens 2 Research Mode as a Tool for Computer Vision Research, (2020).

[22] Criollo-C, S., Guerrero-Arias, A., Samala, A.D., Arif, Y.M. and Luján-Mora, S. Enhancing the Educational Model Using Mixed Reality Technologies With Meta Quest 3: A Usability Analysis Using IBM-CSUQ, IEEE Access 13 (2025) 56930–56945. https://doi.org/10.1109/ACCESS.2025.3555930.

[23] Ma, X., Liu, Y., Wang, J., Wang, D. and Liu, Y. Application of the Virtual Reality Technology in Structural Safety Monitoring of Shield Tunnels, IOP Conf Ser Mater Sci Eng 960 (2020) 022064. https://doi.org/10.1088/1757-899X/960/2/022064.

[24] Janiszewski, M ., Uotinen, L., Rinne, M. and Baghbanan, Digitisation of hard rock tunnel for remote fracture mapping and virtual training environment, in: ISRM International Symposium - EUROCK 2020, ISRM, 2020.

[25] Wu, Z., Chang, Y., Li, Q. and Cai, R. A Novel Method for Tunnel Digital Twin Construction and Virtual-Real Fusion Application, Electronics (Basel) 11 (2022). https://doi.org/10.3390/electronics11091413.

[26] Traxler, C. and Chmelina, K. Storytelling as collaboration support for 3D tunnel monitoring, in: Expanding Underground-Knowledge and Passion to Make a Positive Impact on the World, CRC Press, 2023: pp. 2552–2559.

[27] Kim, M., Hwang, D. and Park, D. Analysis of Maintenance Techniques for a Three-Dimensional Digital Twin-Based Railway Facility with Tunnels, Platforms 1 (2023) 5–17. https://doi.org/10.3390/platforms1010002.

[28] Kinateder, M., Pauli, P., Müller, M., Krieger, J., Heimbecher, F., Rönnau, I., Bergerhausen , U., Vollmann, G., Vogt, P. and Mühlberger, A. Human behaviour in severe tunnel accidents: Effects of information and behavioural training, Transp Res Part F Traffic Psychol Behav 17 (2013) 20–32. https://doi.org/10.1016/j.trf.2012.09.001.

[29] Kirytopoulos, K., Mourelatos, A., Chatzistelios, G., Ntzeremes, P. and Konstantinidou, M. A virtual reality instrument to raise drivers' awareness on safer driving through road tunnels, Proceedings of the Institution of Mechanical Engineers, Part O 238 (2023) 451–463. https://doi.org/10.1177/1748006X231175719.

[30] Cha, M., Han, S., Lee, J. and Choi, B. A virtual reality based fire training simulator integrated with fire dynamics data, Fire Saf J 50 (2012) 12–24. https://doi.org/10.1016/j.firesaf.2012.01.004.

[31] Calandra, D., Pratticò, F.G., Migliorini, M., Verda,V. and Lamberti, F. A multi-role, multi-user, multi-technology virtual reality-based road tunnel fire simulator for training purposes, in: Proc. 16th International Conference on Computer Graphics Theory and Applications (GRAPP 2021), SciTePress, 2021: pp. 96–105.

[32] Yu, X., Yu, P., Wan, C., Wang, D., Shi, W., Shou, W., Wang, J. and Wang, X. Integrating Virtual Reality and Building Information Modeling for Improving Highway Tunnel Emergency Response Training, Buildings 12 (2022). https://doi.org/10.3390/buildings12101523.

[33] Isleyen, E., Duzgun, S. and Nelson, P. Virtual reality for hazard assessment and risk mitigation in tunneling, in: Tunnels and Underground Cities: Engineering and Innovation Meet Archaeology, Architecture and Art, CRC Press, 2020: pp. 4857–4863.

[34] Tender, M., Fuller, P., Vaughan, A., Demian, P., Chow, V., Silva, F., Couto, J.P. and . Santos, R. (2023). Lessons learned from the use of virtual reality for occupational safety and health training in Thames tideway tunnel,

[35] Dolsak, W. and Reich, C. Virtual reality simulator for pipe umbrella installation works, Geomechanics and Tunnelling 15 (2022) 233–238. https://doi.org/10.1002/geot.202100062.

[36] Ren, T., Qiao, M., Roberts, J., Hines, J., Chow, Y.-W., Zong, W., Sugden, A., Shepherd, M., Farrelly, M., Kennedy, G., Hai, F. and Susilo, W. Development and evaluation of an immersive VR-CFD-Based tool for dust exposure mitigation in underground tunnelling operations, Tunnelling and Underground Space Technology 143 (2024) 105496. https://doi.org/10.1016/j.tust.2023.105496.

[37] Kinateder, M., Müller, M., Jost, M., Mühlberger and Pauli, P. Social influence in a virtual tunnel fire – Influence of conflicting information on evacuation behavior, *Appl Ergon* 45 (2014) 1649–1659. https://doi.org/10.1016/j.apergo.2014.05.014.

[38] Ronchi, E., Kinateder, M., Müller, M., Jost, M., Nehfischer, M., Pauli, P. and Mühlberger, A. Evacuation travel paths in virtual reality experiments for tunnel safety analysis, *Fire Saf J* 71 (2015) 257–267. https://doi.org/10.1016/j.firesaf.2014.11.005.

[39] Ronchi, E., Nilsson, D., Kojić, S., Eriksson, J., Lovreglio, R., Modig, H. and Walter, A.l. A Virtual Reality Experiment on Flashing Lights at Emergency Exit Portals for Road Tunnel Evacuation, *Fire Technol* 52 (2016) 623–647. https://doi.org/10.1007/s10694-015-0462-5.

[40] Skjermo, J., Moscoso, C., Nilsson, D., Frantzich, H., Hoem, Å.S., Arnesen, P. and Jenssen, G.D. Analysis of visual and acoustic measures for self-evacuations in road tunnels using virtual reality, Fire Saf J 148 (2024) 104224. https://doi.org/10.1016/j.firesaf.2024.104224.

[41] Zhang, X., Chen, L., Jiang, J., Ji, Y., Han, Zhu, T., Xu, W. and Tang, F. Risk analysis of people evacuation and its path optimization during tunnel fires using virtual reality experiments, Tunnelling and Underground Space Technology 137 (2023) 105133. https://doi.org/10.1016/j.tust.2023.105133.

[42] Schiavi, B., Havard, V., Beddiar, K. and Baudry, D. BIM data flow architecture with AR/VR technologies: Use cases in architecture, engineering and construction, Autom Constr 134 (2022) 104054. https://doi.org/10.1016/j.autcon.2021.104054.

[43] Marino, E., Barbieri, L., Colacino, B., Fleri, A.K. and Bruno, F. An Augmented Reality inspection tool to support workers in Industry 4.0 environments, Comput Ind 127 (2021) 103412. https://doi.org/10.1016/j.compind.2021.103412.

[44] Papachristos, C. and Alexis, K. Augmented reality-enhanced structural inspection using aerial robots, in: 2016 IEEE International Symposium on Intelligent Control (ISIC), 2016: pp. 1–6. https://doi.org/10.1109/ISIC.2016.7579983.

[45] Hsu, S.-H., Hung, H.-T., Lin, Y.-Q. and Chang, C.-M. Defect inspection of indoor components in buildings using deep learning object detection and augmented reality, Earthquake Engineering and Engineering Vibration 22 (2023) 41–54. https://doi.org/10.1007/s11803-023-2152-5.

[46] Mourtzis, D., Angelopoulos, J. and Panopoulos, N. Unmanned Aerial Vehicle (UAV) manipulation assisted by Augmented Reality (AR): The case of a drone, IFAC-PapersOnLine 55 (2022) 983–988. https://doi.org/10.1016/j.ifacol.2022.09.483.

[47] Zhou, Y., Luo, H. and Yang, Y. Implementation of augmented reality for segment displacement inspection during tunneling construction, Autom Constr 82 (2017) 112–121. https://doi.org/10.1016/j.autcon.2017.02.007.

[48] O. Melnyk, O., Huymajer, M., Fenzl, D. Huemer, C. Wenighofer, R., Mazak-Huemer, A. (2024) Augmented reality for enhanced documentation and anchor inspection reporting in conventional tunnelling, Tunnelling and Underground Space Technology 153 106040. https://doi.org/10.1016/j.tust.2024.106040.

[49] Yu, H., Xia, M., Zhu, X., Jin, Y. and Wu, F. Industrial Inspection Auxiliary System Based On AR Technology, in: 2023 International Conference on Image Processing, Computer Vision and Machine Learning (ICICML), 2023: pp. 979–983. https://doi.org/10.1109/ICICML60161.2023.10424734.

Index

About the Editors

Yuxin Zhang is an associate professor in the Department of urban management at Kyoto University and leads the smart safety and underground space engineering group. Before joining Kyoto University, she worked at the Hong Kong Polytechnic University. She obtained her bachelor and doctoral degree majoring in tunnelling and underground space engineering at Tongji University. She has conducted various multi-scale field experiments and VR experiments to study human behavior and applied data analysis and modelling to optimize evacuation plans in tunnels and subway stations. She has published over 40 academic papers in journals and conferences including Nature Cities and presented at several distinguished conferences. She also serves as editorial members, guest editors and reviewers for several top academic journals. In addition, she has been actively advancing the translation of research into real-world applications, with a strong portfolio of patents, and received the Geneva Invention Gold Medal of the Jury (2025) and CES Best of Innovation (2026).

Yi Rui is a Professor at Tongji University. He received a PhD from University of Cambridge, MSc and BEng from Tongji University. Before joining Tongji, Prof. Rui was a lecturer at Aberdeen University, UK. His research background is broad, encompassing several hot research topics in geotechnical engineering: (i) Energy geo-structures; (ii) Offshore geotechnical engineering; (iii) Intelligent structural health monitoring. Prof. Rui has published over 80 journal papers. He is an Editorial member of Underground Space and a Guest Editor of Fractal and Fractional. He has also been honored with the Science and Technology Progress Award from the Chinese Society for Rock Mechanics and Engineering and the Shanghai Science and Technology Progress Award.

Xinyan Huang is an Associate Professor in the Department of Building Environment and Energy Engineering, the Hong Kong Polytechnic University. He received his PhD from Imperial College London and is a Chartered Building Services and Fire Engineer. Prof. Huang has co-authored over 200 journal papers in building safety and smart resilience. He is an Associate Editor of *Fire Technology, and J. Safety Sci. Resil.*, an editorial member of *Fire Safety J and J. Building Engineering*. He serves as a board member of Int. Association of Fire Safety Science (IAFSS) and Society of Fire Protection Engineers (SFPE) Foundation. He is a winner of the NSFC Excellent Young Scientists Fund, IAFSS Early Career Award from IAFSS, and SFPE Engineering Communication Award.

For Product Safety Concerns and Information please contact our
EU representative GPSR@taylorandfrancis.com Taylor & Francis
Verlag GmbH, Kaufingerstraße 24, 80331 München, Germany